Springer Series in Operations Research

Editors:
Peter W. Glynn   Stephen M. Robinson

Springer Series in Operations Research

Altiok: Performance Analysis of Manufacturing Systems
Birge and Louveaux: Introduction to Stochastic Programming
Bonnans and Shapiro: Perturbation Analysis of Optimization Problems
Dantzig and Thapa: Linear Programming 1: Introduction
Dantzig and Thapa: Linear Programming 2: Theory and Extensions
Drezner (Editor): Facility Location: A Survey of Applications and Methods
Facchinei and Pang: Finite-Dimensional Variational Inequalities and Complementarity Problems, Volume I
Facchinei and Pang: Finite-Dimensional Variational Inequalities and Complementarity Problems, Volume II
Fishman: Discrete-Event Simulation: Modeling, Programming, and Analysis
Fishman: Monte Carlo: Concepts, Algorithms, and Applications
Haas: Stochastic Petri Nets: Modeling, Stability, Simulation
Klamroth: Single-Facility Location Problems with Barriers
Nocedal and Wright: Numerical Optimization
Olson: Decision Aids for Selection Problems
Pinedo: Planning and Scheduling in Manufacturing and Services
Simchi-Levi, Chen, and Bramel: The Logic of Logistics: Theory, Algorithms, and Applications for Logistics and Supply Chain Management (second edition)
Whitt: Stochastic-Process Limits: An Introduction to Stochastic-Process Limits and Their Application to Queues
Yao (Editor): Stochastic Modeling and Analysis of Manufacturing Systems
Yao and Zheng: Dynamic Control of Quality in Production-Inventory Systems: Coordination and Optimization

Michael L. Pinedo

Planning and Scheduling in Manufacturing and Services

Includes CD-ROM

 Springer

Michael L. Pinedo
Department of Operations Management
Stern School of Business
New York University
40 West 4th Street, Suite 700
New York, NY 10012-1118
USA
mpinedo@stern.nyu.edu

Series Editors:
Peter W. Glynn
Department of Management Science
  and Engineering
Terman Engineering Center
Stanford University
Stanford, CA 94305-4026
USA
glynn@leland.stanford.edu

Stephen M. Robinson
Department of Industrial Engineering
University of Wisconsin–Madison
1513 University Avenue
Madison, WI 53706-1572
USA
smrobins@facstaff.wisc.edu

Mathematics Subject Classification (2000): 90-xx

Library of Congress Cataloging-in-Publication Data
Pinedo, Michael.
  Planning and scheduling in manufacturing services / Michael L. Pinedo.
    p.   cm.
  Includes bibliographical references and index.
  ISBN 0-387-22198-0
  1. Production scheduling. 2. Production planning. I. Title.
  TS157.5.P558 2005
  658.5'3—dc22
                                                            2004062622

ISBN 0-387-22198-0   Printed on acid-free paper.

© 2005 Springer Science+Business Media, Inc.
All rights reserved. This work may not be translated or copied in whole or in part without the written permission of the publisher (Springer Science+Business Media, Inc., 233 Spring Street, New York, NY 10013, USA), except for brief excerpts in connection with reviews or scholarly analysis. Use in connection with any form of information storage and retrieval, electronic adaptation, computer software, or by similar or dissimilar methodology now known or hereafter developed is forbidden.
The use in this publication of trade names, trademarks, service marks, and similar terms, even if they are not identified as such, is not to be taken as an expression of opinion as to whether or not they are subject to proprietary rights.

Printed in the United States of America.   (AL/EB)

9 8 7 6 5 4 3 2 1   SPIN 10972656

springeronline.com

Preface

This book is an outgrowth of an earlier text that appeared in 1999 under the title "Operations Scheduling with Applications in Manufacturing and Services", coauthored with Xiuli Chao from North Carolina State. This new version has been completely reorganized and expanded in several directions including new application areas and solution methods.

The application areas are divided into two parts: manufacturing applications and services applications. The book covers five areas in manufacturing, namely, project scheduling, job shop scheduling, scheduling of flexible assembly systems, economic lot scheduling, and planning and scheduling in supply chains. It covers four areas in services, namely, reservations and timetabling, tournament scheduling, planning and scheduling in transportation, and workforce scheduling. Of course, this selection does not represent all the applications of planning and scheduling in manufacturing and services. Some areas that have received a fair amount of attention in the literature, e.g., scheduling of robotic cells, have not been included. Scheduling problems in telecommunication and computer science have not been covered either.

It seems to be harder to write a good applications-oriented book than a good theory-oriented book. In the writing of this book one question came up regularly: what should be included and what should not be included? Some difficult decisions had to be made with regard to some of the material covered. For example, should this book discuss Johnson's rule, which minimizes the makespan in a two machine flow shop? Johnson's rule is described in virtually every scheduling book and even in many books on operations management. It is mathematically elegant; but it is not clear how important it is in practice. We finally concluded that it did not deserve so much attention in an applications-oriented book such as this one. However, we did incorporate it as an exercise in the chapter on job shop scheduling and ask the student to compare its performance to that of the well-known shifting bottleneck heuristic (which is one of the better known heuristics used in practice).

The fundamentals concerning the methodologies that are used in the application chapters are covered in the appendixes. They contain the basics of mathematical programming, dynamic programming, heuristics, and constraint programming.

It is not necessary to have a detailed knowledge of computational complexity in order to go through this book. However, at times some complexity terminology is used. That is, a scheduling problem may be referred to as polynomially solvable (i.e., easy) or as NP-hard (i.e., hard). However, we never go into any NP-hardness proofs.

Because of the diversity and the complexity of the models it turned out to be difficult to develop a notation that could be kept uniform throughout the book. A serious attempt has been made to maintain some consistency of notation. However, that has not always been possible (but, of course, within each chapter the notation is consistent). Another issue we had to deal with was the level of the mathematical notation used. We decided that we did have to adopt at times the set notation and use the $\in$ symbol. So $j \in S$ implies that job j belongs to a set of jobs called S and $S_1 \cup S_2$ denotes the union of the two sets S_1 and S_2.

The book comes with a CD-ROM that contains various sets of powerpoint slides. Five sets of slides were developed by instructors who had adopted the earlier version of this book, namely Erwin Hans and Johann Hurink at Twente University of Technology in the Netherlands, Siggi Olafsson at Iowa State, Sanja Petrovic in Nottingham, Sibel Salman at Carnegie-Mellon (Sibel is currently at Koç University in Turkey), and Cees Duin and Erik van der Sluis at the University of Amsterdam. Various collections of slides were also made available by several companies, including Alcan, Carmen Systems, Cybertec, Dash Optimization, Ilog, Multimodal, and SAP. Both Ilog and Dash Optimization provided a substantial amount of additional material in the form of software, minicases, and a movie. The CD-ROM contains also various planning and scheduling systems that have been developed in academia. The LEKIN system has been especially designed for the machine scheduling and job shop models discussed in Chapter 5. Other systems on the CD-ROM include a crew scheduling system, an employee scheduling system, and a timetabling system.

This new version has benefited enormously from numerous comments made by many colleagues. First of all, this text owes a lot to Xiuli Chao from North Carolina State; his comments have always been extremely useful. Many others have also gone through the manuscript and provided constructive criticisms. The list includes Ying-Ju Chen (NYU), Jacques Desrosiers (GERAD, Montreal), Thomas Dong (ILOG), Andreas Drexl (Kiel, Germany), John Fowler (Arizona), Guillermo Gallego (Columbia), Nicholas Hall (Ohio State), Jack Kanet (Clemson), Chung-Yee Lee (HKUST), Joseph Leung (NJIT), Haibing Li (NJIT), Irv Lustig (ILOG), Kirk Moehle (Maersk Line), Detlef Pabst (Arizona), Denis Saure (Universidad de Chile), Erik van der Sluis (University of Amsterdam), Marius Solomon (Northeastern University), Chelliah Sriskandarajah (UT Dallas), Michael Trick (Carnegie-Mellon), Reha Uzsoy (Purdue),

Alkis Vazacopoulos (Dash Optimization), Nitin Verma (Dash Optimization), and Benjamin Yen (Hong Kong University).

The technical production of this book and CD-ROM would not have been possible without the help of Berna Sifonte and Adam Lewenberg. Thanks are also due to the National Science Foundation; without its support this project would not have been completed.

A website for this book will be maintained at

http://www.stern.nyu.edu/~mpinedo

This site will keep an up-to-date list of the instructors who are using the book (including those who used the 1999 version). In addition, the site will contain relevant material that becomes available after the book has gone to press.

New York Michael Pinedo
Fall 2004

Contents

Preface v

Contents of CD-ROM xv

Part I Preliminaries

1 **Introduction** 3
   1.1 Planning and Scheduling: Role and Impact 3
   1.2 Planning and Scheduling Functions in an Enterprise 8
   1.3 Outline of the Book 11

2 **Manufacturing Models** 19
   2.1 Introduction 19
   2.2 Jobs, Machines, and Facilities 21
   2.3 Processing Characteristics and Constraints 24
   2.4 Performance Measures and Objectives 28
   2.5 Discussion 32

3 **Service Models** 37
   3.1 Introduction 37
   3.2 Activities and Resources in Service Settings 40
   3.3 Operational Characteristics and Constraints 41
   3.4 Performance Measures and Objectives 43
   3.5 Discussion 45

Part II Planning and Scheduling in Manufacturing

4 Project Planning and Scheduling 51
  4.1 Introduction 51
  4.2 Critical Path Method (CPM) 54
  4.3 Program Evaluation and Review Technique (PERT) 58
  4.4 Time/Cost Trade-Offs: Linear Costs 61
  4.5 Time/Cost Trade-Offs: Nonlinear Costs 68
  4.6 Project Scheduling with Workforce Constraints............. 69
  4.7 ROMAN: A Project Scheduling System for the Nuclear
      Power Industry... 72
  4.8 Discussion 76

5 Machine Scheduling and Job Shop Scheduling 81
  5.1 Introduction 81
  5.2 Single Machine and Parallel Machine Models................ 82
  5.3 Job Shops and Mathematical Programming 84
  5.4 Job Shops and the Shifting Bottleneck Heuristic............. 87
  5.5 Job Shops and Constraint Programming.................... 93
  5.6 LEKIN: A Generic Job Shop Scheduling System............. 102
  5.7 Discussion 109

6 Scheduling of Flexible Assembly Systems 115
  6.1 Introduction 115
  6.2 Sequencing of Unpaced Assembly Systems 116
  6.3 Sequencing of Paced Assembly Systems 122
  6.4 Scheduling of Flexible Flow Systems with Bypass............ 127
  6.5 Mixed Model Assembly Sequencing at Toyota 132
  6.6 Discussion 135

7 Economic Lot Scheduling 141
  7.1 Introduction 141
  7.2 One Type of Item and the Economic Lot Size 142
  7.3 Different Types of Items and Rotation Schedules 146
  7.4 Different Types of Items and Arbitrary Schedules............ 150
  7.5 More General ELSP Models 159
  7.6 Multiproduct Planning and Scheduling at Owens-Corning
      Fiberglas 162
  7.7 Discussion 164

8 Planning and Scheduling in Supply Chains 171
  8.1 Introduction 171
  8.2 Supply Chain Settings and Configurations 173
  8.3 Frameworks for Planning and Scheduling in Supply Chains ... 178

8.4 A Medium Term Planning Model for a Supply Chain 184
8.5 A Short Term Scheduling Model for a Supply Chain 190
8.6 Carlsberg Denmark: An Example of a System Implementation 193
8.7 Discussion 197

Part III Planning and Scheduling in Services

9 Interval Scheduling, Reservations, and Timetabling 205
    9.1 Introduction 205
    9.2 Reservations without Slack 207
    9.3 Reservations with Slack 210
    9.4 Timetabling with Workforce Constraints 213
    9.5 Timetabling with Operator or Tooling Constraints 216
    9.6 Assigning Classes to Rooms at U.C. Berkeley 221
    9.7 Discussion 224

10 Scheduling and Timetabling in Sports and Entertainment . 229
    10.1 Introduction 229
    10.2 Scheduling and Timetabling in Sport Tournaments 230
    10.3 Tournament Scheduling and Constraint Programming 237
    10.4 Tournament Scheduling and Local Search 240
    10.5 Scheduling Network Television Programs 243
    10.6 Scheduling a College Basketball Conference 245
    10.7 Discussion 248

11 Planning, Scheduling, and Timetabling in Transportation .. 253
    11.1 Introduction 253
    11.2 Tanker Scheduling 254
    11.3 Aircraft Routing and Scheduling 258
    11.4 Train Timetabling 272
    11.5 Carmen Systems: Designs and Implementations 279
    11.6 Discussion 283

12 Workforce Scheduling 289
    12.1 Introduction 289
    12.2 Days-Off Scheduling 290
    12.3 Shift Scheduling 296
    12.4 The Cyclic Staffing Problem 299
    12.5 Applications and Extensions of Cyclic Staffing 301
    12.6 Crew Scheduling 303
    12.7 Operator Scheduling in a Call Center 307
    12.8 Discussion 311

Part IV Systems Development and Implementation

13 Systems Design and Implementation 319
    13.1 Introduction 319
    13.2 Systems Architecture 320
    13.3 Databases, Object Bases, and Knowledge-Bases 322
    13.4 Modules for Generating Plans and Schedules 327
    13.5 User Interfaces and Interactive Optimization 330
    13.6 Generic Systems vs. Application-Specific Systems 336
    13.7 Implementation and Maintenance Issues 339

14 Advanced Concepts in Systems Design 345
    14.1 Introduction 345
    14.2 Robustness and Reactive Decision Making 346
    14.3 Machine Learning Mechanisms 351
    14.4 Design of Planning and Scheduling Engines and Algorithm Libraries 357
    14.5 Reconfigurable Systems 360
    14.6 Web-Based Planning and Scheduling Systems 362
    14.7 Discussion 365

15 What Lies Ahead? 371
    15.1 Introduction 371
    15.2 Planning and Scheduling in Manufacturing 372
    15.3 Planning and Scheduling in Services 373
    15.4 Solution Methods 375
    15.5 Systems Development 377
    15.6 Discussion 378

Appendices

A Mathematical Programming: Formulations and Applications 383
    A.1 Introduction 383
    A.2 Linear Programming Formulations 383
    A.3 Nonlinear Programming Formulations 386
    A.4 Integer Programming Formulations 388
    A.5 Set Partitioning, Set Covering, and Set Packing 390
    A.6 Disjunctive Programming Formulations 391

B	**Exact Optimization Methods** 395	
	B.1 Introduction 395	
	B.2 Dynamic Programming 396	
	B.3 Optimization Methods for Integer Programs 400	
	B.4 Examples of Branch-and-Bound Applications 402	
C	**Heuristic Methods** 413	
	C.1 Introduction 413	
	C.2 Basic Dispatching Rules 414	
	C.3 Composite Dispatching Rules 417	
	C.4 Beam Search 421	
	C.5 Local Search: Simulated Annealing and Tabu-Search 424	
	C.6 Local Search: Genetic Algorithms........................ 431	
	C.7 Discussion 433	
D	**Constraint Programming Methods** 437	
	D.1 Introduction 437	
	D.2 Constraint Satisfaction 438	
	D.3 Constraint Programming 439	
	D.4 OPL: An Example of a Constraint Programming Language ... 441	
	D.5 Constraint Programming vs. Mathematical Programming 444	
E	**Selected Scheduling Systems** 447	
	E.1 Introduction 447	
	E.2 Generic Systems.. 447	
	E.3 Application-Specific Systems 448	
	E.4 Academic Prototypes 448	
F	**The Lekin System User's Guide** 451	
	F.1 Introduction 451	
	F.2 Linking External Algorithms 451	

References 459

Notation 489

Name Index 493

Subject Index 501

Contents of CD-ROM

0. CD Overview

1. Slides from Academia
   (a) Twente University (by Erwin Hans and Johann Hurink)
   (b) Iowa State University (by Siggi Olafsson)
   (c) University of Nottingham (by Sanja Petrovic)
   (d) Carnegie-Mellon University (by Sibel Salman)
   (e) University of Amsterdam (by Eric van der Sluis and Cees Duin)

2. Slides from Corporations
   (a) Alcan
   (b) Carmen Systems
   (c) Cybertec
   (d) Dash Optimization
   (e) Multimodal
   (f) SAP

3. Scheduling Systems
   (a) LEKIN Job Shop Scheduling System
   (b) CSS Crew Scheduling System
   (c) ESS Employee Scheduling System
   (d) TTS Timetabling System

4. **Optimization Software**
   (a) Dash Optimization Software (with Sample Programs for Examples 8.4.1, 11.2.1, 11.3.1)
   (b) ILOG OPL Software (with Sample Programs for Examples 8.4.1 and 11.2.1)

5. **Examples and Exercises**
   (a) Tanker Scheduling (Computational details of Example 11.2.1)
   (b) Aircraft Routing and Scheduling (Computational details of Example 11.3.1)

6. **Mini-cases**
   (a) ILOG
   (b) Dash Optimization

7. **Additional Readings (White Papers)**
   (a) Carmen Systems
   (b) Multimodal Inc.

8. **Movie**
   (a) Saiga - Scheduling at the Paris airports (ILOG)

Part I

Preliminaries

1 Introduction 3
2 Manufacturing Models 19
3 Service Models 37

Chapter 1

Introduction

1.1 Planning and Scheduling: Role and Impact 3
1.2 Planning and Scheduling Functions in an Enterprise 8
1.3 Outline of the Book 11

1.1 Planning and Scheduling: Role and Impact

Planning and scheduling are decision-making processes that are used on a regular basis in many manufacturing and service industries. These forms of decision-making play an important role in procurement and production, in transportation and distribution, and in information processing and communication. The planning and scheduling functions in a company rely on mathematical techniques and heuristic methods to allocate limited resources to the activities that have to be done. This allocation of resources has to be done in such a way that the company optimizes its objectives and achieves its goals. Resources may be machines in a workshop, runways at an airport, crews at a construction site, or processing units in a computing environment. Activities may be operations in a workshop, take-offs and landings at an airport, stages in a construction project, or computer programs that have to be executed. Each activity may have a priority level, an earliest possible starting time and a due date. Objectives can take many different forms, such as minimizing the time to complete all activities, minimizing the number of activities that are completed after the committed due dates, and so on.

The following nine examples illustrate the role of planning and scheduling in real life situations. Each example describes a particular type of planning and scheduling problem. The first example shows the role of planning and scheduling in the management of large construction and installation projects that consist of many stages.

Example 1.1.1 (A System Installation Project). Consider the procurement, installation, and testing of a large computer system. The project involves a number of distinct tasks, including evaluation and selection of hardware, software development, recruitment and training of personnel, system testing, system debugging, and so on. A precedence relationship structure exists among these tasks: some can be done in parallel (concurrently), whereas others can only start when certain predecessors have been completed. The goal is to complete the entire project in minimum time.

Planning and scheduling not only provide a coherent process to manage the project, but also provide a good estimate for its completion time, reveal which tasks are critical and determine the actual duration of the entire project.

The second example is taken from a job shop manufacturing environment, where the importance of planning and scheduling is growing with the increasing diversification and differentiation of products. The number of different items that have to be produced is large and setup costs as well as shipping dates have to be taken into account.

Example 1.1.2 (A Semiconductor Manufacturing Facility). Semiconductors are manufactured in highly specialized facilities. This is the case with memory chips as well as with microprocessors. The production process in these facilities usually consists of four phases: wafer fabrication, wafer probe, assembly or packaging, and final testing.

Wafer fabrication is technologically the most complex phase. Layers of metal and wafer material are built up in patterns on wafers of silicon or gallium arsenide to produce the circuitry. Each layer requires a number of operations, which typically include: (i) cleaning, (ii) oxidation, deposition and metallization, (iii) lithography, (iv) etching, (v) ion implantation, (vi) photoresist stripping, and (vii) inspection and measurement. Because it consists of many layers, each wafer undergoes these operations several times. Thus, there is a significant amount of recirculation in the process. Wafers move through the facility in lots of 24. Some machines may require setups to prepare them for incoming jobs. The setup time often depends on the configurations of the lot just completed and the lot about to start.

The number of orders in the system is often in the hundreds and each has its own release date and committed shipping or due date. The scheduler's objective is to meet as many of the committed shipping dates as possible, while maximizing throughput. The latter goal is achieved by maximizing equipment utilization, especially of the bottleneck machines. Minimization of idle times and setup times is thus also required.

In many manufacturing environments, automated material handling systems dictate the flow of products through the system. Flexible assembly systems fall in this category. The scheduler's job in this kind of environment is to develop the best schedule while satisfying certain timing and sequencing conditions. The scheduler thus has less freedom in constructing the schedule. The next illustration describes a classical example of this type of environment.

1.1 Planning and Scheduling: Role and Impact

Example 1.1.3 (An Automobile Assembly Line). An automobile assembly line typically produces many different models, all belonging to a small number of families of cars. For example, the different models within a family may include a two-door coupe, a four-door sedan, and a stationwagon. There are also a number of different colors and option packages. Some cars have automatic transmissions, while others are manual; some cars have sunroofs, others have not.

In an assembly line there are typically several bottlenecks, where the throughput of a particular machine or process determines the overall production rate. The paint shop is often such a bottleneck; every time the color changes the paint guns have to be cleaned, which is a time consuming process.

One of the objectives is to maximize the throughput by sequencing the cars in such a way that the workload at each station is balanced over time.

The previous examples illustrate some of the detailed and short term aspects of planning and scheduling processes. However, planning and scheduling often deal with medium term and long term issues as well.

Example 1.1.4 (Production Planning in a Paper Mill). The input to a paper mill is wood fiber and pulp; the output is finished rolls of paper. At the heart of the paper mill are its paper machines, which are very large and represent a significant capital investment (between 50 and 100 million dollars each). Each machine produces various types of paper which are characterized by their basis weights, grades and colors.

Master production plans for these machines are typically drawn up on an annual basis. The projected schedules are cyclic with cycle times of two weeks or longer. A particular type of paper may be produced either every cycle, every other cycle, or even less often, depending upon the demand.

Every time the machine switches over from one grade of paper to another there is a setup cost involved. During the changeover the machine keeps on producing paper. Since the paper produced during a changeover does not meet any of the required specifications, it is either sold at a steep discount or considered waste and fed back into the production system.

The production plan tries to maximize production, while minimizing inventory costs. Maximizing production implies minimizing changeover times. This means longer production runs, which in turn result in higher inventory costs. The overall production plan is a trade-off between setup and inventory costs.

Each one of the facilities described in the last three examples may belong to a network of facilities in which raw material or (semi)finished goods move from one facility to another; in a facility the product is either being stored or more value is added. In many industries the planning and scheduling of the supply chains is of crucial importance.

Example 1.1.5 (Planning and Scheduling in a Supply Chain). Consider the paper mill of the previous example. A mill is typically an integral part of a complex network of production facilities that includes timberland (where trees are grown using advanced forest management technology), paper mills where the rolls of paper are produced, converting facilities where the rolls are transformed into paper products (e.g., bags, cartons, or cutsize paper), distribution centers (where inventory is kept) and end-consumers or retailers. Several different modes of transportation are used between the various stages of the supply chain, e.g., trucks, trains, and barges. Each mode has its own characteristics, such as cost, speed, reliability, and so on. Clearly, in each stage of the supply chain more value is added to the product and the further down the supply chain, the more product differentiation. Coordinating the entire network is a daunting process. The overall goal is to minimize the total costs including production costs, transportation costs and inventory holding costs.

In many manufacturing environments customers have close relationships with the manufacturer. The factory establishes its production schedule in collaboration with its customers and may allow them to reserve machines for specific periods of time. Conceptually, the scheduling problem of the manufacturer is similar to the scheduling problems in car rental agencies and hotels, where the cars and rooms correspond to the machines and the objective is to maximize the utilization of these resources.

Example 1.1.6 (A Reservation System). A car rental agency maintains a fleet of various types of cars. It may have full size, midsize, compact, and subcompact cars. Some customers may be flexible with regard to the type of car they are willing to rent, while others may be very specific. A customer typically calls in to make a reservation for certain days and the agency has to decide whether or not to provide him with a car. At times it may be advantageous to deny a customer a reservation that is for a very short period when there is a chance to rent the car out to another customer for a longer period. The agency's objective is to maximize the number of days its cars are rented out.

Scheduling and timetabling play also an important role in sports and entertainment. Sport tournaments have to be scheduled very carefully. The schedule has to be such that all the participating teams are treated fairly and that the preferences of the fans are taken into account. Timetabling plays also an important role in entertainment. For example, television programs have to be scheduled in such a way that the ratings (and therefore the profits) are maximized. After the programs have been assigned to their slots, the commercials have to be scheduled as well.

Example 1.1.7 (Scheduling a Soccer Tournament). Consider a tournament of a soccer league. The games have to be scheduled over a fixed number of rounds. An important consideration in the creation of a schedule is that, ideally, each team should have a schedule that alternates between games at home and games away. However, it often cannot be avoided that a team has to play two consecutive games at home or two consecutive games away. There are many other concerns as well: for example, if a city has two teams participating in the same league, then it is desirable to have in each round one team at home and the other team away. If two teams in a league are very strong, then it would be nice if none of the other teams would have to face these two teams in consecutive rounds.

Planning and scheduling play a very important role in transportation. There are various modes of transportation and different industries focus on different ways of moving either cargo or passengers. The objectives include minimizing total cost as well as maximizing convenience or, equivalently, minimizing penalty costs.

Example 1.1.8 (Routing and Scheduling of Airplanes). The marketing department of an airline usually has a considerable amount of information with regard to customer demand for any given flight (a flight is characterized by its origin and destination and by its scheduled departure time). Based on the demand information, the airline can estimate the profit of assigning a particular type of aircraft to a flight leg under consideration. The airline scheduling problem basically focuses on how to combine the different flight legs into so-called round-trips that can be assigned to a specific plane. A round trip may be subject to many constraints: the turn-around time at an airport must be longer than a given minimum time; a crew cannot be on duty for a duration that is longer than what the Federal Aviation Administration (FAA) allows, and so on.

In many manufacturing and service industries planning and scheduling often have to deal with resources other than machines; the most important resource, besides machines, is usually personnel.

Example 1.1.9 (Scheduling Nurses in a Hospital). Every hospital has staffing requirements that change from day to day. For instance, the number of nurses required on weekdays is usually more than the number required on weekends, while the staffing required during the night shift may be less than that required during the day shift. State and federal regulations and union rules may provide additional scheduling constraints. There are thus different types of shift patterns, all with different costs.

The goal is to develop shift assignments so that all daily requirements are met and the constraints are satisfied at a minimal cost.

From the examples above it is clear that planning and scheduling is important in manufacturing as well as in services. Certain types of scheduling problems are more likely to occur in manufacturing settings (e.g., assembly line scheduling), while others are more likely to occur in service settings (e.g., reservation systems). And certain types of scheduling problems occur in both manufacturing and services; for example, project scheduling is important in the shipbuilding industry as well as in management consulting.

In many environments it may not be immediately clear what impact planning and scheduling has on any given objective. In practice, the choice of schedule usually has a measurable impact on system performance. Indeed, an improvement in a schedule usually can cut direct and indirect costs significantly, especially in a complex production setting.

Unfortunately, planning and scheduling may be difficult to implement. The underlying mathematical difficulties are similar to those encountered in other branches of combinatorial optimization, while the implementation difficulties are often caused by inaccuracies in model representations or by problems encountered in the retrieval of data and the management of information. Resolving these difficulties takes skill and experience, but is often financially and operationally well worth the effort.

1.2 Planning and Scheduling Functions in an Enterprise

Planning and scheduling in either a manufacturing or a service organization must interact with many other functions. These interactions are typically system-dependent and may differ substantially from one setting to another; they often take place within a computer network. There are, of course, also many situations where the exchange of information between planning and scheduling and other decision making functions occurs in meetings or through memos.

Planning and Scheduling in Manufacturing. We first describe a generic manufacturing environment and the role of its planning and scheduling function. Orders that are released in a manufacturing setting have to be translated into jobs with associated due dates. These jobs often have to be processed on the machines in a workcenter in a given order or sequence. The processing of jobs may sometimes be delayed if certain machines are busy. Preemptions may occur when high priority jobs are released which have to be processed at once. Unexpected events on the shopfloor, such as machine breakdowns or longer-than-expected processing times, also have to be taken into account, since they may have a major impact on the schedules. Developing, in such an environment, a detailed schedule of the tasks to be performed helps maintain efficiency and control of operations.

The shopfloor is not the only part of the organization that impacts the scheduling process. The scheduling process also interacts with the production planning process, which handles medium- to long-term planning for the entire organization. This process intends to optimize the firm's overall product mix and long-term resource allocation based on inventory levels, demand forecasts and resource requirements. Decisions made at this higher planning level may impact the more detailed scheduling process directly. Figure 1.1 depicts a diagram of the information flow in a manufacturing system.

1.2 Planning and Scheduling Functions in an Enterprise

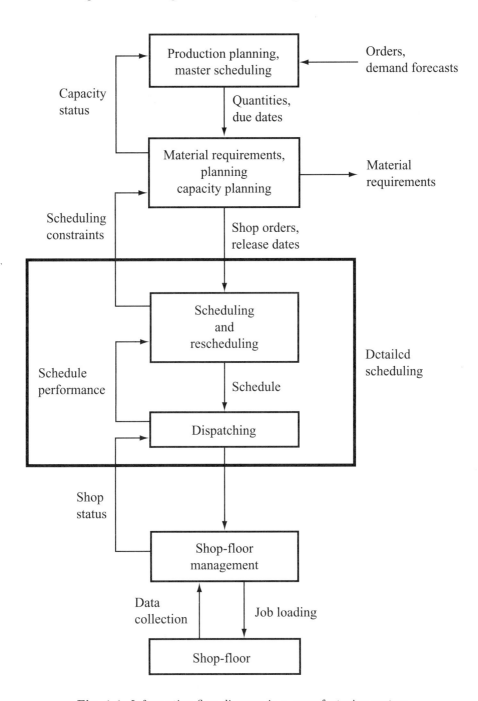

Fig. 1.1. Information flow diagram in a manufacturing system

In manufacturing, planning and scheduling has to interact with other decision making functions in the plant. One popular system that is widely used is the Material Requirements Planning (MRP) system. After a schedule has been set up it is necessary that all the raw materials and resources are available at specified times. The ready dates of the jobs have to be determined by the production planning and scheduling system in conjunction with the MRP system.

MRP systems are normally fairly elaborate. Each job has a Bill Of Materials (BOM) itemizing the parts required for production. The MRP system keeps track of the inventory of each part. Furthermore, it determines the timing of the purchases of each one of the materials. In doing so, it uses techniques such as lot sizing and lot scheduling that are similar to those used in planning and scheduling systems. There are many commercial MRP software packages available. As a result, many manufacturing facilities rely on MRP systems. In the cases where the facility does not have a planning or scheduling system, the MRP system may be used for production planning purposes. However, in a complex setting it is not easy for an MRP system to do the detailed planning and scheduling satisfactorily.

Modern factories often employ elaborate manufacturing information systems involving a computer network and various databases. Local area networks of personal computers, workstations and data entry terminals are connected to a central server, and may be used either to retrieve data from the various databases or to enter new data. Planning and scheduling is usually done on one of these personal computers or workstations. Terminals at key locations may often be connected to the scheduling computer in order to give departments access to current scheduling information. These departments, in turn, may provide the scheduling system with relevant information, such as changes in job status, machine status, or inventory levels.

Companies nowadays rely often on elaborate Enterprise Resource Planning (ERP) systems, that control and coordinate the information in all its divisions and sometimes also at its suppliers and customers. Decision support systems of various different types may be linked to such an ERP system, enabling the company to do long range planning, medium term planning as well as short term scheduling.

Planning and Scheduling in Services. Describing a generic service organization and its planning and scheduling systems is not as easy as describing a generic manufacturing system. The planning and scheduling functions in a service organization may often face many different problems. They may have to deal with the reservation of resources (e.g., trucks, time slots, meeting rooms or other resources), the allocation, assignment, and scheduling of equipment (e.g., planes) or the allocation and scheduling of the workforce (e.g., the assignment of shifts in call centers). The algorithms tend to be completely different from those used in manufacturing settings. Planning and scheduling in a service environment also have to interact with other decision making functions, usually within elaborate information systems, much in the same way as

1.3 Outline of the Book

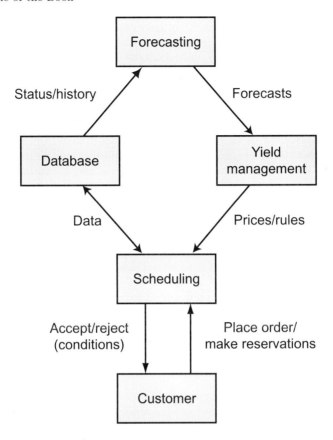

Fig. 1.2. Information flow diagram in a service system

the scheduling function in a manufacturing setting. These information systems typically rely on extensive databases that contain all the relevant information regarding the availability of resources as well as the characteristics of current and potential customers. A planning and scheduling system may interact with a forecasting module; it may also interact with a yield management module (which is a type of module that is not very common in manufacturing settings). On the other hand, in a service environment there is usually no MRP system. Figure 1.2 depicts the information flow in a service organization such as a car rental agency.

1.3 Outline of the Book

This book focuses on planning and scheduling applications. Although thousands of planning and scheduling models and problems have been studied in the literature, only a limited number are considered in this book. The selection is based on the insight the models provide, the methodologies needed for their analyses and their importance with regard to real-world applications.

The book consists of four parts. Part I describes the general characteristics of scheduling models in manufacturing and in services. Part II considers various classes of planning and scheduling models in manufacturing, and Part III discusses several classes of planning and scheduling models in services. Part IV deals with system design, development and implementation issues.

The remainder of Part I consists of Chapters 2 and 3. Chapter 2 discusses the basic characteristics of the manufacturing models that are considered in Part II of this book and Chapter 3 describes the characteristics of the service models that are considered in Part III. These characteristics include machine environments and service settings, processing restrictions and constraints, as well as performance measures and objective functions.

Part II focuses on planning and scheduling in nanufacturing and consists of Chapters 4 to 8. Each one of these chapters focuses on a different class of planning and scheduling models with applications in manufacturing; each chapter corresponds to one of the examples discussed in Section 1.1. At first, it may appear that the various chapters in this part are somewhat unrelated to one another and are selected arbitrarily. However, there is a rationale behind the selection of topics and the chapters are actually closely related to one another.

Chapter 4 focuses on project scheduling. A project scheduling problem usually concerns a single project that consists of a number of separate jobs which are related to one another through precedence constraints. Since only a single project is considered, the basic format of this type of scheduling problem is inherently easy, and therefore the logical one to start out with. Moreover, an immediate generalization of this problem, i.e., the project scheduling problem with workforce constraints, has from a mathematical point of view several important special cases. For example, the job shop scheduling problems discussed in Chapter 5 and the timetabling problems considered in Chapter 9 are such special cases.

Chapter 5 covers the classical single machine, parallel machine, and job shop scheduling models. In a single machine as well as in a parallel machine environment a job consists of a single operation; in a parallel machine environment this operation may be done on any one of the machines available. In a job shop, each job has to undergo multiple operations on the various machines and each job has its own set of processing times and routing characteristics. Several objectives are of interest; the most important one is the makespan, which is the time required to finish all jobs.

Chapter 6 focuses on flexible assembly systems. These systems have some similarities with job shops; however, there are also some differences. In a flexible assembly system there are several different job types; but, in contrast to a job shop, a certain number has to be produced of each type. The routing

1.3 Outline of the Book

constraints in flexible assembly systems are also somewhat different from those in job shops. Because of the presence of a material handling or conveyor system, the starting time of one operation may be a very specific function of the completion time of another operation. (In job shops these dependencies are considerably weaker.)

The problems considered in Chapter 7, lot sizing and scheduling, are somewhat similar to those in Chapter 6. There are again various different job types, and of each type there are a number of identical jobs. However, the variety of job types in this chapter is usually less than the variety of job types in a flexible assembly system. The number to be produced of any particular job type tends to be larger than in a flexible assembly system. This number is called the lot size and its determination is an integral part of the scheduling problem. So, going from Chapter 5 to Chapter 7, the variety in the order portfolio decreases, while the batch or lot sizes increase.

Chapter 8 focuses on planning and scheduling in supply chains. This chapter assumes that the manufacturing environment consists of a network of raw material and parts providers, production facilities, distribution centers, customers, and so on. The product flows from one stage to the next and at each stage more value is added to the product. Each facility can be optimized locally using the procedures that are described in Chapters 5, 6 and 7. However, performing a global optimization that encompasses the entire network requires a special framework. Such a framework has to take now transportation issues also into account; transportation costs appear in the objective function and the quantities to be transported between facilities may be subject to restrictions and constraints.

Part III focuses on planning and scheduling in services and consists of Chapters 9 to 12. Each chapter describes a class of planning and scheduling models in a given service setting, and each chapter corresponds to one of the examples discussed in Section 1.1. The topics covered in Part III are, clearly, different from those in Part II. However, Chapters 9 to 12 also deal with issues and features that may be important in manufacturing settings as well. However, adding these features to the models described in Chapters 4 to 8 would lead to problems that are extremely hard to analyze. That is why these features are considered separately in Part III in relatively simple settings.

Chapter 9 considers reservation systems and timetabling models. These classes of models are basically equivalent to parallel machine models. In reservation models, jobs (i.e., reservations) tend to have release dates and due dates that are tight; the decision-maker has to decide which jobs to process and which jobs not to process. Reservation models are important in hospitality industries such as hotels and carrental agencies. In timetabling models the jobs are subject to constraints with regard to the availability of operators or tools. Timetabling models are also important in the scheduling of meetings, classes, and exams.

Chapter 10 describes scheduling and timetabling in sports and entertainment. The scheduling of tournaments (e.g., basketball, baseball, soccer, and so

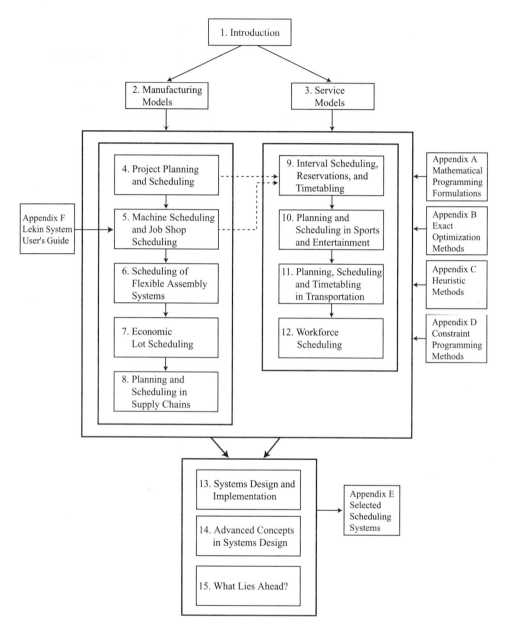

Fig. 1.3. Overview of the Book

1.3 Outline of the Book

on) tends to be very difficult because of the many preferences and constraints concerning the schedules. For example, it is desirable that the sequence of games assigned to any given team alternates between games at home and games away. This chapter discusses an optimization approach, a constraint programming approach, as well as a local search approach for tournament scheduling problems. It also describes how to schedule programs in broadcast television so as to maximize the ratings.

Chapter 11 discusses planning, scheduling and timetabling in transportation settings. The transportation settings include the scheduling of oil tankers, the routing and scheduling of airplanes, and the timetabling of trains. Tankers have to move cargoes from one point to another within given time windows and aircraft have to cover flight legs between certain cities also within given time windows. One important objective is to assign the different trips or flight legs to the given tankers or airplanes in such a way that the total cost is minimized. There are important similarities as well as differences between the scheduling of oil tankers, the routing and scheduling of aircraft, and the timetabling of trains.

Chapter 12 focuses on workforce scheduling. There are various different classes of workforce scheduling models. One class of workforce scheduling models includes the shift scheduling of nurses in hospitals or operators in call centers. A different class of workforce scheduling models includes crew scheduling in airlines or trucking companies.

Part IV concerns systems development and implementation issues and consists of the three remaining chapters. Chapter 13 deals with the design, development, and implementation of planning and scheduling systems. It covers basic issues with regard to system architectures and describes the various types of databases, planning and scheduling engines, and user interfaces. The databases may be conventional relational databases or more sophisticated knowledge-bases; the user interfaces may include Gantt charts, capacity buckets, or throughput diagrams.

Chapter 14 describes more advanced concepts in systems design. The topics discussed include robustness issues, learning mechanisms, systems reconfigurability, as well as Internet related features. These topics have been inspired by the design, development and implementation of planning and scheduling systems during the last decade of the twentieth century; in recent years these concepts have received a fair amount of attention in the academic literature.

Chapter 15, the last chapter of this book, discusses future directions in the research and development of planning and scheduling applications. Issues in manufacturing applications are discussed (which often concern topics in supply chain management), issues in service applications are covered, as well as issues in the design and development of complex systems that consist of many different modules. Important issues concern the connectivity and interface design between the different modules of an integrated system.

There are six appendixes. Appendix A covers mathematical programming formulations. Appendix B focuses on exact optimization methods, including dynamic programming and branch-and-bound. Appendix C describes the most popular heuristic techniques, e.g., dispatching rules, beam search, as well as local search. Appendix D contains a primer on contraint programming. Appendix E presents an overview of selected scheduling systems and Appendix F provides a user's guide to the LEKIN job shop scheduling system.

Figure 1.3 shows the precedence relationships between the various chapters. Some of these precedence relationships are stronger than others. The stronger relationships are depicted by solid arrows and the weaker ones by dotted arrows.

This book is intended for a senior level or masters level course on planning and scheduling in either an engineering or a business school. The selection of chapters to be covered depends, of course, on the instructor. It appears that in most cases Chapters 1, 2 and 3 have to be covered. However, an instructor may not want to go through all of the chapters in Parts II or III; he or she may select, for example, most of Chapters 4 and 5 and a smattering of Chapters 6 to 12. An instructor may decide not to cover Part IV, but still assign those chapters as background reading.

Prerequisite knowledge for this book is an elementary course in Operations Research on the level of Hillier and Lieberman's *Introduction to Operations Research*.

Comments and References

Over the last three decades many scheduling books have appeared, ranging from the elementary to the advanced. Most of these focus on just one or two of the nine model categories covered in Parts II and III of this book.

Several books are completely dedicated to project scheduling; see, for example, Moder and Philips (1970), Kerzner (1994), Kolisch (1995), Neumann, Schwindt, and Zimmermann (2001), and Demeulemeester and Herroelen (2002). There are also various books that have a couple of chapters dedicated to this topic (see, for example, Morton and Pentico (1993)). A survey paper by Brucker, Drexl, Möhring, Neumann, and Pesch (1999) presents a detailed overview of the models as well as the techniques used in project scheduling.

Many books emphasize job shop scheduling (often also referred to as machine scheduling). One of the better known textbooks on this topic is the one by Conway, Maxwell and Miller (1967) (which, even though slightly out of date, is still very interesting). A more recent text by Baker (1974) gives an excellent overview of the many aspects of machine scheduling. An introductory textbook by French (1982) covers most of the techniques that are used in job shop scheduling. The more advanced book by Blazewicz, Cellary, Slowinski and Weglarz (1986) focuses mainly on job shop scheduling with resource constraints and multiple objectives. The book by Blazewicz, Ecker, Schmidt and Weglarz (1993) is also somewhat advanced and deals primarily with computational and complexity aspects of machine scheduling models and their applications to manufacturing. The more applied text by Morton and

Comments and References

Pentico (1993) presents a detailed analysis of a large number of scheduling heuristics that are useful for practitioners. The survey paper by Lawler, Lenstra, Rinnooy Kan and Shmoys (1993) gives a detailed overview of machine scheduling. Recently, a number of books have appeared that focus primarily on machine scheduling and job shop scheduling, see Dauzère-Pérès and Lasserre (1994), Baker (1995), Parker (1995), Pinedo (1995), Ovacik and Uzsoy (1997), Sule (1996), Bagchi (1999), Pinedo (2002), and Brucker (2004). The books edited by Chrétienne, Coffman, Lenstra and Liu (1995) and Lee and Lei (1997) contain a wide variety of papers on machine scheduling. Baptiste, LePape and Nuijten (2001) cover the application of constraint programming techniques to job shop scheduling. The volume edited by Nareyek (2001) contains papers on local search applied to job shop scheduling.

Several books focus on flexible assembly systems, flexible manufacturing systems, or intelligent systems in general; see, for example, Kusiak (1990, 1992). These books usually include one or more chapters that deal with scheduling aspects of these systems. Scholl (1998) focuses in his book on the balancing and sequencing of assembly lines.

Lot sizing and scheduling is closely related to inventory theory. Haase (1994) gives an excellent overview of this field. Brüggemann (1995) and Kimms (1997) provide more advanced treatments. Drexl and Kimms (1997) provide an exhaustive survey with extensions. Lot scheduling has also been covered in a number of survey papers on production scheduling; see, for example, the reviews by Graves (1981) and Rodammer and White (1988).

Planning and scheduling in supply chains have received a fair amount of attention over the last decade. A number of books cover supply chains in general. Some of these emphasize the planning and scheduling aspects; see, for example, Shapiro (2001) and Miller (2002). Stadtler and Kilger (2002) edited a book that has various chapters on planning and scheduling in supply chains.

Interval scheduling, reservation systems and timetabling are fairly new topics that are closely related to one another. Interval scheduling has been considered in a couple of chapters in Dempster, Lenstra and Rinnooy Kan (1982) and timetabling has been discussed in the textbook by Parker (1995). Recently, a series of proceedings of conferences on timetabling have appeared, see Burke and Ross (1996), Burke and Carter (1998), Burke and Erben (2001), and Burke and De Causmaecker (2003).

A fair amount of research has been done on planning and scheduling in sports and entertainment. As far as this author is aware of, no book has appeared that is completely dedicated to this area. However, the book edited by Butenko, Gil-Lafuente and Pardalos (2004) has several chapters that focus on planning and scheduling in sports.

No framework has yet been established for planning and scheduling models in transportation. A series of conferences on computer-aided scheduling of public transport has resulted in a number of interesting proceedings, see Wren and Daduna (1988), Desrochers and Rousseau (1992), Daduna, Branco, and Pinto Paixao (1995), Wilson (1999), and Voss and Daduna (2001). A volume edited by Yu (1998) considers operations research applications in the airline industry; this volume contains several papers on planning and scheduling in the airline industry.

Personnel scheduling has not received the same amount of attention as project scheduling or job shop scheduling. Tien and Kamiyama (1982) present an overview on manpower scheduling algorithms. The text by Nanda and Browne (1992) is completely dedicated to personnel scheduling. The Handbook of Industrial Engineering has a chapter on personnel scheduling by Burgess and Busby (1992). Parker (1995) dedicates a section to staffing problems. Burke, De Causmaecker, Vanden Berghe and Van Landeghem (2004) present an overview of the state of the art in nurse rostering. Several of the books on planning and scheduling in transportation have chapters on crew scheduling.

Recently, a Handbook of Scheduling appeared edited by Leung (2004). This handbook contains numerous chapters covering a wide spectrum that includes job shop scheduling, timetabling, tournament scheduling, and workforce scheduling.

A fair amount has been written on the design, development and implementation of scheduling systems. Most of this research and development has been documented in survey papers. Atabakhsh (1991) presents a survey of constraint based scheduling systems using artificial intelligence techniques and Noronha and Sarma (1991) give an overview of knowledge-based approaches for scheduling problems. Smith (1992) focuses in his survey on the development and implementation of scheduling systems. Two collections of papers, edited by Zweben and Fox (1994) and by Brown and Scherer (1995), describe a number of scheduling systems and their actual implementation.

Chapter 2

Manufacturing Models

    2.1 Introduction 19
    2.2 Jobs, Machines, and Facilities 21
    2.3 Processing Characteristics and Constraints 24
    2.4 Performance Measures and Objectives 28
    2.5 Discussion 32

2.1 Introduction

Manufacturing systems can be characterized by a variety of factors: the number of resources or machines, their characteristics and configuration, the level of automation, the type of material handling system, and so on. The differences in all these characteristics give rise to a large number of different planning and scheduling models. In a manufacturing model, a resource is usually referred to as a "machine"; a task that has to be done on a machine is typically referred to as a "job". In a production process, a job may be a single operation or a collection of operations that have to be done on various different machines. Before describing the main characteristics of the planning and scheduling problems considered in Part II of this book, we give a brief overview of five classes of manufacturing models.

    The first class of models are the project planning and scheduling models. Project planning and scheduling is important whenever a large project, that consists of many stages, has to be carried out. A project, such as the construction of an aircraft carrier or a skyscraper, typically consists of a number of activities or jobs that may be subject to precedence constraints. A job that is subject to precedence constraints cannot be started until certain other jobs have been completed. In project scheduling, it is often assumed that there are an unlimited number of machines or resources, so that a job can start as soon as all its predecessors have been completed. The objective is to minimize the completion time of the last job, commonly referred to as the makespan. It is also important to find the set of jobs that determines the makespan, as these jobs are critical and cannot be delayed without delaying the completion of the entire project. Project scheduling models are also important in the planning and scheduling of services. Consider, for example, the planning and scheduling of a large consulting project.

The second class of models include single machine, parallel machine and job shop models. In a single machine or parallel machine environment, a job consists of one operation that can be done on any one of the machines available. In a full-fledged job shop, a job typically consists of a number of operations that have to be performed on different machines. Each job has its own route that it has to follow through the system. The operations of the jobs in a job shop have to be scheduled to minimize one or more objectives, such as the makespan or the number of late jobs. Job shops are prevalent in industries that make customized industrial hardware. However, they also appear in service industries (e.g., hospitals). A special case of a job shop is a setting where each one of the jobs has to follow the same route through the system (i.e., each job has to be processed first on machine 1, then on machine 2, and so on); such a setting is usually called a flow shop.

The third class of models focuses on production systems with automated material handling. In these settings a job also consists of a number of operations. A material handling or conveyor system controls the movement of the jobs as well as the timing of their processing on the various machines. Examples of such environments are flexible manufacturing systems, flexible assembly systems, and paced assembly lines. The objective is typically to maximize throughput. Such settings are prevalent in the automotive industry and in the consumer electronics industry.

The fourth class of models are known as lot scheduling models. These models are used for medium term and long term production planning. In contrast to the first three classes, the production and demand processes are now continuous. In this class, there are a variety of different products. When a machine switches from one product to another, a changeover cost is incurred. The goal is usually to minimize total inventory and changeover costs. These models are important in the process industries, such as oil refineries and paper mills.

The fifth class of models consists of supply chain planning and scheduling models. These models tend to be hierarchical and are often based on an integration of the lot scheduling models (the fourth class of models) and the job shop scheduling models (the second class of models). The objective functions in supply chain planning and scheduling take into account inventory holding costs at the various stages in the chain as well as costs of transportation between the stages. There are restrictions and constraints on the production quantities as well as on the quantities that have to be transported from one stage to another.

The manufacturing models described above can be classified as either discrete or continuous. Some of the models are discrete, whereas others are continuous. The project scheduling models, job shop models, and flexible assembly systems are discrete models. The lot scheduling models are continuous. The models for supply chain planning and scheduling can be continuous or discrete. A discrete model can usually be formulated as an integer program or disjunctive program, whereas a continuous model can be formulated as a linear or nonlinear program (see Appendix A).

There are similarities as well as differences between the manufacturing models in the second part of this book and the service models in the third part. One class of models, namely the project scheduling models, is important for both manufacturing and services.

2.2 Jobs, Machines, and Facilities

The following terminology and notation is used throughout Part II of the book. The number of jobs is denoted by n and the number of machines by m. The subscripts j and k refer to jobs j and k. The subscripts h and i refer to machines h and i. The following data pertain to job j.

Processing time (p_{ij}) The processing time p_{ij} represents the time job j has to spend on machine i. The subscript i is omitted if the processing time of job j does not depend on the machine or if it only needs processing on one machine. If there are a number of identical jobs that all need a processing time p_j on one machine, then we refer to this set of jobs as items of type j. The machine's production rate of type j items is denoted by $Q_j = 1/p_j$ (number of items per unit time).

Release date (r_j). The release date r_j of job j is also known as the ready date. It is the time the job arrives at the system, i.e., the earliest time at which job j can start its processing.

Due date (d_j). The due date d_j of job j represents the committed shipping or completion date (the date the job is promised to the customer). Completion of a job after its due date *is* allowed, but a penalty is then incurred. When the due date absolutely must be met, it is referred to as a *deadline*.

Weight (w_j). The weight w_j of job j is a priority factor, reflecting the importance of job j relative to other jobs in the system. It may represent the cost of keeping job j in the system for one time unit. The weight can be a holding or inventory cost, or it can be the amount of value already added to the job.

The four pieces of data listed above are *static* data, since they do not depend on the schedule. Conversely, data that are not fixed in advance and that do depend on the schedule are referred to as *dynamic* data. The most important dynamic data are the following:

Starting time S_{ij}. The starting time S_{ij} is the time when job j starts its processing on machine i. If the subscript i is omitted, then S_j refers to the time when job j starts with its first processing in the system.

Completion time C_{ij}. The completion time C_{ij} is the time when job j is completed on machine i. If the subscript i is omitted, then C_j refers to the time when job j leaves the system.

An important characteristic of a scheduling model is its machine configuration. There are several important machine configurations. The remaining part of this section covers the most basic ones.

Single Machine Models. Many production systems give rise to single machine models. For instance, if there is a single bottleneck in a multi-machine environment, then the job sequence at the bottleneck typically determines the performance of the entire system. In such a case, it makes sense to schedule the bottleneck first and all other operations (upstream and downstream) afterward. This implies that the original problem first has to be reduced to a single machine scheduling problem. Single machine models are also important in decomposition methods, when scheduling problems in more complicated machine environments are broken down into a number of smaller single machine scheduling problems.

Single machine models have been thoroughly analyzed under all kinds of conditions and with many different objective functions. The result is a collection of rules that, while easy to identify and apply, often provide optimal solutions in the single machine environment. For example, the *Earliest Due Date first* (*EDD*) rule, which orders the jobs in increasing order of their due dates, has been shown to minimize the maximum lateness among all jobs. The *Shortest Processing Time first* (*SPT*) rule has been shown to minimize the average number of jobs waiting for processing.

Parallel Machine Models. A bank of machines in parallel is a generalization of the single machine model. Many production environments consist of several stages or workcenters, each with a number of machines in parallel. The machines at a workcenter may be identical, so that a job can be processed on any one of the machines available. Parallel machine models are important for the same reason that single machine models are important: If one particular workcenter is a bottleneck, then the schedule at that workcenter will determine the performance of the entire system. That bottleneck can then be modelled as a bank of parallel machines and analyzed separately.

At times, the machines in parallel may not be *exactly* identical. Some machines may be older and operate at a lower speed; or, some machines may be better maintained and capable of doing higher quality work. If that is the case, then some jobs may be processed on any one of the m machines, while other jobs may be processed only on specific subsets of the m machines. When the "machines" are people, then the processing time of an operation may depend on the job as well as on the person or operator. One operator

2.2 Jobs, Machines, and Facilities

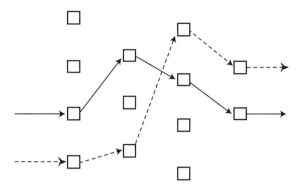

Fig. 2.1. Flexible flow shop

may excel in one type of job while another operator may be more specialized in another type.

Flow Shop Models. In many manufacturing and assembly settings jobs have to undergo multiple operations on a number of different machines. If the routes of all jobs are identical, i.e., all jobs visit the same machines in the same sequence, the environment is referred to as a flow shop. The machines are set up in series and whenever a job completes its processing on one machine it joins the queue at the next. The job sequence may vary from machine to machine, since jobs may be resequenced in between machines. However, if there is a material handling system that transports the jobs from one machine to the next, then the same job sequence is maintained throughout the system.

A generalization of the flow shop is the so-called *flexible flow shop*, which consists of a number of stages in series with at each stage a number of machines in parallel. At each stage a job may be processed on any one of the machines in parallel (see Figure 2.1).

In some (flexible) flow shops, a job may bypass a machine (or stage) if it does not require any processing there, and it may go ahead of the jobs that are being processed there or that are waiting for processing there. Other (flexible) flow shops, however, may not allow bypass.

Job Shop Models. In multi-operation shops, jobs often have different routes. Such an environment is referred to as a job shop, which is a generalization of a flow shop (a flow shop is a job shop in which each and every job has the same route).

The simplest job shop models assume that a job may be processed on a particular machine at most once on its route through the system (see Figure 2.2). In others a job may visit a given machine several times on its route through the system. These shops are said to be subject to recirculation, which increases the complexity of the model considerably.

A generalization of the job shop is the flexible job shop with workcenters that have multiple machines in parallel. From a combinatorial point of view

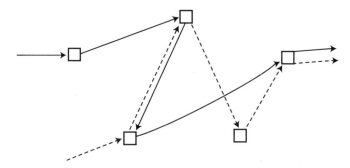

Fig. 2.2. Job shop

the flexible job shop with recirculation is one of the most complex machine environments. It is a very common setting in the semiconductor industry. The wafer fab described in Example 1.1.2 is a classic example of a flexible job shop; the routes of the jobs are order-specific and require recirculation.

Supply Chain Models. More general models assume a production environment that consists of a network of interconnected facilities with each facility being a (flexible) flow shop or a (flexible) job shop. In supply chain management the planning and scheduling of such networks is very important. This planning and scheduling may focus on the actual production in the various facilities as well as on the transportation of the products within the network.

In the real world there are many machine environments that are more complicated. Nonetheless, those described above are so fundamental that their analyses provide insights that are useful for the analyses of more complicated environments.

2.3 Processing Characteristics and Constraints

Job processing has many distinctive characteristics and is often subject to constraints that are peculiar. This section describes some of the most common processing characteristics and constraints.

Precedence Constraints. In scheduling problems a job often can start only after a given set of other jobs has been completed. Such constraints are referred to as precedence constraints and can be described by a precedence constraints graph. A precedence constraints graph may have a specific structure. For example, it may take the form of a set of chains (see Figure 2.3.a), or a tree (see Figure 2.3.b). In the system installation project of Example 1.1.1 some of the activities or jobs are subject to precedence constraints.

Machine Eligibility Constraints. In a parallel machine environment, it may often be the case that job j cannot be assigned to just any one of

2.3 Processing Characteristics and Constraints

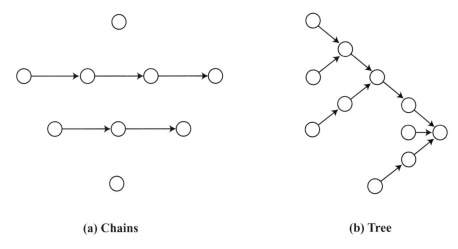

Fig. 2.3. Precedence constraints

the machines available; it can only go on a machine that belongs to a specific subset M_j. As described earlier, this may occur when the m machines in parallel are not all exactly identical.

In the paper mill example, there are a number of machines with different characteristics, e.g., dimensions, speeds, etc. From an operational point of view, it is then advantageous to have the long runs on the fast machines.

Workforce Constraints. A machine often requires one or more specific operators to process a job and a facility may have only a few people who can operate such a machine. Jobs that need processing on such a machine may have to wait until one of these operators becomes available. A workforce may consist of various pools; each pool consists of operators with a specific skill set. The number of operators in pool l is denoted by W_l. If the workforce is homogeneous, i.e., each person has the same qualifications, then the total number is denoted by W. In a parallel machine environment jobs have to be scheduled in such a way that the workforce constraints are satisfied. Because of such constraints, machine scheduling and workforce scheduling often have to be dealt with in an integrated manner.

Routing Constraints. Routing constraints specify the route a job must follow through a system, e.g., a flow shop or job shop. A given job may consist of a number of operations that have to be processed in a given sequence or order. The routing constraints specify the order in which a job must visit the various machines. Routing constraints are common in most manufacturing environments.

Consider again the wafer fab described in Example 1.1.2, in which each order or job must undergo a number of different operations at the various stages. An individual job that does not need processing at a particular stage may be allowed to bypass that stage and go on to the next. The information that specifies the stages a job must visit and the stages it can skip is embedded in the routing constraints.

Material Handling Constraints. Modern assembly systems often have material handling systems that convey the jobs from one workcenter to another. The level of automation of the material handling system depends on the level of automation of the workcenters. If the workcenters are highly automated (e.g., roboticized), then the processing times are deterministic and do not vary. The material handling system may then have to be automated as well. When manual tasks are performed at the workcenters, the pace of the material handling system may be adjustable since the pace may depend on the processing times of the jobs.

A material handling system enforces a strong dependency between the starting time of any given operation and the completion times of its predecessors. Moreover, the presence of a material handling system often limits the amount of buffer space, which in turn limits the amount of Work-In-Process.

The automobile assembly facility described in Example 1.1.3 is a paced assembly line with a material handling system that moves the cars. The cars usually belong to a limited number of different families. However, cars from the same family may be different because of the various option packages that are available. This may cause a certain variability in the processing times of the cars at any given workstation. The processing times at the workstations affect the configuration of the line, its pace, and the number of operators assigned to each station.

Sequence Dependent Setup Times and Costs. Machines often have to be reconfigured or cleaned between jobs. This is known as a changeover or setup. If the length of the setup depends on the job just completed and on the one about to be started, then the setup times are *sequence dependent*. If job j is followed by job k on machine i, then the setup time is denoted by s_{ijk}.

For example, paint operations often require changeovers. Every time a new color is used, the painting devices have to be cleaned. The cleanup time often depends on the color just used and on the color about to be used. In practice it is preferable to go from light to dark colors, because the cleanup process is then easier.

Besides taking valuable machine time, setups also involve costs because of labor and waste of raw material. For example, machines in the process and chemical industries are not stopped when going from one grade of material to another. Instead, a certain amount of the material produced at the start of a new run is typically not of acceptable quality and is discarded or recirculated. If job j is followed by job k on machine i then the setup cost is denoted by c_{ijk}.

Consider the paper machines in Example 1.1.4. A paper machine produces various types of paper, characterized by color, grade and basis weight. When

2.3 Processing Characteristics and Constraints   27

the machine switches from one combination of grade, color and basis weight to another, a certain amount of transition product goes to waste. The machine is not productive for a length of time that depends on the characteristics of the two types of paper.

Storage Space and Waiting Time Constraints. In many production systems, especially those producing bulky items, the amount of space available for Work-In-Process (WIP) storage is limited. This puts an upper bound on the number of jobs waiting for a machine. In flow shops this can cause *blocking*. Suppose the storage space (buffer) between two successive machines is limited. When the buffer is full, the upstream machine cannot release a job that has been completed into the buffer. Instead, that job has to remain on the machine on which it has completed its processing and thus prevents that machine from processing another job.

Make-to-Stock and Make-to-Order. A manufacturing facility may opt to keep items in stock for which there is a steady demand and no risk of obsolescence. This decision to Make-To-Stock affects the scheduling process, because items that have to be produced for inventory do not have tight due dates.

When the demand rates are fixed and constant, the demand rate for items of type j is denoted by D_j. In the case of such a deterministic demand process, the production lot size is determined by a trade-off between setup costs and inventory holding costs. Whenever the inventory drops to zero the facility replenishes its stock. If, in the case of a stochastic demand process, the inventory drops below a certain level, then the facility produces more in order to replenish the in-stock supply. The inventory level that triggers the production process depends on the uncertainty in the demand pattern, while the amount produced depends on the setup costs and inventory holding costs.

Make-to-Order jobs, conversely, have specific due dates, and the amount produced is determined by the customer. Many production facilities operate partly according to Make-to-Stock and partly according to Make-to-Order.

Consider the production plan of the paper machine in Example 1.1.4. The mill keeps a number of standard products in stock, characterized by their combination of grade, color, basis weight and dimensions. The mill has to control the inventory levels of all these combinations in such a way that inventory costs, setup costs, and probabilities of stockouts are minimized.

However, customers may occasionally order a size that is not kept in stock, even though the specified grade, color and basis weights are produced on a regular basis. This item is then made to order. The mill accepts the order through a production reservation system and inserts the order in its cyclic schedule in such a way that setup costs and setup times are minimized.

Preemptions. Sometimes, during the execution of a job, an event occurs that forces the scheduler to interrupt the processing of that job in order to make the machine available for another job. This happens, for instance, when a rush order with a high priority enters the system. The job taken off the machine is said to be *preempted*.

There are various forms of preemptions. According to one form, the processing already done on the preempted job is not lost, and when the job is later put back on the machine, it resumes its processing from where it left off. This form of preemption is referred to as *preemptive-resume*. According to another form of preemption, the processing already done on the preempted job is lost. This form is referred to as *preemptive-repeat*.

Transportation Constraints. If multiple manufacturing facilities are linked to one another in a network, then the planning and scheduling of the supply chain becomes important. The transportation time between any two facilities a and b is known and denoted by τ_{ab}^m (i.e., the time required to *move* the products from facility a to facility b). There may be constraints on the departure times of the trucks as well as on the quantities of goods to be shipped (i.e., there may be upper bounds on the quantities to be shipped because of the capacities of the vehicles).

Of course, the restrictions and constraints described in this section are just a sample of those that occur in practice. There are many other types of processing characteristics and constraints.

2.4 Performance Measures and Objectives

Many different types of objectives are important in manufacturing settings. In practice, the overall objective is often a composite of several basic objectives. The most important of these basic objectives are described below.

Throughput and Makespan Objectives. In many facilities maximizing the throughput is of the utmost importance and managers are often measured by how well they do so. The throughput of a facility, which is equivalent to its output rate, is frequently determined by the bottleneck machines, i.e., those machines in the facility that have the lowest capacity. Thus maximizing a facility's throughput rate is often equivalent to maximizing the throughput rate at these bottlenecks. This can be achieved in a number of ways. First, the scheduler must try to ensure that a bottleneck machine is never idle; this may require having at all times some jobs in the queue waiting for that machine. Second, if there are sequence dependent setup times s_{ijk} on the bottleneck machine, then the scheduler has to sequence the jobs in such a way that the sum of the setup times, or, equivalently, the average setup time, is minimized.

The makespan is important when there are a finite number of jobs. The makespan is denoted by $C_{\max}$ and is defined as the time when the last job leaves the system, i.e.,

$$C_{\max} = \max(C_1, \ldots, C_n),$$

where C_j is the completion time of job j. The makespan objective is closely related to the throughput objective. For example, minimizing the makespan

2.4 Performance Measures and Objectives

in a parallel machine environment with sequence dependent setup times forces the scheduler to balance the load over the various machines and minimize the sum of all the setup times. Heuristics that tend to minimize the makespan in a machine environment with a finite number of jobs also tend to maximize the throughput rate when there is a constant flow of jobs over time.

Consider the paper mill in Example 1.1.4 and suppose there are a number of paper machines in parallel. In this case, the scheduler has two main objectives. The first is to properly balance the production over the machines. The second is to minimize the sum of the sequence dependent setup times in order to maximize the throughput rate.

Due Date Related Objectives. There are several important objectives that are related to due dates. First, the scheduler is often concerned with minimizing the maximum *lateness*. Job lateness is defined as follows. Let d_j denote the due date of job j. The lateness of job j is then

$$L_j = C_j - d_j$$

(see Figure 2.4.a). The maximum lateness is defined as

$$L_{\max} = \max(L_1, \ldots, L_n).$$

Minimizing the maximum lateness is in a sense equivalent to minimizing the worst performance of the schedule.

Another important due date related objective is the number of tardy jobs. This objective does not focus on how tardy a job actually is, but only on whether or not it is tardy. The number of tardy jobs is a statistic that is easy to track in a database, so managers are often measured by the percentage of on-time shipments. However, minimizing the number of tardy jobs may result in schedules with some jobs being very tardy, which may be unacceptable in practice.

A due date related objective that addresses this concern is the total tardiness or, equivalently, the average tardiness. The tardiness of job j is defined as

$$T_j = \max(C_j - d_j, 0)$$

(see Figure 2.4.b) and the objective function is

$$\sum_{j=1}^{n} T_j.$$

Suppose different jobs carry different priority weights, where the weight of job j is w_j. The larger the weight of the job, the more important it is. Then a more general version of the objective function is the total weighted tardiness

$$\sum_{j=1}^{n} w_j T_j.$$

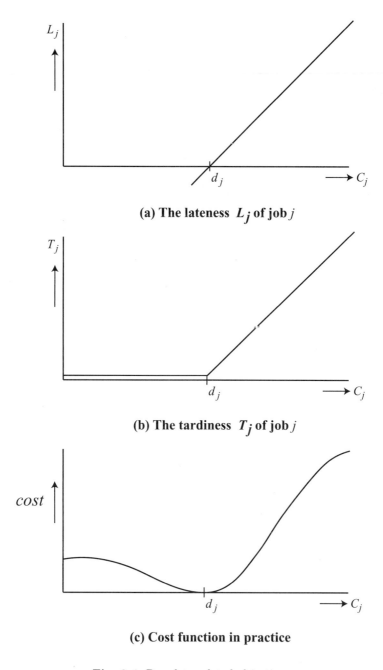

(a) The lateness L_j of job j

(b) The tardiness T_j of job j

(c) Cost function in practice

Fig. 2.4. Due date related objectives

2.4 Performance Measures and Objectives

Consider the semiconductor manufacturing facility described in Example 1.1.2. Each job (or, equivalently, each customer order) has a due date, or committed shipping date. The jobs may also have different priority weights. Minimizing the total weighted tardiness would be a fairly suitable objective in such an environment, even though a production manager may still be concerned with the *number* of late jobs.

None of the due date related objectives described above penalizes the *early* completion of a job. In practice, however, it is usually not advantageous to complete a job early, as this may lead to storage costs and additional handling costs. Cost functions, in practice, may look like the one depicted in Figure 2.4.c.

Setup Costs. It often pays to minimize the setup times when the throughput rate has to be maximized or the makespan has to be minimized. However, there are situations with insignificant setup times s_{ijk} but major setup costs c^s_{ijk} (if setup costs are the only costs in the model, then the superscript s may be omitted). Setup costs are not necessarily proportional with setup times. For example, a setup time on a machine that has ample capacity (lots of idle time) may not be significant, even though such a setup may cause a large amount of material waste.

Work-In-Process Inventory Costs. Another important objective is the minimization of the Work-In-Process (WIP) inventory. WIP ties up capital, and large amounts of it can clog up operations. WIP increases handling costs, and older WIP can easily be damaged or become obsolete. Products are often not inspected until after they have completed their path through the production process. If a defect that is detected during final inspection is caused by a production step at the very beginning of the process, then all the WIP may be affected. This is a very important consideration, especially in the semiconductor industries where the yield may be anywhere in between 60 and 95%. When a defect is detected, the cause has to be determined immediately. This then points to the workstation that is responsible and gives an indication of the proportion of defective WIP. Because of such occurrences it pays to have a low WIP. These kinds of considerations have led manufacturing companies in Japan to the Just-In-Time (JIT) concept.

A performance measure that can be used as a surrogate for WIP is the average *throughput time*. The throughput time is the time it takes a job to traverse the system. Minimizing the average throughput time, given a certain level of output, minimizes the WIP. Minimizing the average throughput time is also closely related to minimizing the sum of the completion times, i.e.,

$$\sum_{j=1}^{n} C_j.$$

This last objective is equivalent to minimizing the average number of jobs in the system. However, at times it may be of more interest to minimize the total value that is tied up as WIP. If that is the case, then it is more appropriate to minimize the sum of the weighted completion times, i.e.,

$$\sum_{j=1}^{n} w_j C_j.$$

Finished Goods Inventory Costs. An important objective is to minimize the inventory costs of the finished goods. The (holding) cost of keeping one item of type j in inventory for one time unit is denoted by h_j.

If the production at a facility is entirely Make-To-Order, then the finished goods inventory costs are equivalent to earliness costs. Conversely, if a facility's mode of production is Make-To-Stock, then it carries inventory on purpose. The production frequency of a given item and the lot size depend on both the inventory carrying cost and the setup time and cost. In this instance, the demand rate as well as the uncertainty in the demand determine the minimum safety stock that has to be kept. But even in such a case, it is important to minimize the costs of the finished goods inventory.

Transportation Costs. In networks that consist of many facilities the transportation costs may represent a significant part of the total production costs. There are various modes of transportation, e.g., truck, rail, air, and sea, and each mode has its own set of characteristics with regard to speed, cost and reliability. In practice, the transportation cost per unit is often increasing concave in the quantity moved. However, in this text the cost of transporting (moving) one unit of product from facility a to facility b is assumed to be independent of the quantity moved and is denoted by c_{ab}^m.

There are many other objectives besides those mentioned above. For example, as Just-In-Time (JIT) concepts have become more entrenched in manufacturing, one objective that is increasing in popularity is the minimization of the sum of the earlinesses. In a JIT system, a job should not be completed until just before its committed shipping date in order to avoid additional inventory and handling costs.

Another goal for the scheduler is to generate a schedule that is as *robust* as possible. If a schedule is robust, then the necessary changes that have to be made in case of a disruption (e.g., machine breakdown, rush order) tend to be minimal. However, the concept of robustness has not been well defined yet, and it is also not clear how robustness is maximized. For more details on this topic, see Chapter 14.

2.5 Discussion

From the previous sections it is clear that planning and scheduling problems in manufacturing have many different aspects. In Part II of this book we consider only some of the most fundamental aspects, but the reader should keep in mind that there are many other important issues.

For example, scheduling problems in practice are never static, because the input data continuously change. For example, the weight of job j, which is hard to measure to begin with, may not be constant; it may be time dependent. A job that is not important today may suddenly become important tomorrow.

Planning and scheduling problems in practice typically have multiple objectives; the overall goal may be a weighted combination of several objectives, e.g., the maximization of the total throughput rate combined with the minimization of the number of tardy jobs. The weights of the individual objectives may change from day to day and may even depend on the scheduler in charge. Also, personnel and operations scheduling problems are closely intertwined. This implies that the two problems often cannot be solved separately; they may have to be solved together.

Another practice that is quite common is job splitting. A job that consists of a collection or batch of items may be partially processed on one machine, and partially on another machine. Job splitting is a concept that is more general than preemption, because a job can be divided into several parts which may be processed concurrently on various machines in parallel. Another form of job splitting can occur when a job is a batch of items that has to go in a flow shop from one machine to the next. After part of a batch has been completed at one stage it can start its processing on the next machine before the entire batch has been completed at the first stage.

There are many more machine configurations that have not been covered in this chapter. For instance, there are environments where each job needs several machines simultaneously, depending upon the size of the job; or, jobs may share a machine. An example is a port with berths for ships; a large ship may take the same space as two small ships, and so on.

Exercises

2.1. A contractor has decided to use project scheduling techniques for the construction of a building. The jobs he has to do are listed in Table 2.1.
(a) Draw the precedence constraints graph.
(b) Compute the makespan of the project.
(c) If it would be possible to shorten one of the jobs by one week, which job should be shortened?

2.2. Define the earliness of job j as

$$E_j = \max(d_j - C_j, 0).$$

In essence the earliness is the dual of the tardiness. Explain how the minimization of the sum of the earlinesses relates to the Just-In-Time concept.

2.3. Consider a flexible flow shop environment. Whenever a job has to wait before it can start its processing at the next stage it is regarded as Work-In-Process (WIP), since it already has received processing on machines upstream.

The costs associated with WIP are storage costs, amount of value already added, cost of capital, and so on. Explain why high WIP costs may postpone the starting times of jobs on certain machines.

2.4. Consider the Central Processing Unit (CPU) of a mainframe computer. Jobs with different priorities (weights) come in from remote terminals. The CPU is capable of processor sharing, i.e., it can process various jobs simultaneously.

(a) Explain how the processor sharing concept can be modeled using preemptions.

(b) Describe an appropriate objective for the CPU to optimize.

2.5. Consider an environment that is prone to machine breakdowns and subject to high inflation and frequent strikes. What are the effects of these variables on

(a) the Material Requirements Planning,

(b) the Work-In-Process levels, and

(c) the finished goods inventory levels?

2.6. An important aspect of a schedule is its *robustness*. If there is a random perturbation in a robust schedule (e.g., machine breakdown, unexpected arrival of a priority job, etc.), then the necessary changes in the schedule are minimal. There is always a desire to have a schedule that is robust.

(a) Define a measure for the robustness of a schedule.

(b) Motivate your definition with a numerical example.

Job	Description of Job	Duration	Immediate Predecessor(s)
1	Excavation	4 weeks	–
2	Foundations	2 weeks	1
3	Floor Joists	3 weeks	2
4	Exterior Plumbing	3 weeks	1
5	Floor	2 weeks	3,4
6	Power On	1 weeks	2
7	Walls	10 weeks	5
8	Wiring	2 weeks	6,7
9	Communication Lines	1 weeks	8
10	Inside Plumbing	5 weeks	7
11	Windows	2 weeks	10
12	Doors	2 weeks	10
13	Sheetrock	3 weeks	9,10
14	interior trim	5 weeks	12,13
15	exterior trim	4 weeks	12
16	Painting	3 weeks	11,14,15
17	Carpeting	1 weeks	16
18	Inspection	1 weeks	17

Table 2.1. Table for Exercise 2.1

Comments and References

Some of the books and articles mentioned in the comments and references of Chapter 1 present fairly elaborate problem classification schemes (including at times a very detailed notation) encompassing many of the issues described in this chapter.

Brucker, Drexl, Möhring, Neumann, and Pesch (1999) present a comprehensive classification scheme for resource constrained project scheduling.

Conway, Maxwell, and Miller (1967) were the first to come up with a classification scheme for machine scheduling and job shop scheduling. Rinnooy Kan (1976), Lawler, Lenstra and Rinnooy Kan (1982), Lawler, Lenstra, Rinnooy Kan and Shmoys (1993), Herrmann, Lee, and Snowdon (1993), and Pinedo (1995) developed more elaborate schemes that contain many of the machine environments, processing characteristics and constraints, and performance measures discussed in this chapter.

MacCarthy and Liu (1993) developed a classification scheme for flexible manufacturing systems. This class of systems contains the class of flexible assembly systems as a subcategory. (Another subcategory of flexible manufacturing systems is the class of general flexible machining systems (GFMS) which is not covered in this book; this subcategory may be regarded as job shops with automated material handling systems.)

Some of the books that cover other model categories, e.g., lot scheduling, also introduce problem classification schemes. However, these other frameworks have not gained that much popularity yet.

Chapter 3

Service Models

    3.1 Introduction 37
    3.2 Activities and Resources in Service Settings 40
    3.3 Operational Characteristics and Constraints 41
    3.4 Performance Measures and Objectives 43
    3.5 Discussion 45

3.1 Introduction

Service industries are in many aspects different from manufacturing industries. A number of these differences affect the planning and the scheduling of the activities involved. One important difference can be attributed to the fact that in manufacturing it is usually possible to inventorize goods (e.g., raw material, Work-In-Process, and finished products), whereas in services there are typically no goods to inventorize. The fact that in manufacturing a job can either wait or be completed early affects the structure of the models in a major way. In service industries, a job tends to be an activity that involves a customer who does not like to wait. Planning and scheduling in service industries is, therefore, often more concerned with capacity management and yield management.

    A second difference is based on the fact that in manufacturing the number of resources (which are typically machines) is usually fixed (at least for the short term), whereas in services the number of resources (e.g., people, rooms, and trucks) may vary over time. This variable may even be a part of the objective function.

    A third difference is due to the fact that denying a customer a service is a more common practice than not delivering a product to a customer in a manufacturing setting. This is one of the reasons why yield management plays such an important role in service industries.

These differences between manufacturing and services affect the processing restrictions and constraints as well as the objectives. Service models tend to be therefore very different from manufacturing models.

Several classes of planning and scheduling models play an important role in service industries. One class of models, which is important in both manufacturing and services, has already been discussed in the previous chapter. This class consists of the project planning and scheduling models. Project planning and scheduling is not only useful in the building of an aircraft carrier but also in the management of a consulting project. Even though the class of project planning and scheduling models is important in service industries, it is in this book only discussed in the manufacturing part (Part II). The remainder of this section provides a short description of four other classes of planning and scheduling models that are important for service industries.

The first class of models considered in the services part (Part III) consists of models for reservation systems and timetabling. These two subclasses are mathematically closely related to one another; however, they have similarities as well as differences. In a model for a reservation system job j has a duration p_j and its starting time and completion time are usually fixed in advance, i.e., there is no slack. For example, in a car rental agency a job is equivalent to the reservation of a car for a given period. The agency may not be able to process all the jobs that present themselves (i.e., they may not be able to confirm all reservations); it may have to decide which jobs to process and which ones not. The objective is typically to process as many jobs as possible. In timetabling (or rostering) job j or activity j may be a meeting or an exam with a duration p_j which has to be scheduled within a given time window. There may be an earliest possible starting time r_j and a latest possible completion time d_j, and there may be a certain amount of slack. In timetabling the starting times and completion times of the jobs are not fixed in advance; but, in order for an activity to take place, certain people have to be present. This implies that two activities which require the same person or operator to be present cannot be done at the same time. There may also be restrictions and constraints with regard to the availability of the operators (an operator may only be available during a given time period). One objective in timetabling may be the minimization of the makespan subject to the constraint that all activities have to be scheduled. An example of a timetabling problem is the exam scheduling problem; the exams are similar to jobs and the rooms are equivalent to machines. Two exams that have to be taken by the same group of students may not overlap in time. Another example of a timetabling problem is the scheduling of operating rooms in hospitals. An operation requires a patient, an operating room, a surgeon, an anastaesiologist and other members of a surgical team. The objective is to find a feasible schedule that includes all operations and satisfies all constraints. Timetabling has some similarities with workforce scheduling as well. A workforce may consist of various pools of different types of operators; a pool of type l consists of W_l operators of type l; the integer W_l is fixed over time.

3.1 Introduction

The second class of service models consists of tournament scheduling models and broadcast television scheduling models. A tournament scheduling model involves a league with a set of teams and a fixed number of games that have to be assigned to given time slots. The games may be subject to many constraints. For example, certain games have to be played within given time windows. The sequence of games a team plays in a tournament may be subject to various constraints; for example, a team should not play more than two consecutive games at home or more than two consecutive games away. A tournament scheduling problem can be compared to a parallel machine scheduling problem in which all the jobs have the same processing time. Moreover, there are constraints that are somewhat similar to the workforce constraints in project scheduling, since a team can play at most one game in any given time slot.

The third class of service models are the transportation scheduling models. Planning and scheduling is important for airlines, railroads, and shipping. A job may be a trip or flight leg that has to be covered by a ship, plane, or vehicle. A ship, plane, or vehicle plays a role that is similar to a machine. A trip or flight leg j has to take place within a given time frame and is subject to given processing restrictions and constraints. The processing restrictions and constraints specify how a particular trip or flight leg can be combined with other trips or flight legs in a feasible round-trip k that can be assigned to a particular vehicle and crew. Round trip k incurs a cost c_k and generates a profit π_k. The objective is either to minimize total cost or maximize total profit.

The fourth class of service models concerns workforce scheduling, which is a very important aspect of planning and scheduling in the service industries. Workforce scheduling models tend to be quite different from machine scheduling models. Workforce scheduling may imply either shift scheduling in a service facility (e.g., a call center) or crew scheduling in a transportation environment. Shift scheduling models are the easier ones to formulate: for each time interval there are requirements with regard to the number of personnel that have to be present. Time interval i requires a staffing of b_i (b_i integer). Personnel can be hired for different shifts and there is a cost associated with each type of hire. The objective is to minimize the total cost. One can argue that shift scheduling is somewhat similar to machine scheduling. During time interval i (which may have unit length) b_i tasks have to be done, each one of unit length. The workforce is in a sense equivalent to a number of machines in parallel which have to process the unit tasks. At least b_i resources have to be available to do these unit tasks in interval i. There are restrictions, constraints and costs associated with keeping these resources available and the objective is to minimize the total cost. Because of the special structures of the shifts the formulation of such a problem turns out to be different from the formulation of a typical machine scheduling problem. The formulations of crew scheduling models tend to be different from the formulations of shift scheduling models as well. In contrast to the shift scheduling models, the crew scheduling models do not assume a steady state pattern. Crew scheduling depends very much on specific tasks that have to be done. Crew scheduling are therefore often intertwined with other scheduling procedures in the organization (e.g., the routing and scheduling of planes or trucks).

From the above, it is clear that there are some similarities between planning and scheduling applications in manufacturing and planning and scheduling applications in services. The areas of interval scheduling, timetabling and reservation models have many similarities with planning and scheduling in manufacturing. Reservation systems are common in transportation and hospitality industries; however, they also play a role in manufacturing settings when customers may reserve specific times on given machines. Planning and scheduling in transportation also have commonalities with machine scheduling. Transportation models are often similar to parallel machine scheduling models: a plane is similar to a machine and a trip or flight leg is similar to a job that has a processing time, release date and due date; a turn-around time of a plane is similar to a sequence dependent setup time.

In practice, workforce scheduling may often be intertwined with other scheduling functions. For example, machine schedules may depend on shift schedules and fleet schedules may depend on crew schedules. However, in this book we do not consider models that integrate workforce scheduling with other scheduling functions.

3.2 Activities and Resources in Service Settings

In manufacturing a job usually represents an activity that transforms a physical component and adds value to it. In services an activity typically involves people. From the previous section it is clear that an activity in a service setting can take many forms. An activity may be, for example,

 (i) a meeting that has to be attended by certain people,
 (ii) a game that has to be played between two teams,
 (iii) a flight leg that has to be covered by a plane, or
 (iv) a personnel position that must be occupied in a given time period.

The characteristics of a job or an activity in the service industries are in some aspects similar to those of a job in a manufacturing environment. The data include

 (i) a duration (processing time p_{ij}),
 (ii) an earliest possible starting time (release date r_j),
 (iii) a latest possible finishing time (due date d_j),
 (iv) a priority level (weight w_j).

The priority level of an activity depends on its profit or cost. An activity may have more parameters which specify, for example, the additional resources that may be required in order to do the activity.

In manufacturing, resources are typically referred to as machines and the configuration of the machines is given. They may be set up in parallel or in series, or the product may follow a more complicated route. In service models, the resources required can take many different forms. A resource can be a classroom, meeting room, hotel room, stadium, rental car, or operating room. In transportation, it may be a plane, ship, truck, train, airport gate, dock, railroad track, person (e.g., a pilot), and so on. In a health care setting, a resource may be an operating room, surgeon, anasthaesiologist, and so on.

In service industries it is often important to synchronize the timing of the use of the different types of resources. It may be the case that, in order to perform a certain activity, it is necessary to have two or three different types of resources available at the same time. To transport a given cargo from one point to another, a truck, a driver and docking areas have to be available. If a train goes from one station to another, then a track between the two stations has to be available while the train is on its way.

Each type of resource may have characteristics or parameters that are important for the planning and scheduling process. For example, if resource i is a room, then it has a capacity A_i (number of seats), a cost c_i (or revenue) per unit time, an availability of certain equipment (e.g., audiovisual), and so on. A plane, ship or truck may have a capacity A_i, a given speed v_i, and a certain range. It has a given operating cost, requires a certain crew, and can be used only on specific routes. If a resource is a person, then he or she may be a specialist (e.g., a surgeon). A specialist usually has a set of characteristics or skills that determines the types of jobs he or she can handle.

3.3 Operational Characteristics and Constraints

The operational characteristics and constraints in service models are more diverse and often more difficult to specify than those in manufacturing models.

Time Windows (Release Dates and Due Dates). Scheduling meetings, exams, or games often must satisfy time window constraints; such time windows are equivalent to release dates and due dates in manufacturing. When an airline schedules its flights, it has to make sure that each flight departs within its time window. A time window is often determined by a marketing department that has estimates of passenger demand as a function of the departure time. The same is true with respect to train timetabling. Each train in a given corridor has an ideal arrival and departure time for each one of the stations in the corridor.

Capacity Requirements and Constraints. Capacity requirements and constraints are important in reservation systems, in timetabling of meetings, as well as in transportation planning and scheduling. A meeting or exam may be scheduled with a number of participants in mind; this number is an important parameter of the meeting, since in a feasible schedule the value of this parameter must be less than the capacity A_i of the meeting room. When an airline assigns its planes (which may be of several different types) to the various flight legs, it has to take into account the number of seats on each plane.

Capacity requirements are similar to the machine eligibility constraints in a manufacturing model. In a parallel machine environment, job j may often not be processed on just any one of the available machines, but rather on a machine that belongs to a specific subset M_j. As described in Chapter 2, this may be the case when the m machines are not all exactly the same.

Capacity constraints are also important in railway scheduling. The fact that there is usually only a single track in each direction between two stations makes it impossible for one train to pass another in between stations. Railway scheduling is particularly difficult because of such track capacity constraints.

Preemptions. Preemptions are less prevalent in services than in manufacturing. Many types of service activities are very difficult to preempt, e.g., an operation in a hospital, a flight leg or a game. However, some types of service jobs can be preempted, but usually such preemptions are not allowed to occur at just any time; preemptions in services tend to occur at specific points in time rather than at arbitrary points in time. The time during which a meeting takes place may be a collection of disjoint periods rather than one contiguous period. A meeting may be split up in segments of one or more hours each.

Setup Times and Turnaround Times. Resources may require setups between consecutive activities. Rooms, planes, and trucks may have to be cleaned and set up for subsequent meetings or trips. This may involve a setup time s_{jk} that may or may not be sequence dependent. Setups may have either fixed or random durations. For example, after a plane arrives at an airport, various activities have to be performed before it can take off again. Arriving passengers have to be deplaned, the plane has to be refueled and cleaned, and departing passengers have to board the plane. This turnaround time is equivalent to a sequence dependent setup time.

In airport models a runway may be considered a resource; the take-offs and landings are the activities that have to be performed and there are sequence dependent setup times. In between two consecutive take-offs it may be necessary to keep the runway idle for one or two minutes in order to allow air turbulence to dissipate. This idle time is longer when a small plane follows a large plane than vice versa.

Operator and Tooling Requirements. In an environment with parallel resources, activities may have to be scheduled in such a way that additional requirements are met (e.g., specific operators or tools have to be available) or, equivalently, the operators and/or tools have to be assigned in such a way that all activities can be done according to the rules. Resources frequently require one or more operators to do the activities. Operators or specialists may be of different types, and of some types there may be a limited availability. The routing and scheduling of planes is therefore often intertwined with the scheduling of the crews.

Operator and tooling requirements play, of course, also a role in manufacturing settings. A machine shop may have only a limited number of operators who can work on a certain machine. Jobs that need processing on that machine have to wait until one of these operators becomes available. In such cases machine scheduling and workforce scheduling become intertwined. If there is only one type of operator and there are W of them, then the W operators constitute a single, homogeneous workforce.

Workforce Scheduling Constraints. Workforce scheduling and shift assignments are usually subject to many constraints. They are often of a form in which people work a given number of consecutive days (e.g., five) and then have a number of consecutive days off (e.g., two). However, there are many different types of shifts as well as many ways of rotating them.

In the hospital described in Example 1.1.9 the personnel requirements for weekends differ from those for weekdays. For instance, only emergency operations are performed over the weekend, whereas all planned operations are performed during the week. This implies that there are different personnel requirements for the operating rooms during these two time periods. Also, fewer beds are occupied over the weekend than during the week; this affects personnel requirements as well.

Union rules may have a significant impact on the working hours of personnel. In call centers even the lunch breaks and coffee breaks have to be scheduled according to rules that are approved by a union. With regard to airline crews the FAA has very tight regulations concerning the maximum number of consecutive hours of flying time and the maximum number of hours that can be flown within a given time period.

3.4 Performance Measures and Objectives

Scheduling objectives in manufacturing are typically a function of the completion times, the due dates, and the deadlines of the jobs. Usually, the number of machines in a manufacturing environment is fixed. In service environments the objectives may be somewhat different. Some objectives in services may indeed also be a function of the completion times, the due dates, and the deadlines; but, objectives in services may have an additional dimension. In contrast to manufacturing, the number of resources in a service environment may be variable (e.g., the number of full-time and part-time people employed). Because of this, there may be a different type of objective which tries to minimize the number of resources used and/or minimize the cost associated with the use of these resources.

An objective in a service environment may thus be a combination of two types of objectives: one concerning the timings of the activities and the other concerning the utilization of the resources. If these two types of objectives are combined in one objective function then appropriate weights have to be determined.

That part of a services objective that is similar to a manufacturing objective tends to be tightly constrained: the time windows in which activities have to be scheduled are often narrow and may, at times, not have any slack at all. A problem may reduce to an assignment problem in which an activity (meeting or game) has to be assigned to a given timeslot and/or to a given room with the assignment having a certain cost (or fit). The objective is to minimize the total cost of all the assignments. In transportation scheduling, certain legs (or trips) have to be combined into feasible round trips that satisfy all constraints. The round trips created must include all the legs that have to be covered at a minimum cost.

Makespan. The makespan $C_{\max}$ is not only important in project planning and scheduling models, it is also important in timetabling and reservation systems. For example, consider a set of exams that has to be squeezed into a given exam period of one week; a primary objective may be to schedule all the exams within that period and a secondary objective may be to minimize the number of rooms used. Since minimizing the makespan is often equivalent to minimizing the sequence dependent setup times, the makespan objective also plays a role in transportation settings when turnaround times have to be minimized.

Setup Costs. Before the take-off of a plane, certain preparations have to be made that may depend on the type of flight. The preparations include cleaning, catering, fueling, and so on. At other times more elaborate setups are required, which may include engine maintenance, etc. Each setup has a cost that may be sequence dependent.

Earliness and Tardiness Costs; Convenience Costs and Penalty Costs. In transportation scheduling, there are often "ideal" departure times (determined by the marketing department). If the departure of a train or plane is shifted too much from its ideal departure time, then a penalty cost is incurred (due to an expected loss of revenue), see Figure 3.1.

Personnel Costs. A cost is associated with the assignment of a given person to a particular shift. This basic cost is usually known in advance. However, overtime may be required, the amount of which often not known in advance. Cost of overtime is typically significantly higher than cost of regular time.

Consider the hospital environment in Example 1.1.9, which has a 24-hour staffing requirement. If the workforce is unionized, then their collective bargaining agreements may contain staffing restrictions and constraints. The scheduler's objective is to develop a variety of shift patterns and assign people to shifts in such a way that all requirements are met. Each shift pattern has a cost associated with it and the objective is to minimize total cost.

Crew scheduling in a transportation environment tends to be more complicated than nurse scheduling. The nurse scheduling problem in a hospital can be solved more or less independently from other optimization problems in a hospital. The crew scheduling problem of an airline, on the other hand, is

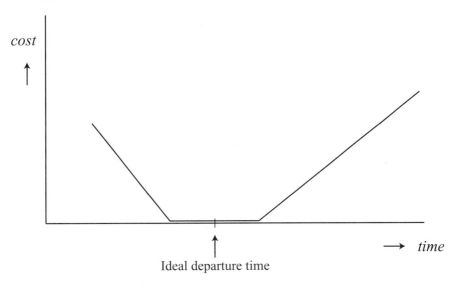

Fig. 3.1. Earliness and Tardiness Penalties

intertwined with other problems that must be dealt with by the airline, e.g., route selection, fleet assignment, etc.

3.5 Discussion

It seems that planning and scheduling research in the past has focused less on services than on manufacturing (in spite of the fact that there are many planning and scheduling applications in services). Planning and scheduling objectives in service industries are often considerably more complicated than planning and scheduling objectives in manufacturing. Service models seem, therefore, harder to categorize than manufacturing models.

In services, it often occurs that various planning and scheduling processes have to interact with one another; for example, in the aviation industry fleet scheduling and crew scheduling are linked to one another, even though the objective functions of the two modules are quite different. In fleet scheduling usually the number of planes utilized has to be minimized, whereas in crew scheduling the total cost of the crews has to be minimized. Other scheduling modules in the aviation industry are not as closely linked to one another. For example, gate assignment and runway scheduling at airports tend to be independent processes.

An important area of research nowadays concerns shift scheduling in call centers. Various types of operators, with different skill sets, have to respond to all kinds of calls. Each time interval requires a number of operators. There are multiple shift patterns and each pattern has its own cost. Each type of operator in each type of shift has its own cost structure. Shift scheduling models may also be combined with machine scheduling models in manufacturing.

Planning and scheduling in manufacturing and in services may have to deal with medium term and long term issues, and also with short term and reactive scheduling issues. Scheduling in services tends to be more short term oriented and more reactive than scheduling in manufacturing.

Exercises

3.1. A consulting company has to install a brand new production planning and scheduling system for a client. This project requires the succesful completion of a number of activities, namely

Activity	Description of Activity	Duration
1	Installation of new computer equipment	8 weeks
2	Testing of computer equipment	5 weeks
3	Development of the software	6 weeks
4	Recruiting of additional systems people	3 weeks
5	Manual testing of software	2 weeks
6	Training of new personnel	5 weeks
7	Orientation of new personnel	2 weeks
8	System testing	4 weeks
9	System training	7 weeks
10	Final debugging	4 weeks
11	System changeover	9 weeks

The precedence relationships between these activities are depicted in Figure 3.2.

(a) Compute the makespan of the project.

(b) If it would be possible to shorten one of the activities by one week, which activity should be shortened?

3.2. Consider an airport terminal with a number of gates. Planes arrive and depart according to a fixed schedule. Upon arrival a plane has to be assigned to a gate in a way that is convenient for the passengers as well as for the airport personnel. Certain gates are only capable of handling narrowbody planes. Model this problem as a machine scheduling problem.

(a) Specify the machines and the processing times.

(b) Describe the processing restrictions and constraints, if any.

(c) Formulate an appropriate objective function.

3.3. Consider the reservation system in a hotel. People call in to make reservations. Model the problem as a machine scheduling problem.

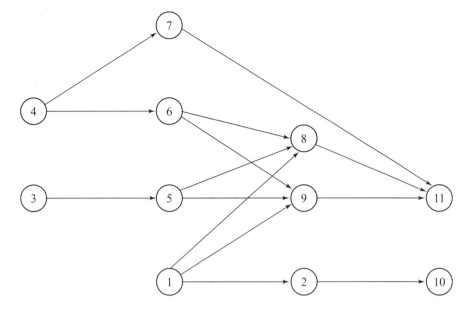

Fig. 3.2. Precedence constraints of system installation

(a) What are the machines?
(b) Are all the machines identical?
(c) What are the jobs and what are their processing times?
(d) Formulate an appropriate objective for the hotel to optimize.

3.4. Consider an exam schedule at a large university. The exams have to be scheduled in such a way that potential conflicts for the students are minimized. There are classroom constraints, because of the number of rooms available and because of their respective sizes.

(a) How can this problem be modeled as a machine scheduling problem?
(b) What are the machines?
(c) Are there operator or tooling constraints? If there are such constraints, what are the operators and what are the tools?

3.5. In the description of a workforce in this chapter it is assumed that a workforce may consist of a number of different pools that do not overlap. Each pool has its own particular skill set and the skill sets of the different pools do not overlap. Give a formal presentation of a framework that assumes a fixed number of skills, say N. Each worker has a subset of all possible skills, However, the skill sets of two different workers may partially overlap, i.e., they may have some skills in common and each worker may have skill(s) that the other one does not have. How does the complexity of this framework compare to the case in which the skill sets do not overlap?

3.6. A company has a central warehouse and a number of clients that have to be supplied from that warehouse. There is a fleet of trucks available and all distances are known, i.e., distances between clients as well as between clients and the warehouse. A truck can supply various clients on a single route. However, the total distance a truck is allowed to cover is bounded from above. The objective is to minimize total distance traveled by all trucks. Show that this problem is equivalent to a parallel machine scheduling problem with setup times.

Comments and References

The number of books on planning and scheduling in services is considerably smaller than the number of books on planning and scheduling in manufacturing.

Models for reservation systems and timetabling are fairly new topics that are closely related to one another. Models for reservation systems (also referred to as interval scheduling models) have been considered in a couple of chapters in Dempster, Lenstra and Rinnooy Kan (1982). Timetabling has received more attention in the literature. The textbook by Parker (1995) discusses this topic and various proceedings of conferences on timetabling have appeared recently, see Burke and Ross (1996), Burke and Carter (1998), Burke and Erben (2001), and Burke and De Causmaecker (2003).

No framework has yet been established for planning and scheduling models in transportation. A series of conferences on computer-aided scheduling of public transport has resulted in a number of very interesting proceedings, see Wren and Daduna (1988), Desrochers and Rousseau (1992), Daduna, Branco, and Pinto Paixao (1995), Wilson (1999), and Voss and Daduna (2001). Christiansen, Fagerholt and Ronen (2004) present a survey of ship routing and scheduling. A volume edited by Yu (1997) considers operations research applications in the airline industry; this volume contains several papers on planning and scheduling in the airline industry.

Personnel scheduling has not received the same amount of attention as project scheduling or job shop scheduling. However, the text by Nanda and Browne (1992) is completely dedicated to personnel scheduling. Parker (1995) devotes a section to staffing problems. Burke, De Causmaecker, Vanden Berghe and Van Landeghem (2004) give an overview of the state of the art of nurse rostering. Several of the books on planning and scheduling in transportation have chapters on crew scheduling.

Part II

Planning and Scheduling in Manufacturing

4   Project Planning and Scheduling 51

5   Machine Scheduling and Job Shop Scheduling 81

6   Scheduling of Flexible Assembly Systems 115

7   Economic Lot Scheduling 141

8   Planning and Scheduling in Supply Chains.............. 171

Chapter 4

Project Planning and Scheduling

4.1	Introduction	51
4.2	Critical Path Method (CPM)	54
4.3	Program Evaluation and Review Technique (PERT)	58
4.4	Time/Cost Trade-Offs: Linear Costs	61
4.5	Time/Cost Trade-Offs: Nonlinear Costs	68
4.6	Project Scheduling with Workforce Constraints	69
4.7	ROMAN: A Project Scheduling System for the Nuclear Power Industry	72
4.8	Discussion	76

4.1 Introduction

This chapter focuses on the planning and scheduling of jobs that are subject to precedence constraints. The setting may be regarded as a parallel machine environment with an unlimited number of machines. The fact that the jobs are subject to precedence constraints implies that a job can start with its processing only when all its predecessors have been completed. The objective is to minimize the makespan while adhering to the precedence constraints. This type of problem is referred to as a project planning and scheduling problem.

A more general version of the project planning and scheduling problem assumes that the processing times of the jobs are not entirely fixed in advance. A project manager has some control on the durations of the processing times of the different jobs through the allocation of additional funds from a budget that he has at his disposal. Since a project may have a deadline and a completion after its deadline may involve a penalty, the project manager has to analyze the trade-off between the costs of completing the project late and the costs of shortening the durations of the individual jobs.

Another more general version of the basic project planning and scheduling problem assumes that a job's processing requires, besides a machine (of which there is always one available), also various operators. The workforce may consist of several different pools of operators; each pool has a fixed number of operators with a specific skill. Because of workforce limitations, it may sometimes occur that two jobs cannot be processed at the same time, even though both are allowed to start as far as the precedence constraints are concerned. This type of problem is in what follows referred to as project scheduling with workforce constraints. (In the literature this type of problem is often also referred to as resource constrained project scheduling.)

The basic project scheduling problem without workforce constraints is from a computational point of view easy. Optimal solutions can be found with little computational effort. On the other hand, project scheduling with workforce constraints is very hard.

These types of planning and scheduling problems occur often in practice when large projects have to be undertaken. Examples of such projects are real estate developments, construction of power generation centers, software developments, and launchings of spacecraft. Other applications include projects in the defense industry, such as the design, development and construction of aircraft carriers and nuclear submarines.

In most of the literature on project scheduling, a job in a project is referred to as an activity. However, in what follows we use the term job rather than activity in order to be consistent with the remaining chapters in this part of the book.

The precedence relationships between the jobs are the basic constraints of the project scheduling problem. The representation of the precedence constraints as a graph may follow either one of two formats. One format is referred to as the "job-on-arc" format and the other as the "job-on-node" format. In the job-on-arc format, the arcs in the precedence graph represent the jobs and the nodes represent the milestones or epochs. For example, if job j is followed by job k, then the precedence graph is as depicted in Figure 4.1.a. The first node represents the starting time of job j, the second the completion time of job j as well as the starting time of job k and the third and last node the completion time of job k.

In the job-on-node format the nodes in the precedence graph represent the jobs and the connecting arcs the precedence relationships between the jobs. If job j is followed by job k, then the precedence graph takes the form depicted in Figure 4.1.b.

Although in practice the first format is more widely used than the second, the second has a number of advantages. A disadvantage of the job-on-arc format is a necessity for so-called "dummy" jobs that are needed to enforce precedence constraints that otherwise would not have been enforcable. The following example illustrates the use of dummy jobs.

4.1 Introduction

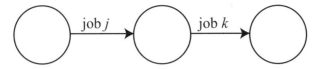

(a) Job-on-arc format

(b) Job-on-node format

Fig. 4.1. Formats of precedence graphs

Example 4.1.1 (Setting up a Production Facility). Consider the problem of setting up a manufacturing facility for a new product. The project consists of eight jobs. The job descriptions and the time requirements are as follows:

Job	Description	Duration (p_j)
1	Design production tooling	4 weeks
2	Prepare manufacturing drawings	6 weeks
3	Prepare production facility for new tools and parts	10 weeks
4	Procure tooling	12 weeks
5	Procure production parts	10 weeks
6	Kit parts	2 weeks
7	Install tools	4 weeks
8	Testing	2 weeks

The precedence constraints are specified below.

Job	Immediate Predecessors	Immediate Successors
1	—	4
2	—	5
3	—	6, 7
4	1	6, 7
5	2	6
6	3, 4, 5	8
7	3, 4	8
8	6, 7	—

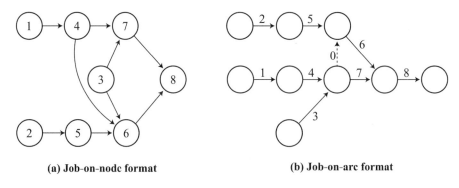

Fig. 4.2. Precedence graphs for Example 4.1.1

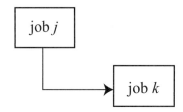

Fig. 4.3. Advantages of depicting nodes as rectangles

The precedence graph is presented in the job-on-node format in Figure 4.2.a and in the job-on-arc format in Figure 4.2.b. From the figure it is clear that there is a need for a dummy job in the job-on-arc format.

The number of dummy jobs required to enforce the proper precedences in a job-on-arc representation of a large project can be substantial and may increase the total number of jobs by as much as 10%.

Another advantage of the job-on-node format is that nodes may be depicted as rectangles and the horizontal sides of the rectangle can be used as a time-axis that corresponds to the processing time of the job. If job k is allowed to start after half of job j has been completed, then the arc establishing this precedence relationship can emanate from the midpoint of a horizontal side of the rectangle, see Figure 4.3.

Precedence constraints are in what follows presented in the job-on-node format. However, nodes are depicted as circles and not as rectangles.

4.2 Critical Path Method (CPM)

Consider n jobs subject to precedence constraints. The processing time of job j is fixed and equal to p_j. There are an unlimited number of machines in

4.2 Critical Path Method (CPM)

parallel and the objective is to minimize the makespan. Besides a machine (and there is always one available) a job does not require any other resource.

The algorithm that yields a schedule with a minimum makespan is relatively simple and can be described in words as follows: Start at time zero with the processing of all jobs that have no predecessors. Every time a job completes its processing, start processing those jobs of which all the predecessors have been completed.

In order to describe the algorithm more formally we need some notation. Let C'_j denote the earliest possible completion time of job j and S'_j the earliest possible starting time. Clearly, $C'_j = S'_j + p_j$. Let the set {all $k \to j$} denote all jobs that are predecessors of job j. This implies that if job k is a predecessor of job j, job k has to be completed before job j can be started.

The following algorithm is referred to as the forward procedure. It is based on one simple fact: a job can start its processing only when all its predecessors have been completed. So the earliest starting time of a job equals the maximum of the earliest completion times of all its predecessors.

Algorithm 4.2.1 (Forward Procedure).

Step 1.

Set time $t = 0$.

Set $S'_j = 0$ and $C'_j = p_j$ for each job j that has no predecessors.

Step 2.

Compute inductively for each job j

$$S'_j = \max_{\{all\ k \to j\}} C'_k,$$

$C'_j = S'_j + p_j.$

Step 3.

The makespan is $C_{\max} = \max(C'_1, \ldots, C'_n)$.

STOP

It can be shown that this procedure yields an optimal schedule and the makespan of this optimal schedule can be computed easily. In the resulting schedule each job starts its processing at its earliest possible starting time and is completed at its earliest possible completion time. However, this may not be necessary in order to minimize the makespan. It may be possible to delay the start of some of the jobs without increasing the makespan.

The next algorithm is used to determine the latest possible starting times and completion times of all the jobs, assuming the makespan is kept at its minimum. This algorithm is referred to as the backward procedure. The algorithm uses the $C_{\max}$, which is an output of the forward procedure, as an input. In order to describe this algorithm some additional notation is needed. Let C_j'' denote the latest possible completion time of job j and S_j'' the latest possible starting time of job j. Let the set $\{j \to \text{all } k\}$ denote all jobs that are successors of job j.

Algorithm 4.2.2 (Backward Procedure).
Step 1.

Set $t = C_{\max}$.

Set $C_j'' = C_{\max}$ and $S_j'' = C_{\max} - p_j$ for each job j that has no successors,

Step 2.

Compute inductively for each job j

$$C_j'' = \min_{\{j \to \text{all } k\}} S_k'',$$

$S_j'' = C_j'' - p_j.$

Step 3.

Verify that $\min(S_1'', \ldots, S_n'') = 0$.

STOP

So the forward procedure determines the earliest possible starting times and completion times as well as the minimum makespan. The backward procedure starts out with the minimum makespan and computes the latest starting times and latest completion times, such that the minimum makespan still can be achieved.

A job of which the earliest starting time is earlier than the latest starting time is referred to as a *slack* job. The difference between a job's latest possible starting time and earliest possible starting time is the amount of slack, also referred to as *float*. A job of which the earliest starting time is equal to the latest starting time is referred to as a *critical* job. The set of critical jobs forms one or more *critical paths*. A critical path is a chain of non-slack jobs, beginning with a job that starts at time zero and ending with a job that completes its processing at $C_{\max}$. There may be more than one critical path and critical paths may partially overlap.

Example 4.2.3 (Application of the Critical Path Method). Consider 14 jobs. The processing times are given below.

Jobs	1	2	3	4	5	6	7	8	9	10	11	12	13	14
p_j	5	6	9	12	7	12	10	6	10	9	7	8	7	5

The precedence constraints are presented in Figure 4.4. The earliest completion time C_j' of job j can be computed using the forward procedure.

4.2 Critical Path Method (CPM)

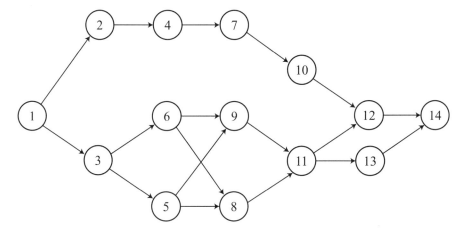

Fig. 4.4. Precedence graph for Example 4.2.3

Jobs	1	2	3	4	5	6	7	8	9	10	11	12	13	14
C'_j	5	11	14	23	21	26	33	32	36	42	43	51	50	56

This implies that the makespan is 56. Assuming that the makespan is 56, the latest possible completion times C''_j can be computed using the backward procedure.

Jobs	1	2	3	4	5	6	7	8	9	10	11	12	13	14
C''_j	5	12	14	24	30	26	34	36	36	43	43	51	51	56

Those jobs of which the earliest possible completion times are equal to the latest possible completion times are critical and constitute the critical path. So the critical path is

$$1 \to 3 \to 6 \to 9 \to 11 \to 12 \to 14.$$

The critical path in this case happens to be unique. The jobs that are not on the critical path are slack.

The Critical Path Method is related to the Critical Path (CP) dispatching rule that is described in Appendix C. According to the CP rule, whenever a machine is freed, the job at the head of the longest chain is given the highest priority. Similarly, in the CPM method one has to make sure that the job at the head of the longest chain is never delayed. However, since there are an unlimited number of machines and no workforce constraints in the environment considered here, other jobs do not have to be delayed either. The method, therefore, does not function as a priority rule. In a more general setting with a limited number of machines or with workforce constraints, the Critical Path rule may actually have to postpone the start of jobs that are at the head of shorter chains.

4.3 Program Evaluation and Review Technique (PERT)

In contrast to the setting in the previous section, the processing times of the n jobs are now random variables. The mean μ_j and the variance σ_j^2 of each of these random variables are either known or can be estimated.

The technique to determine the expected makespan of the project is often referred to as the *Program Evaluation and Review Technique (PERT)*. The algorithm that minimizes the expected makespan is exactly the same as the Critical Path Method. At each job completion, all jobs whose predecessors have all been completed are started. However, the computation of the expected makespan is now more complicated. We assume that we have three pieces of data with regard the processing time of each job, namely

p_j^a = the optimistic processing time of job j,
p_j^m = the most likely processing time of job j,
p_j^b = the pessimistic processing time of job j.

Using these three pieces of data the expected processing time of job j is typically estimated by setting

$$\hat{\mu}_j = \frac{p_j^a + 4p_j^m + p_j^b}{6}.$$

Based on the estimates of the expected processing times an estimate for the expected makespan can be obtained by applying the classical Critical Path Method with fixed processing times that are equal to the estimates for the expected processing times. This application of the Critical Path Method results in one or more critical paths. An estimate of the expected makespan is then obtained by summing the estimates for the expected processing times of all jobs on a critical path. If J_{cp} denotes the set of jobs on a critical path, then an estimate for the expected makespan is

$$\widehat{E}(C_{\max}) = \sum_{j \in J_{cp}} \hat{\mu}_j.$$

To obtain some feeling for the *distribution* of the makespan in the original problem, one proceeds as follows. Compute an estimate for the variance of the processing time of job j by taking

$$\hat{\sigma}_j^2 = \left(\frac{p_j^b - p_j^a}{6}\right)^2.$$

In order to obtain an estimate for the variance of the makespan we focus only on the critical path and disregard all other jobs. Since the jobs on the critical path have to be processed one after another, the variance of the total processing time of all jobs on the critical path can be estimated by taking

$$\widehat{V}(C_{\max}) = \sum_{j \in J_{cp}} \hat{\sigma}_j^2.$$

4.3 Program Evaluation and Review Technique (PERT)

The distribution of the makespan is assumed to be normal, i.e., Gaussian, with mean $\widehat{E}(C_{\max})$ and variance $\widehat{V}(C_{\max})$.

These estimates are, of course, very crude. First of all, there may be more than one critical path. If there are several critical paths, then the actual makespan is the maximum of the total realized processing times of each one of the critical paths. So the expected makespan must be larger than the estimate for the expected makespan obtained by considering a single critical path. Second, the total amount of processing on the critical path is assumed to be normally distributed. If the number of jobs on the critical path is very large, then this assumption may be reasonable (because of the Central Limit Theorem). However, if the number of jobs on the critical path is small (say 4 or 5), then the distribution of the total processing on the critical path may not be that close to Normal (see Exercise 4.7).

Example 4.3.1 (Application of PERT). Consider the 14 jobs of Example 4.2.3. The jobs are subject to the same precedence constraints (see Figure 4.4). However, now the processing times are random, with the following PERT data.

Jobs	1	2	3	4	5	6	7	8	9	10	11	12	13	14
p_j^a	4	4	8	10	6	12	4	5	10	7	6	6	7	2
p_j^m	5	6	8	11	7	12	11	6	10	8	7	8	7	5
p_j^b	6	8	14	18	8	12	12	7	10	15	8	10	7	8

Based on these data the means and the variances of the processing times can be estimated.

Jobs	1	2	3	4	5	6	7	8	9	10	11	12	13	14
$\hat{\mu}_j$	5	6	9	12	7	12	10	6	10	9	7	8	7	5
$\hat{\sigma}_j$	0.33	0.67	1	1.33	0.33	0	1.33	0.33	0	1.33	0.33	0.66	0	1
$\hat{\sigma}_j^2$	0.11	0.44	1	1.78	0.11	0	1.78	0.11	0	1.78	0.11	0.44	0	1

Note that the estimates of the means are equal to the processing times used in Example 4.2.3. So the critical path is the same path as in Example 4.2.3, namely

$$1 \to 3 \to 6 \to 9 \to 11 \to 12 \to 14.$$

The estimate of the makespan is equal to the makespan in Example 4.2.3, i.e., 56. If we compute the estimate of the variance of the makespan, we obtain

$$\widehat{V}(C_{\max}) = \sum_{j \in J_{cp}} \hat{\sigma}_j^2 = 2.66.$$

Assuming that the project duration is normally distributed with the estimated mean and variance of the critical path, then the probability that the project is completed by time 60 is

$$\Phi\left(\frac{60-56}{\sqrt{2.66}}\right) = \Phi(2.449) = 0.993,$$

where $\Phi(x)$ denotes the probability that a normally distributed random variable with mean 0 and variance 1 is less than x. The probability that this project will be done by time 60 is 99.3% and the probability it will be completed after time 60 is 0.7%.

Note that we ignored here the randomness in all the jobs that are not on the critical path. To get an idea of the accuracy of the probability of completing the entire project by time 60, consider the path

$$1 \to 2 \to 4 \to 7 \to 10 \to 12 \to 14.$$

An estimate of the length of this path is 55. This implies that this path is only slightly shorter than the critical path. Estimating the variance of this path yields 7.33. Computing the probability of finishing the project before time 60 by analyzing this path results in

$$\Phi\left(\frac{60-55}{\sqrt{7.33}}\right) = \Phi(1.846) = 0.968.$$

According to this computation the probability that the project is completed after time 60 is 3.2%. This probability is clearly higher than the one obtained by considering the critical path. The reason is clear: the estimate for the variance of the second path (which is not critical) is significantly higher than the estimate for the variance of the critical path.

This example highlights one of the drawbacks of the PERT procedure. The critical path, which is the longest path, may have a relatively low variance, while the second longest path (in expectation) may have a very high variance. If several such paths, not necessarily critical paths, have high variances, then the expected makespan may be significantly larger than the value computed using the PERT method. The next example illustrates an extreme case.

Example 4.3.2 (Drawback of PERT). Consider the precedence graph depicted in Figure 4.5. There are a number of paths in parallel. One path has the longest total expected processing time but its variance is zero. All other paths have a slightly smaller total expected processing time but a high variance. The expected makespan of the project is, in theory, the expected maximum of the $k+1$ paths, i.e.,

$$E(C_{\max}) = E(\max(X_1, X_2, \ldots, X_{k+1})),$$

where X_l, $l = 1, \ldots, k+1$, is the random length of path l. Consider the case where the longest path has a total processing time of 51 with zero variance and the k noncritical paths all have mean 50 and a standard deviation of 20. Assume that the k noncritical paths do not overlap and are therefore

4.4 Time/Cost Trade-Offs: Linear Costs

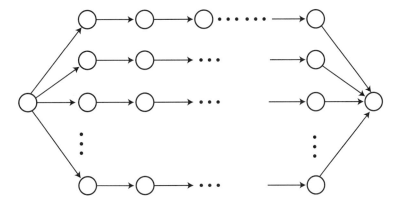

Fig. 4.5. $k+1$ paths in parallel

independent of one another. If the standard PERT procedure is used, then the probability that the makespan is larger than 60 is 0. However, if the k noncritical paths are taken into consideration, then

$$Prob\,(C_{\max} > 60) = 1 - Prob\,(C_{\max} \leq 60).$$

The makespan is less than 60 only if all k paths are less than 60. The probability that one of the paths with mean 50 is less than 60 is

$$\Phi\left(\frac{60-50}{20}\right) = \Phi(0.5) = 0.691.$$

So
$$Prob\,(C_{\max} \leq 60) = (0.691)^k.$$

If k is 5, then
$$Prob\,(C_{\max} \geq 60) = 0.84,$$

i.e., the probability that the makespan is larger than 60 is 84%.

4.4 Time/Cost Trade-Offs: Linear Costs

This section generalizes the deterministic framework presented in Section 4.2. In many situations it may be possible to shorten the processing time of a job by allocating an additional amount of money to the job. We assume in this section and the next that there is a budget that can be used for allocating additional funds to the various jobs.

We assume that the processing time of a job is a linear function of the amount allocated. That is, the more money allocated, the shorter the job, as in Figure 4.6. There is an absolute minimum processing time $p_j^{\min}$ and an absolute maximum processing time $p_j^{\max}$. The cost of processing job j in

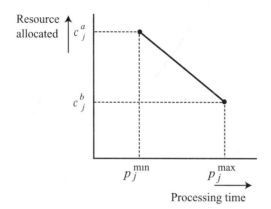

Fig. 4.6. Relationship between job processing time and resource allocated

the minimum amount of time $p_j^{\min}$ is c_j^a and the cost of processing job j in the maximum amount of time $p_j^{\max}$ is c_j^b. Clearly, $c_j^a \geq c_j^b$. Let c_j denote the marginal cost of reducing the processing time of job j by one time unit, i.e.,

$$c_j = \frac{c_j^a - c_j^b}{p_j^{\max} - p_j^{\min}}.$$

So the cost of processing job j in p_j time units, where $p_j^{\min} \leq p_j \leq p_j^{\max}$, is

$$c_j^b + c_j(p_j^{\max} - p_j).$$

Assume there is also a fixed overhead cost c_o that is incurred on a per unit time basis.

In order to determine the minimum cost of the entire project it is necessary to determine the most appropriate processing time for each one of the n jobs. From a complexity point of view this problem is easy and can be solved in a number of ways. We first present a very effective heuristic that often leads to an optimal solution (especially when the problem is small). There are two reasons for presenting this heuristic: first, it is usually the way the problem is dealt with in practice, and, second, it can also be used when costs are nonlinear. After the heuristic we also present a linear programming formulation of the problem, that *always* results in an optimal solution.

Before describing the heuristic it is necessary to introduce some terminology. The precedence graph can be extended by incorporating a source node (with zero processing time) that has an arc emanating to each node that represents a job without predecessors. In the same way each node that represents a job without successors has an arc emanating to a single sink node, which also has zero processing time. A critical path is a longest path from the source to the sink. Let G_{cp} denote the subgraph that consists of the critical path(s)

4.4 Time/Cost Trade-Offs: Linear Costs

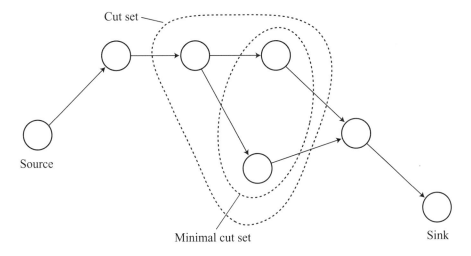

Fig. 4.7. Cut sets in the subgraph of the critical paths

given the current processing times. A *cut set* in this subgraph is a set of nodes whose removal from the subgraph disconnects the source from the sink. A cut set is said to be *minimal* if putting any node back in the graph reestablishes the connection between the source and the sink (see Figure 4.7).

The following heuristic typically leads to a good, but not necessarily optimal allocation.

Algorithm 4.4.1 (Time/Cost Trade-Off Heuristic).

Step 1.

Set all processing times at their maximum.

Determine all critical path(s) with these processing times.

Construct the subgraph G_{cp} of the critical paths.

Step 2.

Determine all minimum cut sets in the current G_{cp}.

Consider only those minimum cut sets of which all processing times are strictly larger than their minimum.

If there is no such set STOP, otherwise go to Step 3.

Step 3.

For each minimum cut set compute the cost of reducing all its processing times by one time unit.

Take the minimum cut set with the lowest cost.

If this lowest cost is less than the overhead cost c_0 per unit time go to Step 4, otherwise STOP.

Step 4.

> *Reduce all the processing times in the minimum cut set by one time unit.*
>
> *Determine the new set of critical paths.*
>
> *Revise graph G_{cp} accordingly and go to Step 2.*

In this heuristic the processing times of the jobs are reduced in each iteration by one time unit. It would be possible to speed up the heuristic by making a larger reduction in the processing times in each iteration. The maximum amount by which the processing times in a minimum cut set can be reduced is determined by two factors. First, while reducing the processing times in a minimum cut set one of these may hit its minimum; when this happens this particular minimum cut set is not relevant any more. Second, while reducing the processing times another path may become critical; when this happens the graph G_{cp} has to be revised and a new collection of minimum cut sets has to be determined.

The reason for presenting the heuristic in the format above is based on the fact that it can also be used when costs are nonlinear (the nonlinear case is considered in the next section).

Example 4.4.2 (Application of Time/Cost Trade-Off Heuristic). Consider again a modification of Example 4.2.3. The processing times in the original example are now the maximum processing times. So the total project time, i.e., the makespan at the maximum processing times, is equal to 56. These processing times can be reduced to some extent at certain costs. The fixed overhead cost per unit time, c_o, is 6. So the total overhead cost over the maximal duration of the project is 336.

Jobs	1	2	3	4	5	6	7	8	9	10	11	12	13	14
$p_j^{\max}$	5	6	9	12	7	12	10	6	10	9	7	8	7	5
$p_j^{\min}$	3	5	7	9	5	9	8	3	7	6	4	5	5	2
c_j^a		20	25	20	15	30	40	35	25	30	20	25	35	20	10
c_j	7	2	4	3	4	3	4	4	4	5	2	2	4	8

Job 12 constitutes a minimum cut set with the lowest cost of processing time reduction. Reducing the processing time of job 12 from 8 to 7 costs 2 and saves 6 in overhead, resulting in a net savings of 4. With the processing time of job 12 equal to 7, a parallel path becomes critical, namely

$$11 \to 13 \to 14.$$

So there are two segments in G_{cp}. The first segment is that part of the original critical path up to job 11; the second segment consists of the two parallel critical paths from job 11 to job 14.

In the first segment there are three cut sets, each one consisting of a single job. Reducing the processing time of job 6 by one time unit results in

4.4 Time/Cost Trade-Offs: Linear Costs

a net savings of 3 (the overhead cost minus the marginal cost of reducing the processing time of job 6). After the processing of job 6 is reduced by one time unit, the path

$$1 \to 2 \to 4 \to 7 \to 10 \to 12$$

becomes critical as well.

There are a number of cut sets in the updated G_{cp}. One cut set is formed by jobs 2 and 11. The total cost of reducing the processing times of these two jobs by one time unit is 4. The net savings is therefore $6 - 4 = 2$. (Job 2 has now hit its minimum processing time.) A cut set that still can be considered is the one that consists of jobs 4 and 11. The makespan can be reduced by shortening the processing times of jobs 4 and 11 by one time unit. The total cost of this reduction is 5 and the net savings is therefore $6 - 5 = 1$. Both jobs 4 and 11 can be reduced one additional time unit, resulting once more in a net savings of 1. Job 11 has now hit its minimum processing time of 4, and cannot be reduced further. The resulting makespan is 51 (see Figure 4.8.a).

However, a number of additional processing time reductions can be accomplished without an increase in total costs. For example, jobs 12 and 13 can both be reduced by two time units without an increase in costs. Jobs 4 and 6 can also be reduced by one time unit each without any increase in costs. The makespan can thus be reduced to 48 without any additional cost (see Figure 4.8.b). This implies that there are a number of optimal solutions, all with a makespan between 48 to 51. The makespan can be reduced even further. However, further tightening increases the total cost.

Although it usually provides a good solution, the heuristic is not guaranteed to yield an optimal solution. The following example illustrates a case where the heuristic yields a suboptimal solution.

Example 4.4.3 (Application of Time/Cost Trade-Off Heuristic). Consider 5 jobs that are subject to the precedence constraints presented in Figure 4.9.

Jobs	1	2	3	4	5
$p_j^{\max}$	5	7	8	4	3
$p_j^{\min}$	4	6	7	3	2
c_j	6	2	2	6	2

The fixed overhead cost per unit time, c_o, is 13. If the processing times of the jobs are at their maximum, the makespan is 12 and there are three critical paths, see Figure 4.9. The total overhead for the duration of the project is therefore 156. The processing time of job j can be reduced at a cost of c_j per unit time.

Applying Algorithm 4.4.1 yields the following result. In the beginning there are various minimum cut sets. The minimum cut set with the smallest cost increase consists of jobs 3, 5, and 2. Reducing jobs 3, 5, and 2 one time unit

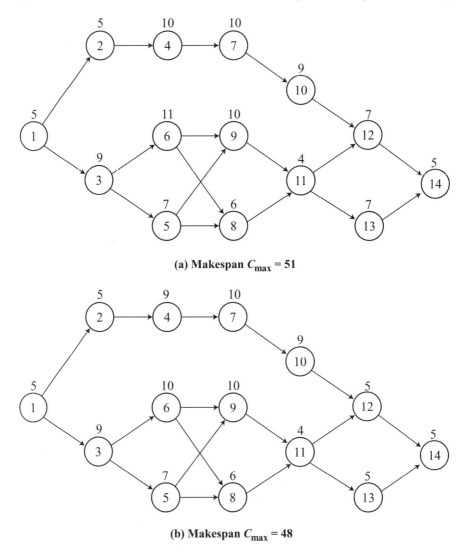

Fig. 4.8. Processing times in Example 4.4.2

costs 6, which is less than the gain of 13 achieved due to reduced overhead. The makespan is now 11 and the critical path graph remains the same. However, jobs 2, 3, and 5 are now processed at their minimum processing times and cannot be reduced further.

At this point there is only one minimum cut set remaining, namely jobs 1 and 4. Reducing both jobs 1 and 4 by one time unit costs 12. This leads to a gain of 1, since the overhead per unit time is 13. The makespan is now 10 and all jobs are processed at their minimum processing times.

4.4 Time/Cost Trade-Offs: Linear Costs

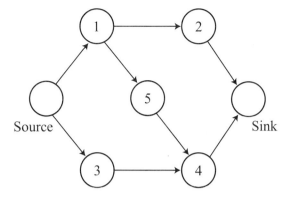

Fig. 4.9. Precedence graph for Example 4.4.3

Algorithm 4.4.1, as presented, stops at this point. However, the solution obtained is not optimal, since job 5 does not have to be processed at its minimum processing time. Extending its processing time to its original processing time does not increase the makespan and reduces the total cost.

When the cost functions are linear, as in Examples 4.4.2 and 4.4.3, the optimization problem can be formulated as a linear program. The processing time of job j, p_j, is a decision variable and not a constant. The earliest possible starting time of job j is a decision variable denoted by x_j, and the makespan $C_{\max}$ is also a decision variable. This implies a total of $2n + 1$ variables. The earliest starting times of the jobs without any predecessors are zero. The duration of each job is subject to two constraints, namely

$$p_j \leq p_j^{\max}$$
$$p_j \geq p_j^{\min}.$$

The total cost of the project, as a function of the processing times $p_1, \ldots, p_n$, is

$$c_o C_{\max} + \sum_{j=1}^{n} \left(c_j^b + c_j(p_j^{\max} - p_j) \right).$$

Since the fixed terms in this cost function do not play a role in the minimization of the objective, the objective is equivalent to

$$c_o C_{\max} - \sum_{j=1}^{n} c_j p_j.$$

If A denotes the set of precedence constraints, then the problem can be formulated as follows.

$$\text{minimize} \quad c_o C_{\max} - \sum_{j=1}^{n} c_j p_j$$

subject to

$$\begin{aligned}
x_k - p_j - x_j &\geq 0 & &\text{for all } j \to k \in A \\
p_j &\leq p_j^{\max} & &\text{for all } j \\
p_j &\geq p_j^{\min} & &\text{for all } j \\
x_j &\geq 0 & &\text{for all } j \\
C_{\max} - x_j - p_j &\geq 0 & &\text{for all } j
\end{aligned}$$

This linear program has $2n+1$ decision variables, i.e., $p_1, \ldots, p_n, x_1, \ldots, x_n$ and $C_{\max}$.

Example 4.4.4 (Linear Programming Formulation). Consider the problem described in Example 4.4.2. This problem can be formulated as a linear program. There are 14 jobs, so there are $2 \times 14 + 1 = 29$ variables. The variable x_1 may be set equal to 0. Actually, this variable does not have to be set equal to zero; the optimization process would force this variable to be zero anyway. The objective function has $14 + 1 = 15$ terms. The first set of constraints consists of 18 constraints, one for each arc in the precedence graph.

In the model under consideration all costs are linear. However, in practice it often occurs that there is a due date associated with the entire project, say d. If the makespan is larger than the due date a penalty is incurred that is proportionate with the tardiness. So instead of the linear overhead cost function considered before, there is a piecewise linear cost function.

The algorithms and the linear programming formulation presented above can be adapted easily to handle this situation. The linear programming formulation can be modified by adding the constraint

$$C_{\max} \geq d.$$

It does not make sense to reduce the makespan to a value that is less than the committed due date, since this would only increase the cost and would not result in any benefit.

4.5 Time/Cost Trade-Offs: Nonlinear Costs

In the previous section the costs of reducing the processing times increase linearly. In practice, the costs of reducing processing times are typically increasing convex (see Figure 4.10.a).

In this section we adopt a discrete time framework. Let $c_j(p_j)$ denote the cost of processing job j in p_j time units, p_j being an integer. We assume that

4.6 Project Scheduling with Workforce Constraints

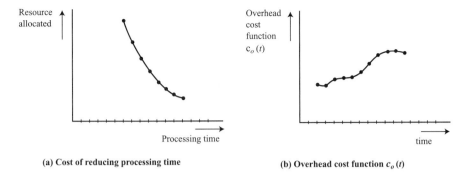

Fig. 4.10. Nonlinear costs

the processing time of job j may take any one of the integer values between $p_j^{\min}$ and $p_j^{\max}$. If the cost is decreasing convex over this range, then

$$c_j(p_j - 1) - c_j(p_j) \geq c_j(p_j) - c_j(p_j + 1).$$

The overhead cost function c_o may also be a function of time. We assume that the overhead cost function is increasing (or nondecreasing) in time (see Figure 4.10.b). Let $c_o(t)$ denote the overhead cost during interval $[t-1, t]$.

An effective heuristic for this problem is similar to Algorithm 4.4.1. Now it is very important to reduce the processing times in a cut set in each iteration by just one time unit. In the linear case it would have been possible to have a larger step size in the reduction of the processing times; in this case it is not advisable since the shape of the cost functions may affect the step size.

A solution for this problem is reached either when no minimum cut sets with reducible processing times remain, or, in case there are such cut sets, the marginal cost of reducing such a cut set is higher than the savings obtained due to the reduced overhead.

It is also possible to formulate this problem in a continuous time setting. This leads to a nonlinear programming formulation that is slightly more general than the linear programming formulation described in the previous section. The objective now is nonlinear, i.e.,

$$\sum_{t=1}^{C_{\max}} c_o(t) + \sum_{j=1}^{n} c_j(p_j).$$

The constraints are the same as those in the linear programming formulation described in the previous section.

4.6 Project Scheduling with Workforce Constraints

In many real world settings there are personnel or workforce constraints. The workforce may consist of various different pools and each pool has a fixed number of operators with a specific skill. Each job requires for its execution a given number from each pool. If the processing of some jobs overlap in time, then the sum of their demands for operators from any given pool may not exceed the total number in that pool. Again, the objective is to minimize the makespan. This problem is in what follows referred to as Project Scheduling with Workforce Constraints. (In the literature this problem is often referred to as Resource Constrained Project Scheduling.)

In order to formulate this problem some additional notation is needed. Let N denote the number of different pools in the workforce. Let W_i denote the total number of operators in pool i and let W_{ij} denote the number of operators job j needs from pool i.

Example 4.6.1 (Workforce Constraints). Consider the following instance with five jobs and two types of operators. There are four of type 1 and eight of type 2. The processing times and workforce requirements are presented in the table below.

$W_1 = 4$

$W_2 = 8$

Jobs	1	2	3	4	5
p_j	8	4	6	4	4
W_{1j}	2	1	3	1	2
W_{2j}	3	0	4	0	3

The precedence constraints are specified below.

Job	Immediate Predecessors	Immediate Successors
1	—	4
2	—	5
3	—	5
4	1	—
5	2,3	—

If there would not have been any workforce constraints, then the critical path is $1 \to 4$ and $C_{\max} = 12$. However, there is no feasible schedule with $C_{\max} = 12$ that satisfies the workforce constraints.

The optimal schedule has $C_{\max} = 18$. According to this schedule jobs 2 and 3 are processed during the interval $[0, 6]$, jobs 1 and 5 are processed during the interval $[6, 14]$ and job 4 is processed during the interval $[14, 18]$.

In contrast to the project scheduling problems described in Sections 4.2 and 4.4, the project scheduling problem with workforce constraints is intrinsically very hard. There is no linear programming formulation for this problem; however, there is an integer programming formulation.

In order to formulate this problem as an integer program, assume that all processing times are fixed and integer. Introduce a dummy job $n+1$ with

4.6 Project Scheduling with Workforce Constraints

zero processing time. Job $n+1$ succeeds all other jobs, i.e., all jobs without successors have an arc emanating to job $n+1$. Let x_{jt} denote a $0-1$ variable that assumes the value 1 if job j is completed exactly at time t and the value 0 otherwise. So the number of operators job j needs from pool i in the interval $[t-1,t]$ is

$$W_{ij} \sum_{u=t}^{t+p_j-1} x_{ju}.$$

Let H denote an upper bound on the makespan. A simple, but not very tight, bound can be obtained by setting

$$H = \sum_{j=1}^{n} p_j.$$

So the completion time of job j can be expressed as

$$\sum_{t=1}^{H} t\, x_{jt}$$

and the makespan as

$$\sum_{t=1}^{H} t\, x_{n+1,t}.$$

The integer program can be formulated as follows:

$$\text{minimize} \quad \sum_{t=1}^{H} t\, x_{n+1,t}$$

subject to

$$\sum_{t=1}^{H} t\, x_{jt} + p_k - \sum_{t=1}^{H} t\, x_{kt} \leq 0 \qquad \text{for all } j \to k \in A$$

$$\sum_{j=1}^{n} \left(W_{ij} \sum_{u=t}^{t+p_j-1} x_{ju} \right) \leq W_i \qquad \text{for all } i \text{ and } t$$

$$\sum_{t=1}^{H} x_{jt} = 1 \qquad \text{for all } j$$

The objective of the integer program is to minimize the makespan. The first set of constraints ensure that the precedence constraints are enforced, i.e., if job j is followed by job k, then the completion time of job k has to be greater than or equal to the completion of job j plus p_k. The second set of constraints ensure that the total demand for pool i at time t does not surpass the availability of pool i. The third set of constraints ensure that each job is processed.

Since this integer program is very hard to solve when the number of jobs is large and the time horizon is long, it is typically solved via heuristics. It turns out that even special cases of this problem are quite hard. However, for a number of important special cases heuristics have been developed that have proven to be quite effective.

The next chapter focuses on a very important special case of this problem, namely the job shop scheduling problem with the makespan objective. A very effective heuristic for this particular special case is described in Section 5.4. Other special cases of the project scheduling problem with workforce constraints are the timetabling problems described in Sections 9.4 and 9.5. Several heuristics are presented there as well.

4.7 ROMAN: A Project Scheduling System for the Nuclear Power Industry

Nuclear power plants have to be shut down periodically for refueling, maintenance and repair. These shutdown periods are referred to as *outages*. Managing an outage is a daunting project that may involve anywhere from 10,000 to 45,000 jobs. A good schedule is not only crucial for nuclear safety reasons but also for economic reasons. The cost per day of shutdown is in the order of $1,000,000.

The outage scheduling problem can be described as follows: Given a set of outage jobs, resources and precedence constraints, the system has to assign resources to jobs for specific time periods so that the makespan is minimized and the jobs are performed safely. There are four types of jobs that have to be done during an outage:

   (i) refueling jobs,
   (ii) repairs,
   (iii) plant modifications, and
   (iv) maintenance jobs.

An outage must be planned and managed to minimize shutdown risks by taking appropriate preventive measures. The main safety functions and systems components that must be monitored during an outage are:

   (i) the AC power control system,
   (ii) the primary and secondary containment,
   (iii) the fuel pool cooling system,
   (iv) the inventory control,
   (v) the reactivity control,
   (vi) the shutdown cooling, and
   (vii) the vital support systems.

4.7 ROMAN: A Project Scheduling System for the Nuclear Power Industry

Current automated outage management approaches are based on conventional project scheduling techniques such as CPM and PERT. The systems used in practice are typically commercial software packages that are rather generic (see Appendix E). Safety and risk assessments are usually done manually. This requires experienced personnel to take the decisions with regard to the various scheduling alternatives. The final schedules are validated using a simulator developed by the Electric Power Research Institute (EPRI).

Generic systems based on CPM and PERT techniques suffer from major limitations with regard to this specific application because they generally cannot take safety considerations into account. The Rome Laboratory of the U.S. Air Force, in collaboration with EPRI, Kaman Science and Kestrel Insitute, designed and developed a system specifically for scheduling outages. The system is referred to as *ROMAN* (Rome Laboratory Outage MANager).

A key concept in enforcing safety constraints is the *state of the plant,* which is measured in colors: green, yellow, orange and red, in increasing order of risk. The state of the plant is computed from complex decision trees regarding safety levels; see, for example, the safety function of the AC power control depicted in Figure 4.11. If there is a job being processed that may cause AC power loss, then, for the plant to be in a yellow state, two off-site AC power sources and three operable emergency safeguard buses have to be available.

ROMAN uses two measures associated with the processing time of a job: the *definite period* and the *potential period.* The definite period of a job corresponds to that period of time during which the job is definitely being worked on; it is the time interval between the latest start time of the job (LST) and its earliest finish time (EFT). The potential period of a job corresponds to the period of time during which it may be processed, i.e., the time period between its earliest start time (EST) and its latest finish time (LFT). In a sense, the definite period represents a lower bound on the processing time of a job and the potential period an upper bound.

With regard to the safety considerations, two concepts are defined, namely, the *definite* state of the plant and the *potential* state of the plant. The definite state of the plant is associated with the definite periods and represents the state of the plant with regard to a given safety function, assuming that all jobs are done within their definite periods. The potential state of the plant is associated with the potential periods. The risk associated with the potential state of the plant is always higher than the risk associated with the definite state of the plant, since the processing times of the jobs in the potential state are longer than those in the definite state.

The schedule generation method adopted in ROMAN is based on a constraint programming approach, i.e., a global search method combined with constraint propagation. Initially, the search assumes that the processing times are at their minimum, i.e., the processing takes place during the definite periods. The search method attempts to generate a feasi-

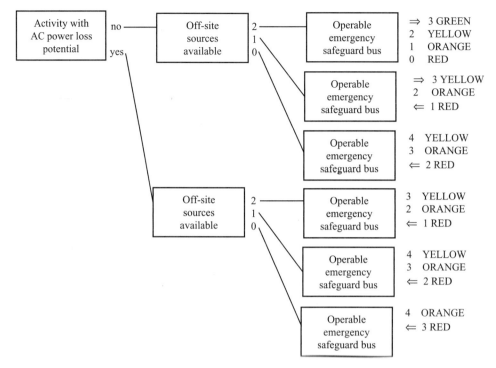

Fig. 4.11. Example of a decision tree for the safety function of the AC power control

ble schedule in a constructive manner. At every step, a job that has not been scheduled yet (with a given time window) is appended to the partial schedule. This is followed by a constraint propagation procedure, which recomputes the time windows of all jobs, enforcing the precedence constraints as well as the safety constraints regarding the definite periods. After the contraint propagation step has been completed, the process repeats itself.

If the schedule obtained by this initial global search does not satisfy the safety requirements, the time windows have to be adjusted in order to satisfy the safety constraints over the potential periods of all the jobs. This phase involves solving another global search problem, this time considering the potential periods of the jobs.

ROMAN has proven to be successful since it extends the functionality offered by existing software tools for outage management. All the technological constraints currently used for automatic schedule generation are incorporated into the system. In addition, ROMAN generates schedules that satisfy the safety constraints. The latest version of ROMAN schedules up to 2000 jobs in approximately 1 minute on a Sparc 2. The schedules produced are often better than those obtained manually since many new possibilities are explored.

4.7 ROMAN: A Project Scheduling System for the Nuclear Power Industry

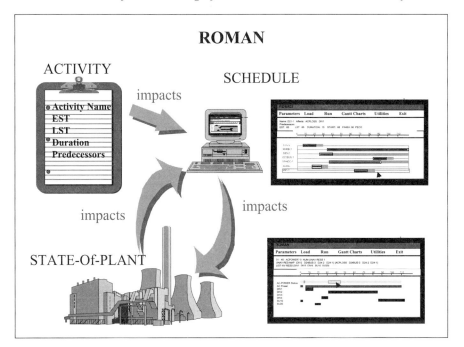

Fig. 4.12. State of plant with respect to AC power

Human schedulers tend to aggregate jobs and schedule them as blocks rather than explore interesting possibilities that occur when the jobs are scheduled one at a time.

A key feature of ROMAN that utility personnel find attractive is the robustness of the schedules that are generated. The current scheduler generates a schedule that includes start time windows for each job. Choosing the start time of a job within its window allows for a feasible execution of the schedule. The window provides information about how critical the start time of a job is; if a predecessor job is delayed, a user can decide whether there still is enough freedom in the start time window to allow on-time completion, or whether it is time to reschedule parts of the overall operation.

ROMAN comes configured with a Graphics User Interface (GUI) that displays an interactive Gantt chart for jobs, showing their start time window, duration, job description and predecessors. Another Gantt chart shows the history of the state of the plant with respect to AC power (see Figure 4.12).

The outage management problem is somewhat similar to the project scheduling problem with workforce constraints discussed in Section 4.6. However, it would have been difficult to formulate the outage management problem as an integer program like the one presented in Section 4.6. There are many constraints that are nonlinear, subjective, and hard to formulate.

4.8 Discussion

Project scheduling is probably the area in scheduling that has received the most attention from practitioners. The more important deterministic project scheduling problems that are not subject to workforce constraints are well solved.

The problems with workforce constraints are obviously harder than the problems without workforce constraints. The workforce constrained project scheduling problem is an important problem in the scheduling literature since many other well-known problems are special cases. One of the most important special cases of this problem is the job shop scheduling problem with the makespan objective that is considered in Chapter 5. The project scheduling problem with workforce constraints becomes a job shop scheduling problem when each pool consists of a single operator with a specific skill. An operator is then equivalent to a machine and the precedence constraints in the project scheduling problem are equivalent to the routing constraints in the job shop scheduling problem.

The workforce structure described in this chapter is somewhat special. All the operators in a given pool are assumed to have the same skill and capable of doing a task that cannot be done by an operator from any other pool. A more general structure is based on the following assumptions: The workforce has a number of skill sets. Each operator has a subset of these skill sets. Two operators may have skill sets that partially overlap, i.e., the first operator has skills A and B, and the second operator has skills A and C. When operators have partially overlapping skill sets the project scheduling problem with workforce constraints becomes more complicated.

When the processing times are random, then the problems become even harder. Even without workforce constraints, problems with random processing times are hard; the PERT procedure may not always give a satisfactory solution. Workforce constrained project scheduling problems with random processing times have not received much attention in the literature. This is an area in planning and scheduling that clearly deserves more attention.

Exercises

4.1. Consider the job-on-node representation depicted in Example 4.2.3. Transform this job-on-node representation into a job-on-arc representation. Compare the number of nodes and arcs in the first representation with the number of nodes and arcs in the second representation. Can you make any general statement with regard to such a comparison?

4.2. Construct the smallest precedence constraints graph in a job-on-node format that requires a dummy job in the corresponding job-on-arc format (*Hint:* You need 4 jobs).

4.3. Consider the setup of the production facility described in Example 4.1.1. The durations of the 8 jobs are tabulated below.

Jobs	1	2	3	4	5	6	7	8
p_j	4	6	10	12	10	2	4	2

(a) Compute the makespan and determine the critical path(s).

(b) Suppose that the duration of job 7 can be reduced by 3 weeks to 1 week. Compute the new makespan and determine the new critical path(s).

4.4. Consider the installation of the production planning and scheduling system described in Exercise 3.1.

Jobs	1	2	3	4	5	6	7	8	9	10	11
p_j	8	5	6	3	2	5	2	4	7	4	9

(a) Compute the makespan and determine the critical path.

Suppose that each job can be shortened at a certain expense. The overhead cost is 6 per week (in tens of thousands of dollars). The cost functions are linear. The minimum and maximum processing times as well as the marginal costs are tabulated below.

Jobs	1	2	3	4	5	6	7	8	9	10	11
p_j^{max}	8	5	6	3	2	5	2	4	7	4	9
p_j^{min}	5	3	4	2	2	3	2	3	5	3	7
c_j^a	30	25	20	15	30	40	35	25	30	20	30
c_j	7	2	2	1	2	3	4	4	4	5	4

(b) Apply Algorithm 4.4.1 to this instance.

(c) Verify whether the solution obtained in (b) is optimal.

4.5. Consider Example 4.2.3. Assume that there are an unlimited number of machines in parallel and each machine can do any one of the 14 jobs.

(a) What is the minimum number of machines needed to achieve the minimum makespan? Call this minimum number m_{min}.

(b) If the number of available machines is $m_{min} - 1$, by how much does the makespan go up?

(c) Plot the relationship between the makespan and the number of machines available, $m = 1, \ldots, m_{min}$.

4.6. Consider again Example 4.1.1. Suppose it is possible to add resources to the various jobs on the critical paths in order to reduce the makespan. The overhead cost c_o per unit time is 6. The costs are linear and the marginal costs are tabulated below.

Jobs	1	2	3	4	5	6	7	8
p_j^{max}	4	6	10	12	10	2	4	2
p_j^{min}	2	5	7	10	8	1	2	1
c_j^a	20	25	20	15	30	40	35	25
c_j	4	3	4	2	3	2	4	4

(a) Determine the processing times and the makespan of the solution with the total minimum cost. Is the solution unique?

(b) Plot the relationship between the total cost of the project and the makespan.

4.7. Consider two independent random variables with the following three point distribution.

$$Prob\,(X = 1) = 0.25$$
$$Prob\,(X = 2) = 0.50$$
$$Prob\,(X = 3) = 0.25$$

Consider the convolution or sum of the two independent random variables with this three point distribution. This new distribution has mass on the points 2, 3, 4, 5, and 6.

(a) Compute the expectation and the variance of this distribution.

(b) Plot the distribution function and plot the distribution function of the normal distribution with the same mean and variance.

(c) Do the same with the convolution of three independent random variables with the original three point distribution.

4.8. Consider the following PERT version of the installation of the production planning and scheduling system described in Exercises 3.1 and 4.4.

Jobs	1	2	3	4	5	6	7	8	9	10	11
p_j^a	2	1	5	2	1	4	1	16	1		8
p_j^m	8	5	6	3	2	5	2	4	7	4	9
p_j^b	14	9	7	4	3	6	3	7	8	7	10

(a) Rank the paths according to the means of their total processing time.

(b) Rank the paths according to the variance in their total processing times. Which path has the highest variance? Which path has the lowest variance?

(c) Compute the probability that the makespan is larger than 27 following the standard PERT procedure.

(d) Compute the probability that the makespan is larger than 27 by considering only the path with the largest variance.

4.9. Formulate the resource allocation problem described in Example 4.4.3 as a linear program.

4.10. Formulate the following project scheduling problem with workforce constraints as an integer program.

Jobs	1	2	3
p_j	4	6	2
W_{1j}	2	3	1
W_{2j}	3	4	0

Both workforce pools have four operators. The precedence constraints are specified below.

Job	Immediate Predecessors	Immediate Successors
1	–	–
2	–	3
3	2	–

Comments and References

The Critical Path Method was first presented in an industry report of DuPont and Remington Rand by Walker and Sayer (1959) and the PERT method was first described in a U.S. Navy report under the title PERT (1958).

Since then these two subjects have been the focus of many books; see, for example, Moder and Philips (1970), Wiest and Levy (1977), Kerzner (1994), Neumann, Schwindt, and Zimmermann (2001), and Demeulemeester and Herroelen (2002). Project scheduling is often also covered in broader based scheduling and production planning and control books; see, for example, Baker (1974), and Morton and Pentico (1993).

PERT has received a significant amount of attention from the research community as well; see, for example, Fulkerson (1962), Elmaghraby (1967), and Sasieni (1986).

Project scheduling with workforce constraints is in the literature usually referred to as resource constrained scheduling. This area also has received an enormous amount of attention from the research community; see, for example, Balas (1970), Talbot (1982), Patterson (1984), Kolisch (1995), and Brucker, Drexl, Möring, Neumann, and Pesch (1999).

The ROMAN system is described in detail by Alguire and Gomes (1996) and Gomes, Smith, and Westfold (1996).

Chapter 5

Machine Scheduling and Job Shop Scheduling

5.1	Introduction	81
5.2	Single Machine and Parallel Machine Models	82
5.3	Job Shops and Mathematical Programming	84
5.4	Job Shops and the Shifting Bottleneck Heuristic	87
5.5	Job Shops and Constraint Programming	93
5.6	LEKIN: A Generic Job Shop Scheduling System	102
5.7	Discussion	109

5.1 Introduction

This chapter focuses on job shops. There are n jobs and each job visits a number of machines following a predetermined route. In some models a job may visit any given machine at most once and in other models a job may visit each machine more than once. In the latter case it is said that the job shop is subject to recirculation. A generalization of the basic job shop is a so-called flexible job shop. A flexible job shop consists of a collection of workcenters and each workcenter consists of a number of identical machines in parallel. Each job follows a predetermined route visiting a number of workcenters; when a job visits a workcenter, it may be processed on any one of the machines at that workcenter.

Job shops are prevalent in industries where each customer order is unique and has its own parameters. Wafer fabs in the semiconductor industry often function as job shops; an order usually implies a batch of a certain type of item and the batch has to go through the facility following a certain route with given processing times. Another classical example of a job shop is a hospital. The patients in a hospital are the jobs. Each patient has to follow a given route and has to be treated at a number of different stations while he or she goes through the system.

Some of the job shop problems considered in this chapter are, from a mathematical point of view, special cases of the project scheduling problem with workforce constraints described in Section 4.6. These job shop problems are NP-hard and cannot be formulated as linear programs. However, they can be formulated either as integer programs or as disjunctive programs.

This chapter is organized as follows. The second section focuses on job shops that consist of a single workcenter; these job shops are basically equivalent to either a single machine or a number of machines in parallel. Various objective functions are considered. The third section presents a mathematical programming formulation of a job shop with a single machine at each workcenter and the makespan objective. The fourth section describes the well-known shifting bottleneck heuristic which is tailor-made for the job shop and can be adapted to the flexible job shop as well. The fifth section focuses on an entirely different approach for the job shop with the makespan objective, namely the constraint programming approach. This approach has gained popularity over the last decade with real world scheduling problems. The subsequent section describes the LEKIN scheduling system which is specifically designed for job shops and flexible job shops. The discussion section lists some other popular solution techniques that have been used in practice for job shop scheduling.

All the models considered in this chapter assume that jobs are not allowed to be preempted. A job, once started on a machine, remains on that machine until it is completed.

5.2 Single Machine and Parallel Machine Models

A single machine is the simplest case of a job shop and a parallel machine environment is equivalent to a flexible job shop that consists of a single workcenter. Decomposition techniques for more complicated job shops often have to consider single machine and parallel machine subproblems within their framework.

Consider a single machine and n jobs. Job j has a processing time p_j, a release date r_j and a due date d_j. If $r_j = 0$ and $d_j = \infty$, then the processing of job j is basically unconstrained. It is clear that the makespan $C_{\max}$ in a single machine environment does not depend on the schedule. For various other objectives certain priority rules generate optimal schedules. If the objective to be minimized is the total weighted completion time, i.e., $\sum w_j C_j$, and the processing of the jobs is unconstrained, then the *Weighted Shortest Processing Time first (WSPT)* rule, which schedules the jobs in decreasing order of w_j/p_j, is optimal. If the objective is the maximum lateness $L_{\max}$ and the jobs are all released at time 0, then the *Earliest Due Date first (EDD)* rule, which schedules the jobs in increasing order of d_j, is optimal. Both the WSPT rule and the EDD rule are examples of *static* priority rules. A rule that is somewhat related to the EDD rule is the so-called *Minimum Slack first (MS)* rule which selects, when the machine becomes available at time t, the job with the

5.2 Single Machine and Parallel Machine Models

least slack; the slack of job j at time t is defined as $\max(d_j - p_j - t, 0)$. This rule does not operate in exactly the same way as the EDD rule, but will result in schedules that are somewhat similar to EDD schedules. However, the MS rule is an example of a *dynamic* priority rule, in which the priority of each job is a function of time.

Other objectives, such as the total tardiness $\sum T_j$ and the total weighted tardiness $\sum w_j T_j$, are much harder to optimize than the total weighted completion time or the maximum lateness. A heuristic for the total weighted tardiness objective is described in Appendix C. This heuristic is referred to as the *Apparent Tardiness Cost first (ATC)* rule. If the machine is freed at time t, the ATC rule selects among the remaining jobs the job with the highest value of

$$I_j(t) = \frac{w_j}{p_j} \exp\left(-\frac{\max(d_j - p_j - t, 0)}{K\bar{p}}\right),$$

where K is a so-called scaling parameter and $\bar{p}$ is the average of the processing times of the jobs that remain to be scheduled. The ATC priority rule is actually a weighted mixture of the WSPT and MS priority rules mentioned above. By adjusting the scaling parameter K the rule can be made to operate either more like the WSPT rule or more like the MS rule.

When the jobs in a single machine problem have different release dates r_j, then the problems tend to become significantly more complicated. One famous problem that has received a significant amount of attention is the nonpreemptive single machine scheduling problem in which the jobs have different release dates and the maximum lateness $L_{\max}$ has to be minimized. This problem is NP-Hard, which implies that, unfortunately, no efficient (polynomial time) algorithm exists for this problem. The problem can be solved either by branch-and-bound or by dynamic programming (a branch-and-bound method for this problem is described in Appendix B). It turns out that a procedure for solving this single machine problem is an important step within the well-known shifting bottleneck heuristic for general job shops.

When there are m machines in parallel the makespan $C_{\max}$ is schedule dependent. The makespan objective, which plays an important role when the loads of the various machines have to be balanced, gives rise to another priority rule, namely the *Longest Processing Time first (LPT)* rule. According to this rule, whenever one of the machines is freed, the longest job among those waiting for processing is selected to go next. The intuition behind this rule is clear: One would like to process the smaller jobs towards the end of the schedule since that makes it easier to balance the machines. If the smaller jobs go last, then the longer jobs have to be scheduled more in the beginning. The LPT rule, unfortunately, does not guarantee an optimal solution (it can be shown that this rule guarantees a solution that is within 33% of the optimal solution, see Exercise 5.3).

The SPT and the WSPT rules are important in the parallel machine setting as well. If all n jobs are available at $t = 0$, then the nonpreemptive SPT rule minimizes the total completion time $\sum C_j$; the nonpreemptive SPT rule remains optimal even when preemptions are allowed. Unfortunately, minimizing the total weighted completion time $\sum w_j C_j$ in a parallel machine setting when all jobs are available at $t = 0$ is NP-Hard, so the WSPT rule does not minimize the $\sum w_j C_j$ in this case. However, the WSPT rule still can be used as a heuristic and it is guaranteed to generate a solution that is within 22% of optimality.

The more general parallel machine problem with the total weighted tardiness objective ($\sum w_j T_j$) is even harder. The ATC rule described above is applicable to the parallel machine setting as well; however, the quality of the solutions may at times leave something to be desired.

5.3 Job Shops and Mathematical Programming

Consider a job shop with n jobs and m machines. Each job has to be processed by a number of machines in a given order and there is no recirculation. The processing of job j on machine i is referred to as operation (i, j) and its duration is p_{ij}. The objective is to minimize the makespan $C_{\max}$.

The problem of minimizing the makespan in a job shop without recirculation can be represented by a so-called disjunctive graph. Consider a directed graph G with a set of nodes N and two sets of arcs A and B. The nodes N correspond to all of the operations (i, j) that must be performed on the n jobs. The so-called *conjunctive* (solid) arcs A represent the routes of the jobs. If arc $(i, j) \to (h, j)$ is part of A, then job j has to be processed on machine i before it is processed on machine h, i.e., operation (i, j) precedes operation (h, j). Two operations that belong to two different jobs and which have to be processed on the same machine are connected to one another by two so-called *disjunctive* (broken) arcs going in opposite directions. The disjunctive arcs B form m cliques of double arcs, one clique for each machine. (A clique is a term in graph theory that refers to a graph in which any two nodes are connected to one another; in this case each connection within a clique is a pair of disjunctive arcs.) All operations (nodes) in the same clique have to be done on the same machine. All arcs emanating from a node, conjunctive as well as disjunctive, have as length the processing time of the operation that is represented by that node. In addition there is a source U and a sink V, which are dummy nodes. The source node U has n conjunctive arcs emanating to the first operations of the n jobs and the sink node V has n conjunctive arcs coming in from all the last operations. The arcs emanating from the source have length zero, see Figure 5.1. We denote this graph by $G = (N, A, B)$.

A feasible schedule corresponds to a *selection* of one disjunctive arc from each pair such that the resulting directed graph is acyclic. This implies that each selection of arcs from within a clique must be acyclic. Such a selection determines the sequence in which the operations are to be performed on that machine. That a selection from a clique has to be acyclic can be argued as

5.3 Job Shops and Mathematical Programming

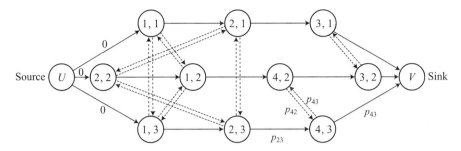

Fig. 5.1. Directed graph for job shop with the makespan as the objective

follows: If there were a cycle within a clique, a feasible sequence of the operations on the corresponding machine would not have been possible. It may not be immediately obvious why there should not be any cycle formed by conjunctive arcs and disjunctive arcs from different cliques. Such a cycle would also correspond to a situation that is infeasible. For example, let (h,j) and (i,j) denote two consecutive operations that belong to job j and let (i,k) and (h,k) denote two consecutive operations that belong to job k. If under a given schedule operation (i,j) precedes operation (i,k) on machine i and operation (h,k) precedes operation (h,j) on machine h, then the graph contains a cycle with four arcs, two conjunctive arcs and two disjunctive arcs from different cliques. Such a schedule is physically impossible. Summarizing, if D denotes the subset of the selected disjunctive arcs and the graph $G(D)$ is defined by the set of conjunctive arcs and the subset D, then D corresponds to a feasible schedule if and only if $G(D)$ contains no directed cycles.

The makespan of a feasible schedule is determined by the longest path in $G(D)$ from the source U to the sink V. This longest path consists of a set of operations of which the first starts at time 0 and the last finishes at the time of the makespan. Each operation on this path is immediately followed either by the next operation on the same machine or by the next operation of the same job on another machine. The problem of minimizing the makespan is reduced to finding a selection of disjunctive arcs that minimizes the length of the longest path (i.e., the *critical* path).

There are several mathematical programming formulations for the job shop without recirculation, including a number of integer programming formulations, see Section 4.6 and Exercise 5.4. However, the formulation most often used is the so-called disjunctive programming formulation, see Appendix A. This disjunctive programming formulation is closely related to the disjunctive graph representation of the job shop.

To present the disjunctive programming formulation, let the variable y_{ij} denote the starting time of operation (i,j). Recall that set N denotes the set of all operations (i,j), and set A the set of all routing constraints $(i,j) \to (h,j)$

which require job j to be processed on machine i before it is processed on machine h. The following mathematical program minimizes the makespan.

$$\text{minimize } C_{\max}$$

subject to

$$y_{hj} - y_{ij} \geq p_{ij} \qquad \text{for all } (i,j) \to (h,j) \in A$$
$$C_{\max} - y_{ij} \geq p_{ij} \qquad \text{for all } (i,j) \in N$$
$$y_{ij} - y_{ik} \geq p_{ik} \text{ or } y_{ik} - y_{ij} \geq p_{ij} \qquad \text{for all } (i,k) \text{ and } (i,j),\ i=1,\ldots,m$$
$$y_{ij} \geq 0 \qquad \text{for all } (i,j) \in N$$

In this formulation, the first set of constraints ensure that operation (h,j) cannot start before operation (i,j) is completed. The third set of constraints are called the disjunctive constraints; they ensure that some ordering exists among operations of different jobs that have to be processed on the same machine. Because of these constraints this formulation is referred to as the disjunctive programming formulation.

Example 5.3.1 (Disjunctive Programming Formulation). Consider the following example with four machines and three jobs. The route, i.e., the machine sequence, as well as the processing times are presented in the table below.

jobs	machine sequence	processing times
1	1, 2, 3	$p_{11} = 10,\quad p_{21} = 8,\quad p_{31} = 4$
2	2, 1, 4, 3	$p_{22} = 8,\quad p_{12} = 3,\quad p_{42} = 5,\quad p_{32} = 6$
3	1, 2, 4	$p_{13} = 4,\quad p_{23} = 7,\quad p_{43} = 3$

The objective consists of the single variable $C_{\max}$. The first set of constraints consists of seven constraints: two for job 1, three for job 2 and two for job 3. For example, one of these is

$$y_{21} - y_{11} \geq 10 \ (= p_{11}).$$

The second set consists of ten constraints, one for each operation. An example is

$$C_{\max} - y_{11} \geq 10 \ (= p_{11}).$$

The set of disjunctive constraints contains eight constraints: three each for machines 1 and 2 and one each for machines 3 and 4 (there are three operations to be performed on machines 1 and 2 and two operations on machines 3 and 4). An example of a disjunctive constraint is

$$y_{11} - y_{12} \geq 3 \ (= p_{12}) \quad \text{or} \quad y_{12} - y_{11} \geq 10 \ (= p_{11}).$$

The last set includes ten nonnegativity constraints, one for each starting time.

That a scheduling problem can be formulated as a disjunctive program does not imply that there is a standard solution procedure available that always will work satisfactorily. Minimizing the makespan in a job shop is a very hard problem and solution procedures are either based on enumeration or on heuristics. To find optimal solutions branch-and-bound methods have to be used. Appendix B provides an example of a branch-and-bound procedure applied to a job shop.

5.4 Job Shops and the Shifting Bottleneck Heuristic

Since it has turned out to be very hard to solve job shop problems with large numbers of jobs to optimality, many heuristic procedures have been designed. One of the most successful procedures for minimizing the makespan in a job shop is the *Shifting Bottleneck* heuristic.

In the following overview of the Shifting Bottleneck heuristic M denotes the set of all m machines. In the description of an iteration of the heuristic it is assumed that in previous iterations a selection of disjunctive arcs has been fixed for a subset M_0 of machines. So for each one of the machines in M_0 a sequence of operations has already been determined.

An iteration results in a selection of a machine from $M - M_0$ for inclusion in set M_0. The sequence in which the operations on this machine are to be processed is also generated in this iteration. To determine which machine should be included next in M_0, an attempt is made to determine which unscheduled machine causes in one sense or another the severest disruption. To determine this, the original directed graph is modified by deleting *all* disjunctive arcs of the machines still to be scheduled (i.e., the machines in set $M - M_0$) and keeping only the relevant disjunctive arcs of the machines in set M_0 (one from every pair). Call this graph G'. Deleting all disjunctive arcs of a specific machine implies that all operations on this machine, which originally were supposed to be done on this machine one after another, now may be done in parallel (as if the machine has infinite capacity, or equivalently, each one of these operations has the machine for itself). The graph G' has one or more critical paths that determine the corresponding makespan. Call this makespan $C_{\max}(M_0)$.

Suppose that operation (i,j), $i \in \{M - M_0\}$, has to be processed in a time window of which the release date and due date are determined by the critical (longest) paths in G', i.e., the release date is equal to the longest path in G' from the source to node (i,j) and the due date is equal to $C_{\max}(M_0)$, minus the longest path from node (i,j) to the sink, plus p_{ij}. Consider each of the machines in $M - M_0$ as a separate nonpreemptive single machine problem with release dates and due dates and with the maximum lateness to be minimized. As stated in the previous section, this problem is NP-hard; however, procedures have been developed that perform reasonably well. The minimum

$L_{\max}$ of the single machine problem corresponding to machine i is denoted by $L_{\max}(i)$ and is a measure of the criticality of machine i.

After solving all these single machine problems, the machine i with the *largest* maximum lateness $L_{\max}(i)$ is chosen. Among the remaining machines, this machine is in a sense the most critical or the "bottleneck" and therefore the one to be included next in M_0. Label this machine h, call its maximum lateness $L_{\max}(h)$ and schedule it according to the optimal solution obtained for the single machine problem associated with this machine. If the disjunctive arcs that specify the sequence of operations on machine h are inserted in graph G', then the makespan of the current partial schedule increases by at least $L_{\max}(h)$, that is,

$$C_{\max}(M_0 \cup h) \geq C_{\max}(M_0) + L_{\max}(h).$$

Before starting the next iteration and determining the next machine to be scheduled, an additional step has to be done within the current iteration. In this additional step all the machines in the original set M_0 are resequenced one by one in order to see if the makespan can be reduced. That is, a machine, say machine l, is taken out of set M_0 and a graph G'' is constructed by modifying graph G' through the inclusion of the disjunctive arcs that specify the sequence of operations on machine h and the exclusion of the disjunctive arcs associated with machine l. Machine l is resequenced by solving the corresponding single machine maximum lateness problem with the release and due dates determined by the critical paths in graph G'''. Resequencing each of the machines in the original set M_0 completes the iteration.

In the next iteration the entire procedure is repeated and another machine is added to the current set $M_0 \cup h$. The shifting bottleneck heuristic can be summarized as follows.

Algorithm 5.4.1 (Shifting Bottleneck Heuristic).

Step 1. *(Initial conditions)*

Set $M_0 = \emptyset$.

Graph G is the graph with all the conjunctive arcs and no disjunctive arcs.

Set $C_{\max}(M_0)$ equal to the longest path in graph G.

Step 2. *(Analysis of machines still to be scheduled)*

Do for each machine i in set $M - M_0$ the following: formulate a single machine problem with all operations subject to release dates and due dates (the release date of operation (i,j) is determined by the longest path in graph G from the source to node (i,j); the due date of operation (i,j) is determined by considering the longest path in graph G from node (i,j) to the sink and subtracting p_{ij}).

Minimize the $L_{\max}$ in each one of these single machine subproblems.

Let $L_{\max}(i)$ denote the minimum $L_{\max}$ in the subproblem corresponding to machine i.

5.4 Job Shops and the Shifting Bottleneck Heuristic

Step 3. *(Bottleneck selection and sequencing)*

Let
$$L_{\max}(h) = \max_{i \in \{M - M_0\}} (L_{\max}(i))$$

Sequence machine h according to the sequence generated for it in Step 2.

Insert all the corresponding disjunctive arcs in graph G.

Insert machine h in M_0.

Step 4. *(Resequencing of all machines scheduled earlier)*

Do for each machine $l \in \{M_0 - h\}$ the following:

Delete the corresponding disjunctive arcs from G; formulate a single machine subproblem for machine l with release dates and due dates of the operations determined by longest path calculations in G.

Find the sequence that minimizes $L_{\max}(l)$ and insert the corresponding disjunctive arcs in graph G.

Step 5. *(Stopping criterion)*

If $M_0 = M$ then STOP, otherwise go to Step 2.

The structure of the shifting bottleneck heuristic shows the relationship between the bottleneck concept and the more combinatorial concepts such as critical (longest) path and maximum lateness. A critical path indicates the location and the timing of a bottleneck. The maximum lateness gives an indication of the amount by which the makespan increases if a machine is added to the set of machines already scheduled. The following example illustrates the use of the shifting bottleneck heuristic.

Example 5.4.2 (Application of Shifting Bottleneck Heuristic). Consider the instance with four machines and three jobs described in Examples 5.3.1. The routing, i.e., the machine sequences, and the processing times are given in the following table:

jobs	machine sequence	processing times
1	1,2,3	$p_{11} = 10$, $p_{21} = 8$, $p_{31} = 4$
2	2,1,4,3	$p_{22} = 8$, $p_{12} = 3$, $p_{42} = 5$, $p_{32} = 6$
3	1,2,4	$p_{13} = 4$, $p_{23} = 7$, $p_{43} = 3$

Iteration 1: Initially, set M_0 is empty and graph G' contains only conjunctive arcs and no disjunctive arcs. The critical path and the makespan $C_{\max}(\emptyset)$ can be determined easily: this makespan is equal to the maximum total processing time required for any job. The maximum of 22 is achieved in this case by both job 1 and job 2. To determine which machine to schedule first, each machine is considered as a nonpreemptive single machine maximum lateness

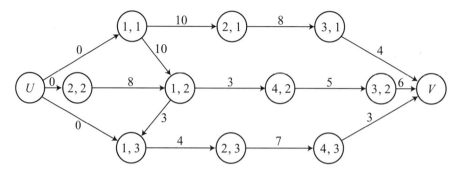

Fig. 5.2. Iteration 1 of shifting bottleneck heuristic (Example 5.4.2)

problem with the release dates and due dates determined by the longest paths in G' (assuming a makespan of 22).

The data for the nonpreemptive single machine maximum lateness problem corresponding to machine 1 is determined as follows.

jobs	1	2	3
p_{1j}	10	3	4
r_{1j}	0	8	0
d_{1j}	10	11	12

The optimal sequence turns out to be $1, 2, 3$ with $L_{\max}(1) = 5$.

The data for the subproblem associated with machine 2 are:

jobs	1	2	3
p_{2j}	8	8	7
r_{2j}	10	0	4
d_{2j}	18	8	19

The optimal sequence for this problem is $2, 3, 1$ with $L_{\max}(2) = 5$. Similarly, it can be shown that

$$L_{\max}(3) = 4$$

and

$$L_{\max}(4) = 0.$$

From this it follows that either machine 1 or machine 2 may be considered a bottleneck. Breaking the tie arbitrarily, machine 1 is selected to be included in M_0. The graph G'' is obtained by fixing the disjunctive arcs corresponding to the sequence of the jobs on machine 1 (see Figure 5.2). It is clear that

$$C_{\max}(\{1\}) = C_{\max}(\emptyset) + L_{\max}(1) = 22 + 5 = 27.$$

5.4 Job Shops and the Shifting Bottleneck Heuristic

Iteration 2: Given that the makespan corresponding to G''' is 27, the critical paths in the graph can be determined. The three remaining machines have to be analyzed separately as nonpreemptive single machine problems. The data for the problem concerning machine 2 are:

jobs	1	2	3
p_{2j}	8	8	7
r_{2j}	10	0	17
d_{2j}	23	10	24

The optimal schedule is $2, 1, 3$ and the resulting $L_{\max}(2) = 1$. The data for the problem corresponding to machine 3 are:

jobs	1	2
p_{3j}	4	6
r_{3j}	18	18
d_{3j}	27	27

Both sequences are optimal and $L_{\max}(3) = 1$. Machine 4 can be analyzed in the same way and the resulting $L_{\max}(4) = 0$. Again, there is a tie and machine 2 is selected to be included in M_0. So $M_0 = \{1, 2\}$ and

$$C_{\max}(\{1,2\}) = C_{\max}(\{1\}) + L_{\max}(2) = 27 + 1 = 28.$$

The disjunctive arcs corresponding to the job sequence on machine 2 are added to G''' and graph G'''' is obtained. At this point, still as a part of iteration 2, an attempt may be made to decrease $C_{\max}(\{1,2\})$ by resequencing machine 1. It can be checked that resequencing machine 1 does not give any improvement.

Iteration 3: The critical path in G'''' can be determined and machines 3 and 4 remain to be analyzed. These two problems turn out to be very simple with both having a zero maximum lateness. Neither of the machines constitutes a bottleneck in any way.

The final schedule is determined by the following machine sequences: the job sequence $1, 2, 3$ on machine 1; the job sequence $2, 1, 3$ on machine 2; the job sequence $2, 1$ on machine 3 and the job sequence $2, 3$ on machine 4. The makespan is 28.

The single machine maximum lateness problem that has to be solved repeatedly within each iteration of the shifting bottleneck heuristic may actually be slightly different and more complicated than the single machine maximum lateness problem described in Section 5.2 and in Example B.4.1. In the single machine subproblem that has to be solved in Step 2 of the shifting bottleneck heuristic, the operations on the given machine may be subject to a special type of precedence constraints. It may be that an operation on the machine to be scheduled *must* be processed after certain other operations have been completed on that machine; these precedence constraints must be satisfied because of the sequences of operations on the machines that already have been scheduled in previous iterations. Moreover, it may even be the case that in between the processing of two operations subject to these precedence constraints a certain minimum amount of time (i.e., a delay) has to elapse. The lengths of the delays are also determined by the sequences of the operations on the machines already scheduled in previous iterations. These precedence constraints are therefore referred to as *delayed* precedence constraints.

It is easy to construct examples that show the need for delayed precedence constraints in the single machine subproblem. Without these constraints the shifting bottleneck heuristic may end up in a situation with a cycle in the disjunctive graph and the corresponding schedule being infeasible.

Extensive numerical research has shown that the Shifting Bottleneck heuristic is extremely effective. When applied to a famous test problem with 10 machines and 10 jobs that had remained unsolved for more than 20 years, the heuristic found a very good solution very fast. This solution actually turned out to be optimal after a branch-and-bound procedure found the same result and verified its optimality. The branch-and-bound approach, in contrast to the heuristic, needed many hours of CPU time. The disadvantage of the heuristic is that there is never a guarantee that the solution generated is optimal.

The Shifting Bottleneck heuristic can be extended to models that are more general than the basic job shop considered above. It can be applied to more general machine environments and processing characteristics as well as to other objective functions.

The disjunctive graph formulation for the job shop problem without recirculation also extends to the job shop problem with recirculation. The set of disjunctive arcs for a machine that is subject to recirculation is now not a clique. If two operations of the same job have to be performed on the same machine a precedence relationship is given. These two operations are not connected by a pair of disjunctive arcs, since they are already connected by conjunctive arcs. The Shifting Bottleneck heuristic described can still be implemented. The jobs in the single machine subproblem are now subject to precedence constraints, which are imposed by the fact that different operations of the same job may require processing on the same machine.

The most general machine environment that can be dealt with is the flexible job shop. Each workcenter consists of a set of machines in parallel. The disjunctive graph representation can be extended to this machine setting as follows: again, all pairs of operations that have to be performed at a workcenter are linked by pairs of disjunctive arcs. If two operations are scheduled for processing on the same machine, then the appropriate disjunctive arc is selected. If two operations are assigned to different machines in the workcenter, then both disjunctive arcs are deleted from the graph, since the two operations do not affect each other's processing. The subproblem that then has to be solved in Step 2 of the Shifting Bottleneck Heuristic (wich was in the original version of the Shifting Bottleneck heuristic a single machine maximum lateness problem with jobs having different release dates) now becomes a parallel machine maximum lateness problem with jobs that have different release dates.

There are variations of the shifting bottleneck heuristic that can be applied to job shop problems with the total weighted tardiness objective. The disjunctive graph formulation for the job shop with the total weighted tardiness objective is slightly different from the disjunctive graph formulation for the makespan objective and the objective function of the subproblem is a more complicated objective than the $L_{\max}$ objective in the subproblem described above.

The technique described above can also be adapted to flexible job shops with multiple objectives. We have already seen that the makespan objective in the original job shop problem leads to a maximum lateness objective in its single machine subproblems. The total weighted tardiness in the original job shop problem leads to a more complicated due date related objective function in the single machine subproblems. Multiple objectives in the original problem lead to multiple objectives in the single machine (or parallel machines) subproblems; the weights of the various objectives in the original problem have, of course, an impact on the weights of the objectives in the subproblems.

The heuristic approaches described above can be linked to local search procedures as described in Appendix C. For example, the shifting bottleneck heuristic can be used first to obtain a schedule with a reasonably good makespan and this solution is then fed into a local search procedure which may yield an even better solution.

5.5 Job Shops and Constraint Programming

Constraint programming is a technique that originated in the Artificial Intelligence (AI) community. In recent years, it has often been implemented in conjunction with Operations Research (OR) techniques in order to improve its overall effectiveness (see Appendix D).

Constraint programming, in its original design, only tries to find a good solution that is feasible and satisfies all the given constraints. However, the solutions found may not necessarily be optimal. The constraints may include release dates and due dates for the jobs. It is possible to embed such a technique in a framework that is designed to minimize any due date related objective function.

Constraint programming can be applied to the basic job shop with the makespan objective as follows. Suppose that in a job shop a schedule has to be found with a makespan $C_{\max}$ that is less than or equal to a given deadline $\bar{d}$. A constraint satisfaction algorithm has to produce for each machine a sequence of the operations in such a way that the overall schedule has a makespan that is less than or equal to $\bar{d}$.

Before the actual procedure starts, an initialization step has to be done. For each operation a computation is done to determine its earliest possible starting time and latest possible completion time on the machine in question. After all the time windows have been computed, the time windows of all the operations on each machine are compared with one another. When the time windows of two operations on any given machine do not overlap, a precedence relationship between the two operations can be imposed; in any feasible schedule the operation with the earlier time window must precede the operation with the later time window. Actually, a precedence relationship may even be inferred when the time windows do overlap. Let S'_{ij} (S''_{ij}) denote the earliest (latest) possible starting time of operation (i,j) and C'_{ij} (C''_{ij}) the earliest (latest) possible completion time of operation (i,j) under the current set of precedence constraints. Note that the earliest possible starting time of operation (i,j), i.e., S'_{ij}, may be regarded as a local release date of the operation and may be denoted by r_{ij}, whereas the latest possible completion time, i.e., C''_{ij}, may be considered a local due date denoted by d_{ij}. Define the slack between the processing of operations (i,j) and (i,k) on machine i as

$$\sigma_{(i,j)\to(i,k)} = S''_{ik} - C'_{ij}$$
$$= C''_{ik} - S'_{ij} - p_{ij} - p_{ik}$$
$$= d_{ik} - r_{ij} - p_{ij} - p_{ik}.$$

If

$$\sigma_{(i,j)\to(i,k)} < 0$$

then, under the current set of precedence constraints, no feasible schedule exists in which operation (i,j) precedes operation (i,k) on machine i. So a precedence relationship can be imposed which requires operation (i,k) to appear before operation (i,j). In the initialization step of the procedure all pairs of time windows are compared with one another and all implied precedence relationships are inserted in the disjunctive graph. Because of these additional precedence constraints the time windows of each one of the operations can be adjusted (narrowed), i.e., this involves a recomputation of the release date and the due date of each operation.

Constraint satisfaction techniques often rely on constraint propagation. A constraint satisfaction technique typically attempts, in each step, to insert new precedence constraints (disjunctive arcs) that are implied by the precedence constraints inserted before. With the new precedence constraints in place the technique recomputes the time windows of all operations. For each pair of operations that have to be processed on the same machine it has to be verified which one of the following four cases holds:

Case 1:
If $\sigma_{(i,j)\to(i,k)} \geq 0$ and $\sigma_{(i,k)\to(i,j)} < 0$,
then the precedence constraint $(i,j) \to (i,k)$ has to be imposed.

5.5 Job Shops and Constraint Programming

Case 2:
If $\sigma_{(i,k)\to(i,j)} \geq 0$ and $\sigma_{(i,j)\to(i,k)} < 0$,
then the precedence constraint $(i,k) \to (i,j)$ has to be imposed.

Case 3:
If $\sigma_{(i,j)\to(i,k)} < 0$ and $\sigma_{(i,k)\to(i,j)} < 0$,
then there is no schedule that satisfies the precedence constraints already in place.

Case 4:
If $\sigma_{(i,j)\to(i,k)} \geq 0$ and $\sigma_{(i,k)\to(i,j)} \geq 0$,
then either ordering between the two operations is still possible.

In one step of the algorithm that is described later on in this section, a pair of operations has to be considered that satisfy the conditions of Case 4, i.e., either ordering between the operations is still possible. It may be the case that in this step of the algorithm many pairs of operations still satisfy Case 4. If there is more than one pair of operations that satisfy the conditions of Case 4, then a selection heuristic has to be applied. The selection of a pair is based on the sequencing flexibility this pair still provides. The pair with the lowest flexibility is selected. The reasoning behind this approach is straightforward: if a pair with low flexibility is not scheduled early on in the process, then later on in the process that pair may not be schedulable at all. So it makes sense to give priority to those pairs with a low flexibility and postpone pairs with a high flexibility. Clearly, the flexibility depends on the amounts of slack in the two orderings. One simple estimate of the sequencing flexibility of a pair of operations, $\phi((i,j)(i,k))$, is the minimum of the two slacks, i.e.,

$$\phi((i,j)(i,k)) = \min\left(\sigma_{(i,j)\to(i,k)}, \sigma_{(i,k)\to(i,j)}\right).$$

However, relying on this minimum may lead to problems. For example, suppose one pair of operations has slack values 3 and 100, whereas another pair has slack values 4 and 4. In this case, there may be only limited possibilities for scheduling the second pair and postponing a decision with regard to the second pair may well eliminate them. A feasible ordering of the first pair may not really be in jeopardy. Instead of using $\phi((i,j)(i,k))$ the following measure of sequencing flexibility has proven to be more effective:

$$\phi'((i,j)(i,k)) = \sqrt{\min(\sigma_{(i,j)\to(i,k)}, \sigma_{(i,k)\to(i,j)}) \times \max(\sigma_{(i,j)\to(i,k)}, \sigma_{(i,k)\to(i,j)})}.$$

So if the max is large, then the flexibility of a pair of operations increases and the urgency to order the pair goes down. After the pair of operations with the least flexibility $\phi'((i,j)(i,k))$ has been selected, the precedence constraint that retains the most flexibility is imposed, i.e., if

$$\sigma_{(i,j)\to(i,k)} \geq \sigma_{(i,k)\to(i,j)}$$

operation (i,j) must precede operation (i,k).

In one of the steps of the algorithm it also can happen that a pair of operations satisfies Case 3. When this is the case the partial schedule that is under construction cannot be completed and the algorithm has to backtrack. Backtracking can take any one of several forms. Backtracking may imply that either (i) one or more of the ordering decisions made in earlier iterations has to be annulled, or (ii) there does not exist a feasible solution for the problem in the way it was formulated and one or more of the original constraints in the problem have to be relaxed.

The constraint satisfaction procedure can be summarized as follows.

Algorithm 5.5.1 (Constraint Satisfaction Procedure).

Step 1.

Compute for each unordered pair of operations the slacks $\sigma_{(i,j)\to(i,k)}$ and $\sigma_{(i,k)\to(i,j)}$.

Step 2.

Check dominance conditions and classify remaining ordering decisions.

If any ordering decision is of Case 3, then BACKTRACK.

If any ordering decision is either of Case 1 or Case 2 go to Step 3;

otherwise go to Step 4.

Step 3.

Insert new precedence constraint and go to Step 1.

Step 4.

If no ordering decision is of Case 4, then solution is found. STOP.

Otherwise go to Step 5.

Step 5.

Compute $\phi'((i,j)(i,k))$ for each pair of operations not yet ordered.

Select the pair with the minimum $\phi'((i,j)(i,k))$.

If $\sigma_{(i,j)\to(i,k)} \geq \sigma_{(i,k)\to(i,j)}$, then operation (i,k) must follow (i,j);

otherwise operation (i,j) must follow operation (i,k).

Go to Step 3.

In order to apply the constraint satisfaction procedure to a job shop problem with the makespan objective, it has to be embedded in the following framework. First, an upper bound d_u and a lower bound d_l have to be found for the makespan.

5.5 Job Shops and Constraint Programming

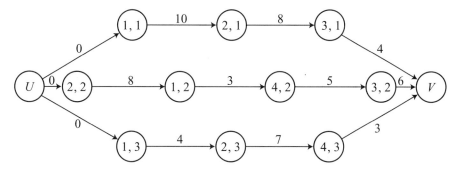

Fig. 5.3. Disjunctive graph without disjunctive arcs

Algorithm 5.5.2 (Framework for Constraint Programming).
Step 1.
  Set $d = (d_l + d_u)/2$.
  Apply Algorithm 5.5.1.
Step 2.
  If $C_{\max} < d$, set $d_u = d$.
  If $C_{\max} > d$, set $d_l = d$.
Step 3.
  If $d_u - d_l > 1$ return to Step 1.
  Otherwise STOP.

The following example illustrates the use of the contraint satisfaction technique.

Example 5.5.3 (Application of Constraint Programming to a Job Shop). Consider the instance of the job shop problem described in Example 5.3.1.

jobs	machine sequence	processing times			
1	1, 2, 3	$p_{11} = 10$,	$p_{21} = 8$,	$p_{31} = 4$	
2	2, 1, 4, 3	$p_{22} = 8$,	$p_{12} = 3$,	$p_{42} = 5$,	$p_{32} = 6$
3	1, 2, 4	$p_{13} = 4$,	$p_{23} = 7$,	$p_{43} = 3$	

Consider a due date $d = 32$ when all jobs have to be completed. Consider again the disjunctive graph but disregard all disjunctive arcs, see Figure 5.3. By doing all longest path computations, the local release dates and local due dates for all operations can be established.

operations	r_{ij}	d_{ij}
(1,1)	0	20
(2,1)	10	28
(3,1)	18	32
(2,2)	0	18
(1,2)	8	21
(4,2)	11	26
(3,2)	16	32
(1,3)	0	22
(2,3)	4	29
(4,3)	11	32

Given these time windows for all the operations, it has to be verified whether these constraints already imply any additional precedence constraints. Consider, for example, the pair of operations (2,2) and (2,3) which both have to go on machine 2. Computing the slack yields

$$\sigma_{(2,3)\to(2,2)} = d_{22} - r_{23} - p_{22} - p_{23}$$
$$= 18 - 4 - 8 - 7$$
$$= -1,$$

which implies that the ordering $(2,3) \to (2,2)$ is not feasible. So the disjunctive arc $(2,2) \to (2,3)$ has to be inserted. In the same way, it can be shown that the disjunctive arcs $(2,2) \to (2,1)$ and $(1,1) \to (1,2)$ have to be inserted as well.

Given these additional precedence constraints the release and due dates of all operations have to be updated. The updated release and due dates are presented in the table below.

operations	r_{ij}	d_{ij}
(1,1)	0	18
(2,1)	10	28
(3,1)	18	32
(2,2)	0	18
(1,2)	10	21
(4,2)	13	26
(3,2)	18	32
(1,3)	0	22
(2,3)	8	29
(4,3)	15	32

These updated release and due dates do not imply any additional precedence constraints. Going through Step 5 of the algorithm requires the computation of the factor $\phi'((i,j)(i,k))$ for every unordered pair of operations on each machine.

5.5 Job Shops and Constraint Programming

pair	$\phi'((i,j)(i,k))$
(1,1)(1,3)	$\sqrt{4 \times 8} = 5.65$
(1,2)(1,3)	$\sqrt{5 \times 14} = 8.36$
(2,1)(2,3)	$\sqrt{4 \times 5} = 4.47$
(3,1)(3,2)	$\sqrt{4 \times 4} = 4.00$
(4,2)(4,3)	$\sqrt{3 \times 11} = 5.74$

The pair with the least flexibility is $(3,1)(3,2)$. Since the slacks are such that

$$\sigma_{(3,2)\to(3,1)} = \sigma_{(3,1)\to(3,2)} = 4,$$

either precedence constraint can be inserted. Suppose the precedence constraint $(3,2) \to (3,1)$ is inserted. This precedence constraint causes significant changes in the time windows during which the operations have to be processed.

operations	r_{ij}	d_{ij}
(1,1)	0	14
(2,1)	10	28
(3,1)	24	32
(2,2)	0	14
(1,2)	10	17
(4,2)	13	22
(3,2)	18	28
(1,3)	0	22
(2,3)	8	29
(4,3)	15	32

However, this new set of time windows imposes an additional precedence constraint, namely $(4,2) \to (4,3)$. This new precedence constraint causes the following changes in the release dates and due dates of the operations.

operations	r_{ij}	d_{ij}
(1,1)	0	14
(2,1)	10	28
(3,1)	24	32
(2,2)	0	14
(1,2)	10	17
(4,2)	13	22
(3,2)	18	28
(1,3)	0	22
(2,3)	8	29
(4,3)	18	32

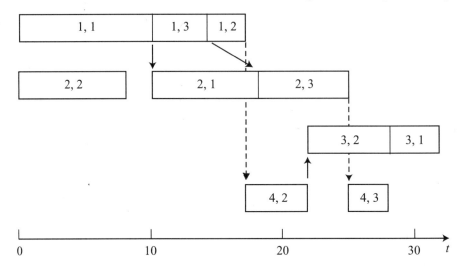

Fig. 5.4. Final schedule in Example 5.5.3.

These updated release and due dates do not imply additional precedence constraints. Going through Step 5 of the algorithm requires the computation of the factor $\phi'((i,j)(i,k))$ for every unordered pair of operations on each machine.

pair	$\phi'((i,j)(i,k))$
(1,1)(1,3)	$\sqrt{0 \times 8} = 0.00$
(1,2)(1,3)	$\sqrt{5 \times 10} = 7.07$
(2,1)(2,3)	$\sqrt{4 \times 5} = 4.47$

The pair with the least flexibility is $(1,1)(1,3)$ and the precedence constraint $(1,1) \to (1,3)$ has to be inserted.

Inserting this last precedence constraint enforces one more constraint, namely $(2,1) \to (2,3)$. Now only one unordered pair of operations remains, namely pair $(1,3)(1,2)$. These two operations can be ordered in either way without violating any due dates. A feasible ordering is

$$(1,3) \to (1,2).$$

The resulting schedule with a makespan of 32 is depicted in Figure 5.4. This schedule meets the due date originally set but is not optimal.

When the pair $(3,1)(3,2)$ had to be ordered, it could have been ordered in either direction because the two slack values were equal. Suppose at that point the opposite ordering was selected, i.e., $(3,1) \to (3,2)$. Restarting the process at that point yields the following release and due dates.

5.5 Job Shops and Constraint Programming

operations	r_{ij}	d_{ij}
(1,1)	0	14
(2,1)	10	22
(3,1)	18	26
(2,2)	0	18
(1,2)	10	21
(4,2)	13	26
(3,2)	18	32
(1,3)	0	22
(2,3)	8	29
(4,3)	15	32

These release and due dates enforce a precedence constraint on the pair of operations $(2,1)(2,3)$ and the constraint is $(2,1) \rightarrow (2,3)$. This additional constraint has the following effect on the release and due dates:

operations	r_{ij}	d_{ij}
(1,1)	0	14
(2,1)	10	22
(3,1)	18	26
(2,2)	0	18
(1,2)	10	21
(4,2)	13	26
(3,2)	22	32
(1,3)	0	22
(2,3)	18	29
(4,3)	25	32

These new release and due dates have an effect on the pair $(4,2)(4,3)$ and the arc $(4,2) \rightarrow (4,3)$ has to be included. This additional arc does not cause any additional changes in the release and due dates. At this point only two pairs of operations remain unordered, namely the pair $(1,1)(1,3)$ and the pair $(1,2)(1,3)$.

pair	$\phi'((i,j)(i,k))$
(1,1)(1,3)	$\sqrt{0 \times 8} = 0.00$
(1,2)(1,3)	$\sqrt{5 \times 14} = 8.36$

So the pair $(1,1)(1,3)$ is more critical and has to be ordered $(1,1) \rightarrow (1,3)$. It turns out that the last pair to be ordered, $(1,2)(1,3)$, can be ordered either way.

The resulting schedule turns out to be optimal and has a makespan of 28, see Figure 5.5.

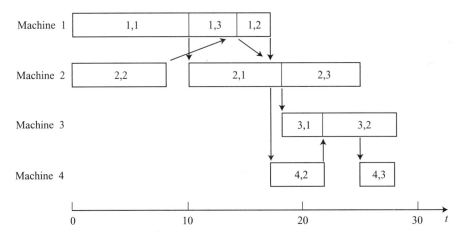

Fig. 5.5. Alternative schedule on Example 5.5.3

As stated before, constraint satisfaction is not only suitable for makespan minimization. It can also be applied to problems with due date related objectives and also when each job has its own release date.

5.6 LEKIN: A Generic Job Shop Scheduling System

Since the job shop is a very common machine environment, many scheduling systems have been developed for this environment. One example of such a system is the LEKIN system. Originally, this system was designed as a tool for teaching and research. Even though the original version was designed with academic goals in mind, several offshoots are being used in real world implementations. The academic version of the system is available on the CD-ROM that is attached to this book.

The system contains a number of scheduling algorithms and heuristics and is designed to allow the user to link and test his or her own heuristics and compare their performances with the heuristics and algorithms that are embedded in the system. The LEKIN system can accomodate various machine environments, namely:

(i) single machine
(ii) parallel machines
(iii) flow shop
(iv) flexible flow shop
(v) job shop
(vi) flexible job shop

Furthermore, it is capable of dealing with sequence dependent setup times in all the environments listed above.

5.6 LEKIN: A Generic Job Shop Scheduling System

In the main menu the user can select a machine environment. After the user has selected an environment, he has to enter all the necessary machine data and job data manually. However, in the main menu the user also has the option of opening an existing data file. An existing file contains data with regard to one of the machine environments and a specific set of jobs. The user can open an existing file, make changes in the file and work with the modified file. At the end of the session the user can save the modified file under a new name.

If the user wants to enter a data set that is completely new, he first must select a machine environment, and then a dialog box appears where he has to enter the most basic information, i.e., the number of workcenters and the number of jobs to be scheduled. After the user has done this, a second dialog box appears and he has to enter the more detailed workcenter information, i.e., the number of machines at the workcenter, their availability, and the details needed to determine the setup times on each machine (if there are setup times). In the third dialog box the user has to enter the detailed information with regard to the jobs, i.e., release dates, due dates, priorities or weights, routing, and the processing times of the various operations. If the jobs require sequence dependent setup times, then the machine settings that are required for the processing have to be entered. In the LEKIN system the required machine setting is equivalent to a setup time parameter; a sequence dependent setup time is then a function of the parameter of the job just completed and the parameter of the job about to be started.

After all the data have been entered four windows appear simultaneously, namely,

(i) the machine park window,
(ii) the job pool window,
(iii) the sequence window, and
(iv) the Gantt chart window,

(see Figure 5.6).

The machine park window displays all the information regarding the workcenters and the machines. This information is organized in the format of a tree. This window first shows a list of all the workcenters. If the user clicks on a workcenter, the individual machines of that workcenter appear.

The job pool window contains the starting time, completion time, and more information with regard to each job. The information with regard to the jobs is also organized in the form of a tree. First, the jobs are listed. If the user clicks on a specific job, then immediately a list of the various operations that belong to that job appear.

The sequence window contains the lists of jobs in the order in which they are processed on each one of the various machines. The presentation here also has a tree structure. First, all the machines are listed. If the user clicks on a machine, then all the operations that are processed on that machine appear in the sequence in which they are processed. This window is equivalent to the dispatch list interface described in Chapter 13. At the bottom of this

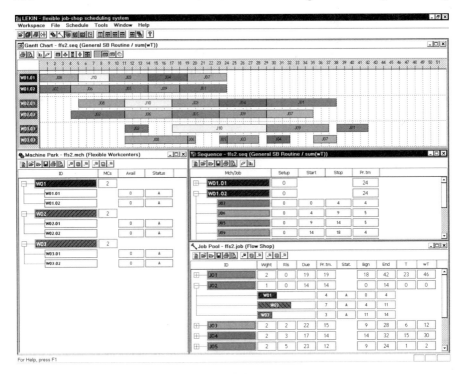

Fig. 5.6. Four windows of the LEKIN system

sequence window there is a summary of the various performance measures of the current schedule.

The Gantt chart window contains a conventional Gantt chart. This Gantt chart window enables the user to do a number of things. For example, the user can click on an operation and a window pops up displaying the detailed information with regard to the corresponding job (see Figure 5.7). The Gantt chart window also has a button that activates a window where the user can see the current values of all the objectives.

The windows described above can be displayed simultaneously on the screen in a number of ways, e.g., in a quadrant style, tiled horizontally, or tiled vertically (see Figure 5.6). Besides these four windows there are two other windows, which will be described in more detail later. These two windows are the logbook window and the objective chart window. The user can print out the windows separately or all together by selecting the print option in the appropriate window.

The data set of a particular scheduling problem can be modified in a number of ways. First, information with regard to the workcenters can be modified in the machine park window. When the user double-clicks on the workcenter the relevant information appears. Machine information can be ac-

5.6 LEKIN: A Generic Job Shop Scheduling System

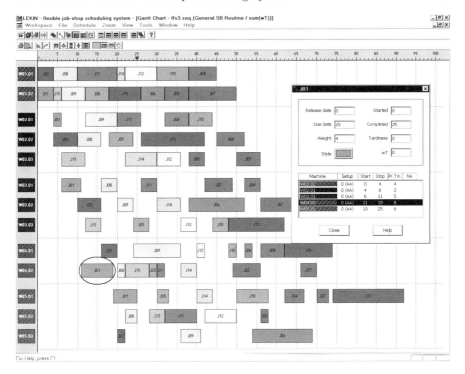

Fig. 5.7. Gantt chart window

cessed in a similar manner. Jobs can be added, modified, or deleted in the job pool window. A double click on a job displays all the relevant information.

After the user has entered the data set, all the information is displayed in the machine park window and job pool window. However, the sequence window and the Gantt chart window remain empty. If the user in the beginning had opened an existing file, then the sequence window and the Gantt chart window may display information pertaining to a sequence that had been generated during an earlier session.

The user can select a schedule from any window. Typically the user would do so either from the sequence window or from the Gantt chart window by clicking on schedule and selecting a heuristic or algorithm from the dropdown menu. A schedule is then generated and displayed in both the sequence window and the Gantt chart window. The schedule generated and displayed in the Gantt chart is a so-called semi-active schedule. A semi-active schedule is characterized by the fact that the start time of any operation of any given job on any given machine is equal to either the completion time of the preceding operation of the same job on a different machine or the completion time of an operation of a different job on the same machine.

The system contains a number of algorithms for several of the machine environments and objective functions. These algorithms include

(i) dispatching rules,
(ii) heuristics of the shifting bottleneck type,
(iii) local search techniques, and
(iv) a heuristic for the flexible flow shop with the total weighted tardiness as objective (SB-LS).

The dispatching rules include EDD and WSPT. The way these dispatching rules are applied in a single machine environment and in a parallel machine environment is standard. However, they can also be applied in the more complicated machine environments such as the flexible flow shop and the flexible job shop. They are then applied as follows: each time a machine is freed the system checks which job should go next on that machine. The system then uses the following data for its priority rules: the due date of a candidate job is then the due date at which the job has to leave the system. The processing time that is plugged in the WSPT rule is the sum of the processing times of all the remaining operations of that job.

The system also has a general purpose routine of the shifting bottleneck type that can be applied to each one of the machine environments and every objective function. Since this routine is quite generic and designed for many different machine environments and objective functions, it cannot compete against a shifting bottleneck heuristic that is tailor-made for a specific machine environment and a specific objective function.

The system also contains a neighbourhood search routine that is applicable to the flow shop and job shop (but not to the flexible flow shop or flexible job shop) with either the makespan or the total weighted tardiness as objective. If the user selects the shifting bottleneck or the local search option, then he must select also the objective he wants to minimize. When the user selects the local search option and the objective, a window appears in which the user has to enter the number of seconds he wants the local search to run.

The system has also a specialized routine for the flexible flow shop with the total weighted tardiness as objective; this routine is a combination of a shifting bottleneck routine and a local search (SB-LS). This routine tends to yield schedules that are of reasonably high quality.

If the user wants to construct a schedule manually, he can do so in one of two ways. First, he can modify an existing schedule in the Gantt chart window as much as he wants by clicking, dragging and dropping operations. After such modifications the resulting schedule is, again, semi-active. However, the user can also construct this way a schedule that is *not* semi-active. To do this he has to activate the Gantt chart, hold down the shift button and move the operation to the desired position. When the user releases the shift button, the operation remains fixed.

A second way of creating a schedule manually is the following. After clicking on schedule, the user must select "manual entry". The user then has to

5.6 LEKIN: A Generic Job Shop Scheduling System

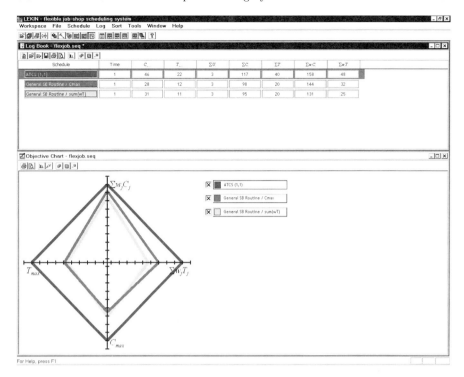

Fig. 5.8. Logbook and comparison of different schedules

enter for each machine a job sequence. The jobs in a sequence are separated from one another by a ";".

Whenever the user generates a schedule for a particular data set, the schedule is stored in a logbook. The system automatically assigns an appropriate name to every schedule generated. The system can store and retrieve a number of different schedules, see Figure 5.8. If the user wants to compare the different schedules he has to click on "logbook". The user can change the name of each schedule and give each schedule a different name for future reference.

The schedules stored in the logbook can be compared with one another by clicking on the "performance measures" button. The user may then select one or more objectives. If the user selects a single objective, a bar chart appears that compares the schedules stored in the logbook with respect to the objective selected. If the user wants to compare the schedules with regard to two objectives, an (x, y) coordinate system appears and each schedule is represented by a dot. If the user selects three or more objectives, then a multi-dimensional graph appears that displays the performance measures of the schedules stored in the logbook.

Some users may want to incorporate the concept of setup times. If there are setup times, then the relevant data have to be entered together with all other

108   5 Machine Scheduling and Job Shop Scheduling

Fig. 5.9. Setup time matrix

job and machine data at the very beginning of a session. (However, if at the beginning of a session an existing file is opened, then such a file may already contain setup data.) The setup times are based on the following structure. Each operation has a single parameter or attribute, which is represented by a letter, e.g., A, B, and so on. This parameter represents the machine setting required for processing that operation. When the user enters the data for each machine, he has to fill in a setup time matrix for that machine. The setup time matrix for a machine specifies the time that it takes to change that machine from one setting to another, i.e., from B to E, see Figure 5.9. The setup time matrices of all the machines at any given workcenter have to be the same (the machines at a workcenter are assumed to be identical). This setup time structure does not allow the user to implement *arbitrary* setup time matrices.

A more advanced user can link his own algorithms to the system. This feature allows the developer of a new algorithm to test his algorithm using the interactive Gantt chart features of the system. The process of making such an external program recognizable by the system consists of two steps, namely, the preparation of the code (programming, compiling and debugging), and the linking of the code to the system.

Exercises

The linking of an external program is done by clicking on "Tools" and selecting "Options". A window appears with a button for a New Folder and a button for a New Item. Clicking on New folder creates a new submenu. Clicking on a New Item creates a placeholder for a new algorithm. After the user has clicked on a New Item, he has to enter all the data with respect to the New Item. Under "Executable" he has to enter the full path to the executable file of the program. The development of the code can be done in any environment under Win3.2. Appendix F provides examples of the format of a file that contains the workcenter information and the format of a file that contains the information pertaining to the job. After a new algorithm has been added, it is included as one of the heuristics in the schedule option menu.

5.7 Discussion

An enormous amount of research on machine scheduling and job shop scheduling has resulted in many books. This chapter just gives a flavor of the types of results that have appeared in the literature. It focuses mainly on the type of research that has proven to be useful in practice, i.e., decomposition techniques such as the shifting bottleneck heuristic and constraint programming techniques.

A significant amount of more theoretical research has appeared in the literature focusing on exact optimization techniques based on integer programming and disjunctive programming formulations. Some of these formulations are presented in Appendix A. The techniques developed include branch-and-bound as well as branch-and-cut. An example of a branch-and-bound application is presented in Appendix B.

Some of the more applied techniques have not been included in this chapter either. A fair amount of research has been done on local search techniques applied to machine scheduling and job shop scheduling. These techniques include simulated annealing, tabu-search as well as genetic algorithms. Examples of such applications are given in Appendix C.

Appendix D contains a constraint programming formulation of the job shop problem with the makespan objective in the OPL language.

Exercises

5.1. What does the ATC rule reduce to
  (a) when K goes to ∞, and
  (b) when K is very close to zero?

5.2. Consider an instance with two machines in parallel and the total weighted tardiness as objective to be minimized. There are 5 jobs.

jobs	1	2	3	4	5
p_j	13	9	13	10	8
d_j	6	18	10	11	13
w_j	2	4	2	5	4

(a) Apply the ATC heuristic on this instance with the look-ahead parameter $K = 1$.

(b) Apply the ATC heuristic on this instance with the look-ahead parameter $K = 5$.

5.3. Consider 6 machines in parallel and 13 jobs. The processing times of the 13 jobs are tabulated below.

jobs	1	2	3	4	5	6	7	8	9	10	11	12	13
p_j	6	6	6	7	7	8	8	9	9	10	10	11	11

(a) Compute the makespan under LPT.
(b) Find the optimal schedule.

5.4. Explain why the problem discussed in Section 5.3 is a special case of the problem described in Section 4.6. (*Hint:* Consider an arbitrary job shop scheduling problem and describe it as a workforce constrained project scheduling problem. Consider a machine in the job shop scheduling problem as a workforce pool in the project scheduling problem and the available number in a pool being equal to 1.)

5.5. Consider the following instance of the job shop problem with no recirculation and the makespan as objective.

jobs	machine sequence	processing times
1	1,2,3	$p_{11} = 9$, $p_{21} = 8$, $p_{31} = 4$
2	1,2,4	$p_{12} = 5$, $p_{22} = 6$, $p_{42} = 3$
3	3,1,2	$p_{33} = 10$, $p_{13} = 4$, $p_{23} = 9$

Give an integer programming formulation of this instance (*Hint:* Consider the integer programming formulation of the project scheduling problem with workforce constraints described in Section 4.6 and Exercise 5.4.)

5.6. Consider the following heuristic for the job shop problem with no recirculation and the makespan objective. Each time a machine is freed, select the job (among those immediately available for processing on the machine) with the longest *total* remaining processing (including its processing on the machine freed). If at any point in time more than one machine is freed, consider first the machine with the largest remaining workload. Apply this heuristic to the instance in Example 5.3.1.

5.7. Apply the heuristic described in Exercise 5.6 to the following instance with the makespan objective.

jobs	machine sequence	processing times			
1	1,2,3,4	$p_{11}=9,$	$p_{21}=8,$	$p_{31}=4,$	$p_{41}=4$
2	1,2,4,3	$p_{12}=5,$	$p_{22}=6,$	$p_{42}=3,$	$p_{32}=6$
3	3,1,2,4	$p_{33}=10,$	$p_{13}=4,$	$p_{23}=9,$	$p_{43}=2$

5.8. Consider the instance in Exercise 5.7.

(a) Apply the Shifting Bottleneck heuristic to this instance (doing the computation by hand).

(b) Compare your result with the result of the shifting bottleneck routine in the LEKIN system.

5.9. Consider the following two machine job shop with 10 jobs. All jobs have to be processed first on machine 1 and then on machine 2. (This implies that the two machine job shop is actually a two machine flow shop).

jobs	1	2	3	4	5	6	7	8	9	10	11
p_{1j}	3	6	4	3	4	2	7	5	5	6	12
p_{2j}	4	5	5	2	3	3	6	6	4	7	2

(a) Apply the heuristic described in Exercise 5.6 to this two machine job shop.

(b) Construct now a schedule as follows. The jobs have to go through the second machine in the same sequence as they go through the first machine. A job whose processing time on machine 1 is shorter than its processing time on machine 2 must precede each job whose processing time on machine 1 is longer than its processing time on machine 2. The jobs with a shorter processing time on machine 1 are sequenced in increasing order of their processing times on machine 1. The jobs with a shorter processing time on machine 2 follow in decreasing order of their processing times on machine 2. (This rule is usually referred to as *Johnson's Rule*.)

(c) Compare the schedule obtained with Johnson's rule to the schedule obtained under (a).

5.10. Apply the shifting bottleneck heuristic to the two machine flow shop instance in Exercise 5.9. Compare the resulting schedule with the schedules obtained in Exercise 5.9.

Comments and References

Many of the basic priority rules originated in the fifties and early sixties. For example, Smith (1956) introduced the WSPT rule, Jackson (1955) introduced the EDD rule, and Hu (1961) the CP rule. Conway (1965a, 1965b) was one of the first to make a comparative study of priority rules. A fairly complete list of the most common priority or dispatching rules is given by Panwalkar and Iskander (1977) and a detailed description of composite dispatching rules by Bhaskaran and Pinedo (1992). Special examples of composite dispatching rules are the COVERT rule developed by Carroll (1965) and the ATC rule developed by Vepsalainen and Morton (1987). Ow and Morton (1989) describe a rule for scheduling problems with earliness and tardiness penalties. The ATCS rule is due to Lee, Bhaskaran and Pinedo (1997).

The total weighted tardiness has been the focus of a number of studies in the more basic machine environments, e.g., the single machine. There are many researchers who have dealt with the single machine total weighted tardiness problem; see, for example, Fisher (1976), Potts and Van Wassenhove (1982, 1985, 1987), and Rachamadugu (1987). Abdul-Razaq, Potts, and Van Wassenhove (1990) present a survey of algorithms for the single machine total weighted tardiness problem. Vepsalainen and Morton (1987) developed priority rule based heuristics for job shops with the total weighted tardiness objective.

Job shop scheduling has received an enormous amount of attention in the research literature as well as in books. The special case of the job shop where all the jobs have the same route, i.e., the flow shop, also has received considerable attention. For results on the flow shop, see, for example, Johnson (1954), Palmer (1965), Campbell, Dudek and Smith (1970), Szwarc (1971, 1973, 1978), Gupta (1972), Baker (1975), Smith, Panwalkar and Dudek (1975, 1976), Dannenbring (1977), Muth (1979), Papadimitriou and Kannelakis (1980), Ow (1985), Pinedo (1985), Widmer and Hertz (1989), and Taillard (1990).

An algorithm for the minimization of the makespan in a two machine job shop without recirculation is due to Jackson (1956) and the disjunctive programming formulation described in Section 5.3 is due to Roy and Sussmann (1964).

Branch-and-bound techniques have often been applied to minimize the makespan in job shops; see, for example, Lomnicki (1965), Brown and Lomnicki (1966), McMahon and Florian (1975), Barker and McMahon (1985), Carlier and Pinson (1989), Applegate and Cook (1991), and Hoitomt, Luh and Pattipati (1993). For an overview of branch-and-bound techniques, see Pinson (1995). Singer and Pinedo (1998) developed a branch-and-bound approach for the job shop with the total weighted tardiness objective.

The famous shifting bottleneck heuristic is due to Adams, Balas and Zawack (1988). Their algorithm makes use of a single machine scheduling algorithm developed by Carlier (1982). Earlier work on this particular single machine scheduling problem was done by McMahon and Florian (1975). Nowicki and Zdrzalka (1986), Dauzère-Pérès and Lasserre (1993, 1994) and Balas, Lenstra and Vazacopoulos (1995) all developed more sophisticated versions of the Carlier algorithm. The monograph by Ovacik and Uzsoy (1996) presents an excellent treatise of the application of decomposition methods and shifting bottleneck techniques to large scale job shops with the makespan and the maximum lateness objectives. This monograph is based on a number of papers by the two authors; see, for example, Uzsoy (1993) for the application of decomposition methods to flexible job shops. Pinedo and Singer (1998) developed a shifting bottleneck approach for the job shop problem with the total weighted tardiness objective and Yang, Kreipl, and Pinedo (2000) developed a similar approach for the flexible flow shop with the total weighted tardiness objective.

A fair amount of research has focused on constraint programming approaches for the job shop scheduling problem with the makespan objective; Section 5.5 is based on the work by Cheng and Smith (1997). Several other researchers have also considered this problem; see, for example, Nuijten and Aarts (1996), and Baptiste, LePape, and Nuijten (2001).

The LEKIN system is due to Asadathorn (1997) and Feldman and Pinedo (1998). The general-purpose routine of the shifting bottleneck type that is embedded in the system is also due to Asadathorn (1997). The local search routines that are applicable to the flow shop and job shop are due to Kreipl (1998). The more specialized SB-LS routine for the flexible flow shop is due to Yang, Kreipl, and Pinedo (1998).

In addition to the procedures discussed in this chapter flow shop and job shop problems have been tackled with local search procedures; see, for example, Matsuo, Suh, and Sullivan (1988), Dell'Amico and Trubian (1991), Della Croce, Tadei and Volta (1992), Storer, Wu and Vaccari (1992), Nowicki and Smutnicki (1996), and Kreipl (1998).

For a broader view of the job shop scheduling problem, see Wein and Chevelier (1992). For an interesting special case of the job shop, i.e., a flow shop with reentry, see Graves, Meal, Stefek and Zeghmi (1983). For results on a generalization of the flow shop, i.e., the flexible flow shop, see Hodgson and McDonald (1981a, 1981b, 1981c).

Of course, many applications of job shop scheduling problems in industry have been discussed in the literature. For job shop scheduling problems and solutions in the micro-electronics industry, see, for example, Wein (1988), Lee, Uzsoy and Martin-Vega (1992) and Uzsoy, Lee and Martin-Vega (1992).

Chapter 6

Scheduling of Flexible Assembly Systems

6.1 Introduction 115
6.2 Sequencing of Unpaced Assembly Systems 116
6.3 Sequencing of Paced Assembly Systems 122
6.4 Scheduling of Flexible Flow Systems with Bypass 127
6.5 Mixed Model Assembly Sequencing at Toyota .. 132
6.6 Discussion 135

6.1 Introduction

Flexible assembly systems differ in a number of ways from the job shops considered in the previous chapter. In a job shop, each job has its own identity and may be different from all other jobs. In a flexible assembly system, there are typically a limited number of different product types and the system has to produce a given quantity of each product type. So two units of the same product type are identical.

The movements of jobs in a flexible assembly system are often controlled by a material handling system, which imposes constraints on the starting times of the jobs at the various machines or workstations. The starting time of a job at a machine is a function of its completion time on the previous machine on its route. A material handling system usually also limits the number of jobs waiting in buffers between machines.

In this chapter we analyze three different models for flexible assembly systems. The machine environments in the three models are similar to the machine environments of a flow shop or a flexible flow shop. However, the models tend to be more complicated than the models considered in Chapter 5. This is mainly because of the additional constraints that are imposed by the material handling systems.

The first model represents a flow line with a number of machines or workstations in series. The line is unpaced, i.e., a machine can spend as much time as needed on any job. There are finite buffers between successive machines which may cause blocking and starving. A number of different product types have to be produced in given quantities and the goal is to maximize the throughput. This type of environment occurs, for example, in the assembly of copiers. Different types of copiers are often assembled on the same line. The different models are usually from the same family and may have many common characteristics. However, they differ with regard to the options. Some types have an automatic document feeder while others have not; some have more elaborate optics than others. The fact that different copiers have different options implies that the processing times at certain stations may vary.

The second model is a paced assembly system with a conveyor system that moves at a constant speed. The units that have to be assembled are moved from one workstation to the next at a fixed speed. Each workstation has its own capacity and constraints. Again, a number of different product types have to be assembled. The goal is to sequence the jobs so that no workstation is overloaded and setup costs are minimized. Paced assembly systems are very common in the automotive industry, where different models have to be assembled on the same line. The cars may have different colors and different option packages. The sequencing of the cars has to take setup costs as well as workload balancing into account.

The third model is a flexible flow system with limited buffers and bypass. In contrast to the first two models, there are a number of machines in parallel at each workcenter. A job may be processed on any one of the parallel machines or it may bypass a workcenter altogether. The objective is to maximize the throughput. This type of system is used, for example, in the manufacturing of Printed Circuit Boards (PCBs). PCBs are produced in batches with each batch requiring its own set of operations.

6.2 Sequencing of Unpaced Assembly Systems

Consider a number of machines in series with a limited buffer between successive machines. The material handling system that moves a job from one machine to the next is unpaced. So whenever a machine finishes processing a job it can release the job to the buffer of the next machine provided that buffer is not full. If that buffer is full, then the machine cannot release the job and is blocked. The material handling system does not permit bypassing, i.e., each machine serves the jobs according to the *First In First Out* discipline. This type of serial processing is common in the assembly of bulky items, such as television sets or copiers; their size makes it difficult to keep many units waiting in front of a machine.

This machine environment with limited buffers is in a mathematical sense equivalent to a system of machines in series with *no* buffers in between ma-

6.2 Sequencing of Unpaced Assembly Systems

chines. This happens to be the case because a buffer space between two machines may be regarded as a "machine" where the processing times of all jobs happen to be zero. So any number of machines with limited buffers between them can be transformed into an equivalent system that consists of a (larger) number of machines with *zero* buffers in between them. Transforming a model with buffers into an equivalent model without buffers is advantageous, since a model without buffers is easier to describe and to formulate mathematically than a model with buffers. In this section we focus, therefore, on models without buffers; of course, we may have machines where all processing times are zero (playing the role of buffers).

Schedules used in flow lines with blocking are often periodic or cyclic. Such schedules are generated as follows. First, a given set of jobs are scheduled in a certain order. This set contains jobs of all the product types and there may be more than one job of the same product type. This set is followed by a second set that is identical to the first one and scheduled in the same way. This process is repeated over and over again; the resulting schedule is a cyclic schedule.

Cyclic schedules are generally not practical when there are significant sequence dependent setup times between different product types. It is then more efficient to have long runs of the same product type. However, if setup times are negligible, then cyclic schedules have important advantages. For example, if the demand for each product type remains constant over time, then a cyclic schedule results in inventory holding costs that are significantly less than the costs incurred with schedules that have long runs of each product type. Cyclic schedules also have an advantage in that they are easy to keep track of and impose a certain discipline. However, it is not necessarily true that a cyclic schedule has the maximum throughput; often, an acyclic schedule has the maximum throughput. Nevertheless, in practice, a scheduler often adopts a cyclic schedule from which he allows minor deviations, dependent upon current demand.

Suppose there are l different product types. Let $\mathcal{N}_k$ denote the number of jobs of product type k in the overall production target. The production target may be for a period of six months or one year and the numbers may be large. If z is the greatest common divisor of the integers $\mathcal{N}_1, \ldots, \mathcal{N}_l$, then the vector

$$\mathcal{N}^* = \Big(\frac{\mathcal{N}_1}{z}, \ldots, \frac{\mathcal{N}_l}{z}\Big)$$

represents the smallest set having the same proportions of the different product types as the long range production target. This set is usually referred to as the *Minimum Part Set (MPS)*. Given the vector $\mathcal{N}^*$, the elements in an MPS may be regarded, without loss of generality, as n jobs, where

$$n = \frac{1}{z}\sum_{k=1}^{l}\mathcal{N}_k.$$

Let p_{ij} denote the processing time of job j, $j = 1, \ldots, n$, on machine i. Cyclic schedules are specified by the sequence of the n jobs in the MPS. The fact that various jobs within an MPS may correspond to the same product type and have identical processing requirements does not affect the approach described below. Maximizing system throughput is basically equivalent to minimizing the cycle time of an MPS in a steady state. In this case the MPS cycle time can be defined as the time between the first jobs of two consecutive MPS's entering the system. The following example illustrates the MPS cycle time concept.

Example 6.2.1 (MPS Cycle Time). Consider an assembly system with four machines and no buffers between the machines. There are three different product types that have to be produced in equal amounts, i.e.,

$$\mathcal{N}^* = (1, 1, 1).$$

The processing times of the three jobs in the MPS are:

Jobs	1	2	3
p_{1j}	0	1	0
p_{2j}	0	0	0
p_{3j}	1	0	1
p_{4j}	1	1	0

The second "machine" (that is, the second row of processing times), with zero processing times for all three jobs, functions as a buffer between the first and third machines. The Gantt charts for this example under the two different sequences are shown in Figure 6.1. Under both sequences a steady state is reached during the second cycle. Under sequence $1, 2, 3$ the MPS cycle time is three, while under sequence $1, 3, 2$ the MPS cycle time is two.

For the minimization of the MPS cycle time the so-called *Profile Fitting (PF)* heuristic can be used. This heuristic works as follows: one job is selected to go first. The selection of the first job in the MPS sequence may be done arbitrarily or according to some scheme. For example, one can choose the job with the largest total amount of processing as the one to go first. This first job generates a *profile*. For the time being, we assume that the job does not encounter any blocking and proceeds smoothly from one machine to the next (in a steady state the first job in an MPS may be blocked by the last job from the previous MPS). The profile is determined by the departure time of this first job, say job j_1, from machine i. If X_{i,j_1} denotes the departure (or exit) time of job j_1 from machine i, then

$$X_{i,j_1} = \sum_{h=1}^{i} p_{h,j_1}$$

6.2 Sequencing of Unpaced Assembly Systems

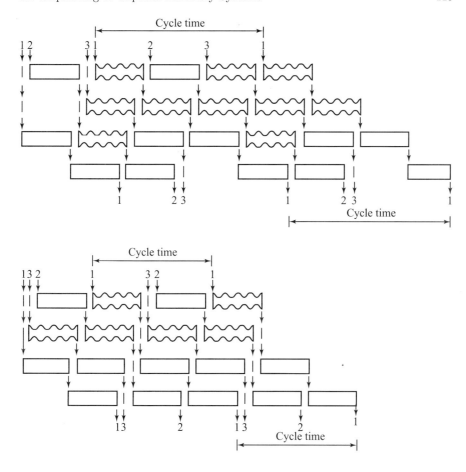

Fig. 6.1. Gantt charts for Example 6.2.1

In order to determine which is the most appropriate job to go second, every one of the remaining jobs in the MPS is tried out. For each candidate job, the amounts of time the machines are idle and the amounts of time the job is blocked at a machine are computed. The departure times of a candidate job for the second position, say job c, can be computed recursively as follows:

$$X_{1,j_2} = \max(X_{1,j_1} + p_{1c}, X_{2,j_1})$$
$$X_{i,j_2} = \max(X_{i-1,j_2} + p_{ic}, X_{i+1,j_1}), \qquad i = 2,\ldots,m-1$$
$$X_{m,j_2} = X_{m-1,j_2} + p_{mc}$$

The nonproductive time on machine i, either from being idle or from being blocked, if candidate c is put into the second position, is $X_{i,j_2} - X_{i,j_1} - p_{ic}$. The sum of these idle and blocked times over all m machines is computed for candidate c, i.e.,

$$\sum_{i=1}^{m}\left(X_{i,j_2} - X_{i,j_1} - p_{ic}\right).$$

This procedure is repeated for all remaining jobs in the MPS. The candidate with the smallest total amount of nonproductive time is then selected for the second position.

After the best fitting job is added to the partial sequence, the new profile (the departure times of this job from all the machines) is computed and the procedure is repeated. From the remaining jobs in the MPS again the one that fits best is selected. This process continues until all the jobs are scheduled. The PF heuristic functions, in a sense, as a dynamic dispatching rule.

The PF heuristic can be summarized as follows.

Algorithm 6.2.2 (Profile Fitting Heuristic).

Step 1. *(Initial Condition)*

Select the job with the longest total processing time as the first job in the MPS.

Step 2. *(Analysis of Remaining Jobs to be Scheduled)*

For each job that has not yet been scheduled do the following: Consider that job as the next one in the partial sequence and compute the total nonproductive time on all m machines (machine idle time as well as machine blocking time).

Step 3. *(Selection of the Next Job in Partial Schedule)*

Of all jobs analyzed in Step 2, select the job with the smallest total nonproductive time as the next one in the partial sequence.

Step 4. *(Stopping Criterion)*

If all jobs in the MPS have been scheduled, then STOP.

Otherwise, go to Step 2.

Observe that in Example 6.2.1, after job 1, the Profile Fitting heuristic would select job 3 to go next, as this would cause only one unit of blocking time (on machine 2) and no idle times. If job 2 were selected to go after job 1, one unit of idle time would be incurred at machine 2 and one unit of blocking on machine 3, resulting in two units of nonproductive time. So in this example the heuristic would yield an optimal sequence.

Experiments have shown that the PF heuristic results in good schedules. However, it can be refined to perform even better. In the description above, the goodness of fit of a particular job was measured by summing all the nonproductive times on the m machines. Each machine was considered equally important. Suppose one machine is a bottleneck which has more processing to do than any other machine. It is intuitive that lost time on a bottleneck machine is worse than lost time on a machine that, on average, does not have much processing to do. When measuring the total amount of lost

6.2 Sequencing of Unpaced Assembly Systems

time, it may be appropriate to weight each inactive time period by a factor that is proportional to the degree of congestion at the corresponding machine. The higher the degree of congestion at a machine, the larger the weight. One measure for the degree of congestion of a machine is easy to calculate; simply determine the total amount of processing to be done on all jobs in an MPS at that machine. In the numerical example presented above the third and the fourth machines are more heavily used than the first and second machines (the second machine was not used at all and basically functioned as a buffer). Nonproductive time on the third and fourth machines is therefore less desirable than nonproductive time on the first and second machines. Experiments have shown that such a weighted version of the PF heuristic performs significantly better than the unweighted version.

Example 6.2.3 (Application of Weighted Profile Fitting Heuristic). Consider three machines and an MPS of four jobs. There are no buffers between machines. The processing times of the four jobs on the three machines are as follows.

Jobs	1	2	3	4
p_{1j}	2	4	2	3
p_{2j}	4	4	0	2
p_{3j}	2	0	2	0

All three machines are actual machines and none are buffers. The workloads on the three machines are not entirely balanced. The workload on machine 1 is 11, on machine 2 it is 10 and on machine 3 it is 4. So, time lost on machine 3 is less detrimental than time lost on the other two machines. To apply a weighted version of the Profile Fitting heuristic, the weight given to nonproductive time on machine 3 should be smaller than the weights given to nonproductive times on the other two machines. In this example we set the weights for machines 1 and 2 equal to 1 and the weight for machine 3 equal to 0.

Assume job 1 is the first job in the MPS. The profile can be determined easily. If job 2 is selected to go second, then there will be no idle time or blocking on machines 1 and 2; however machine 3 will be idle for 2 time units. If job 3 is selected to go second, machines 1 and 2 are both blocked for two units of time. If job 4 is selected to go second, machine 1 will be blocked for one unit of time. As the weight of machine 1 is 1 and of machine 3 is 0, the weighted PF selects job 2 to go second. It can be verified that the weighted PF results in the cycle $1, 2, 4, 3$, with an MPS cycle time of 12. If the unweighted version of the PF heuristic were applied and job 1 again was selected to go first, then job 4 would be selected to go second. The unweighted PF heuristic would result in the sequence $1, 4, 3, 2$ with an MPS cycle time of 14.

6.3 Sequencing of Paced Assembly Systems

In a paced assembly line, a conveyor system moves the jobs (e.g., cars) from one workstation to the next at a constant speed. The jobs maintain a fixed distance from one another. If the assembly at a particular station is done manually, the workers walk alongside the moving line while performing their operation; after its completion, they walk back towards that point in the line that corresponds to the beginning of their station or section. The time it takes to walk back is often negligible in comparison with the time it takes to perform the operation. The operations to be done at the various stations are, of course, different and their processing times may not be identical. The space along the line reserved for a particular operation is in proportion to the amount of time needed for that operation. In this type of assembly line a station is basically equivalent to a designated section of the assembly line. At any point in time there may be more than one job at a given station; if that is the case, then several jobs may be processed at that station at the same time. In this environment bypass is not allowed.

An important characteristic of a paced assembly system is its unit cycle time, which is defined as the time between two successive jobs coming off the line. The unit cycle time is the reciprocal of the production rate.

Paced assembly systems are very common in the automotive industry. Assembly lines in the automotive industry have, of course, many other characteristics that are not included in the model described above. One of these is, for example, the point in the line where the engine and the body come together.

In the automotive industry some jobs have due dates. These jobs are made to order and to provide good customer service it is important to assemble these in a timely fashion. However, these committed shipping dates are not as stringent as those in other industries (a job may have to be shipped in a given week, not necessarily on a given day). Other (more standard) jobs are made for inventory and sent to dealers. The target inventory levels of the different models with the various different option packages are determined by the marketing department. This implies that production is a mixture of Make-To-Order and Make-To-Stock. The Make-To-Stock component of the production provides some flexibility for the scheduling system, since the timing of the production of these jobs is somewhat flexible.

Each job that goes through the line has a number of attributes or characteristics such as color, options (e.g., sunroof, power windows), and destination. From a production point of view it is advantageous to sequence jobs that have certain attributes in common one after another. For example, it pays to have a number of consecutive jobs with the same color, since cleaning the paint guns in the paint shop involves a changeover cost. Also, if a number of jobs have to go on the same trailer to the same destination, then these jobs should all appear within the same segment of the sequence. If the first and the last job for one particular trailer are sequenced far apart, then there is a waiting

6.3 Sequencing of Paced Assembly Systems

cost involved. (The destinations that cause problems are usually the points of low demand where only a single trailer goes.) Because of the characteristics that involve changeover costs, a sequencing heuristic has to go through a grouping phase where it attempts to keep jobs with similar attributes together. Grouping may occur with regard to all operations that have significant setup or changeover costs. This does not include only color and destination, it also includes intricate manual assembly, where an immediate repetition has a positive impact on the quality.

There is another class of attributes that have a different (actually, an opposite) effect on the sequencing. Consider the installation of optional equipment such as a sunroof or the special parts that are required in a station wagon. A given percentage of the jobs, say 10%, have to undergo such a special operation. Often, such an operation takes more time than an operation that is common to all jobs; the section of the line assigned to such an operation must therefore be longer than the sections of the line assigned to other operations. As the number of workers assigned to such an operation (e.g., sunroof installation) is proportional to the *average* workload, the jobs that need that operation must be spaced out more or less evenly over the sequence. A heuristic, therefore, has to go through a spacing step that schedules such jobs at regular intervals.

Example 6.3.1 (Sunroof Installation). Suppose the installation of a sunroof lasts four times longer than a typical operation that has to be done on all jobs (e.g., attaching the hood). The section of the line corresponding to the longer operation has to be at least four times longer than the section of the line assigned to the shorter operation. If the section of the line corresponding to the shorter operation contains, on the average, a single job, then the section of the line corresponding to the longer operation contains, on the average, four jobs. If only 10% of the jobs need to undergo the long operation, then, in a perfectly balanced sequence, every tenth job on the line has to undergo the long operation. However, if two consecutive jobs require the installation of a sunroof, then the workers may not be able to complete the work on the second job in time. The reason is the following: they complete the work on the first job when it is about to leave their section and then turn towards the next job and start working on that one. However, this second job is already relatively close to the end of their section and it may not be possible to complete the operation before it leaves their section. If these two particular jobs were spaced more than four positions apart in the sequence, then there would not have been any problem.

This type of operation is usually referred to as a *capacity constrained* operation. For each capacity constrained operation a criticality index can be computed. This criticality index is the minimum number of positions between two jobs requiring the operation divided by the average number of positions between two such jobs. In Example 6.3.1 the index is 4/10. The higher the index, the more critical the operation. Capacity constrained operations require a careful balancing of the sequence, since proper spacing has a significant impact on quality. This spacing must satisfy the so-called capacity violation constraints; these are basically upper bounds on the frequencies with which jobs may end up in positions where the capacity constrained operations cannot be completed in time.

Based on the considerations described above there are four objectives that have to be taken into account in the sequencing process. One important objective is the minimization of the total setup cost. If in the paint shop a job with color j is followed by a job with color k there is a setup cost c_{jk}. If two consecutive jobs have the same color, then the setup cost is zero. The first objective is to minimize the total setup cost over the entire production horizon.

A second objective concerns the due dates of the Make-To-Order jobs. A job's due date can be translated into a certain position in the sequence; if the job would appear after its "due date position" it would be considered late. The objective to minimize is similar to the total weighted tardiness objective $\sum w_j T_j$. If job j is tardy, its tardiness T_j is measured by the number of positions in between its assigned position and its due date position. In some plants this objective is very important while in other plants it is of no importance (e.g., Cadillacs are more often made to order than Ford Escorts).

A third objective concerns the spacing with regard to the capacity constrained operations. Let $\psi_i(\ell)$ denote the penalty incurred if workstation i has to do work on two jobs that are ℓ positions apart. The penalty $\psi_i(\ell)$ is monotonically decreasing in ℓ and is zero when ℓ is larger than a given $\ell_i^{\min}$. The objective is to minimize the sum of all $\psi_i(\ell)$. This third objective could be considered as part of a more general objective that is described next.

A fourth objective concerns the rate of consumption of material and parts at the workstations. The objective is to keep the consumption rates of all the parts at all the workstations as regular as possible.

The relative weights of the different objectives depend on the specific assembly system and product line. In some assembly lines all jobs are made to stock while in others most jobs are made to order. Some lines have no capacity constrained operations (such as sunroof installation) and the main costs are setup costs (e.g., cleaning paint guns).

One procedure for sequencing paced assembly lines is the so-called *Grouping and Spacing (GS)* heuristic, which has been specifically designed for the paced assembly lines that are common in the automotive industry. In a car assembly facility the number of different models is usually fairly large and there are many different options or option packages that a car can have. Before the heuristic can be applied a detailed analysis has to be done of each operation. Operations with significant setup costs have to be ranked in decreasing order of setup costs and operations with capacity constraints have to be ranked in decreasing order of their criticality indices. The output of this analysis is an input to the heuristic.

6.3 Sequencing of Paced Assembly Systems

The GS heuristic consists of four phases.
(i) Determining the total number of jobs to be scheduled.
(ii) Grouping jobs with regard to operations that have high setup-costs.
(iii) Ordering of the subgroups taking shipping dates into account.
(iv) Spacing jobs within subgroups taking capacity constrained operations into account.

The first phase determines the total number of jobs to be scheduled. This number is usually decided by the scheduler based on his knowledge of the likelihood and the timing of exogenous disruptions that require rescheduling. This number is a trade-off between two factors. A high number allows the scheduler to find a sequence with a lower cost. However, the probability of a disruption is high and if there are disruptions, then the production of some jobs may be postponed unduly. The number to be scheduled typically ranges from the equivalent of one day of production to one week of production.

The second phase forms subgroups taking operations with high setup costs into account. The run lengths of the different colors are determined. These run lengths may be different for different colors. A frequent color, e.g., grey, may have a run length of up to 50, whereas an infrequent color, e.g., purple, may have a run length of 1 or 2. The grouping with regard to destinations is slightly different since the constraints here are less strict. Jobs with the same destination have to be close to one another in the sequence. However, jobs that have to go to other destinations may also be scattered throughout that segment of the sequence. The run lengths determined in this phase are the results of trade-offs between the minimization of setup-costs and the minimization of the holding costs of finished jobs. They also depend somewhat on the committed shipping dates of the Make-To-Order jobs. As such, determining a run length is in a sense similar to computing an Economic Order Quantity (see Chapter 7).

The third phase orders the different subgroups. The order of the different subgroups is mainly determined by the urgency with which the jobs in a group have to be shipped. This urgency is determined by the committed shipping dates of the Make-To-Order jobs as well as the current inventory levels of the Make-To-Stock jobs.

The fourth phase does the internal sequencing within the subgroups considering the capacity constrained operations. It first considers the most critical operation and spaces the jobs that have to undergo this operation as uniformly as possible. Assuming the positions of these jobs as fixed, it then considers the second most capacity constrained operation and spaces out these jobs as uniformly as possible.

Before presenting a numerical example, we describe a simple mathematical model that has many of the characteristics of a paced assembly system. Consider a single machine and n jobs. All the processing times are equal to 1 (since the assembly line moves at a constant speed). Job j has a due date d_j and a weight w_j. Some due dates may be infinite. Job j has l parameters

$a_{j1}, a_{j2}, \ldots, a_{jl}$. The first parameter represents the color, the second one is equal to 1 when the job has a sunroof and 0 otherwise, the third one represents the destination, and so on. If job j is followed by job k and $a_{j1} \neq a_{k1}$ (i.e., the jobs have different colors), then a setup cost c_{jk} is incurred, which is a function of a_{j1} and a_{k1}. If both jobs j and k have sunroofs, i.e., $a_{j2} = a_{k2} = 1$, and they are spaced ℓ positions apart, a penalty $\psi_2(\ell)$ is incurred, which is decreasing in ℓ. If job j is completed after its due date, a weighted tardiness penalty is incurred. The objective is to minimize the total cost, including the setup costs, spacing costs and tardiness costs.

Example 6.3.2 (Application of Grouping and Spacing Heuristic). Consider a single machine and 10 jobs. Each job has unit processing time. Job j has two parameters a_{j1} and a_{j2}. If two consecutive jobs j and k have $a_{j1} \neq a_{k1}$, then a setup cost

$$c_{jk} = |a_{j1} - a_{k1}|$$

is incurred. If two jobs j and k have $a_{j2} = a_{k2} = 1$ and they are spaced ℓ positions apart (i.e., there are $\ell - 1$ jobs scheduled in between) a penalty cost

$$\psi_2(\ell) = \max(3 - \ell, 0)$$

is incurred. Some jobs have due dates. If job j, with due date d_j, is completed after its due date, a penalty $w_j T_j$ is incurred.

Jobs	1	2	3	4	5	6	7	8	9	10
a_{j1}	1	1	1	3	3	3	5	5	5	5
a_{j2}	0	1	1	0	1	1	1	0	0	0
d_j	∞	2	∞	∞	∞	∞	6	∞	∞	∞
w_j	0	4	0	0	0	0	4	0	0	0

From the data it is clear that there are three groups with regard to the operation with setup costs (e.g., there may be three colors: color 1, color 3, and color 5). The objective is to find the sequence with the minimum total cost.

The first phase of the GS heuristic is not applicable here. There are three groups with regard to attribute 1 (color). Group A consists of jobs 1, 2, and 3, group B consists of jobs 4, 5, and 6, and group C consists of jobs 7, 8, 9, and 10. The best order with regard to setup costs would be A,B,C. However, group C contains a job with due date 6 that would not be completed in time if the groups are ordered that way. Since the tardiness penalty is somewhat high, it may be better to order the groups A,C,B, because in this way job 7 can be completed in time. So the result of the third phase of the GS heuristic is that the groups are ordered in the sequence A,C,B.

The last phase of the GS heuristic considers the capacity constrained operations which are embodied in attribute 2 of this model. The jobs with attribute 2 equal to 1 have to be spaced out as much as possible. It can be verified that the following sequence minimizes the penalties with regard to the capacity constrained operation: Group A in the order 2, 1, 3, followed by Group C in the order 8, 7, 9, 10, and Group B in the order 5, 4, 6. The sequence of attribute 2 values is then 1, 0, 1, 0, 1, 0, 0, 1, 0, 1. The total cost with regards to the capacity constrained operation is 3.

Thus, the total cost associated with this sequence is $6 + 3 = 9$ (6 because of setup cost and 3 because of the capacity constrained operation).

The example presented above is clearly very simple. It was not necessary to determine the sizes of each group, since it appeared that there would be just a single group for each value of attribute 1. In larger instances with more jobs, the question of group sizes becomes, of course, a more important issue.

It is not easy to formulate a precise model and algorithm for the paced assembly line sequencing problem in general. The difficulty lies in the various objective functions. The structure of an algorithm depends on the relative importance of each one of the objectives.

Most paced assembly systems in the real world are significantly more complicated than those described in this section. For example, paced assembly lines in the automotive industry may have a resequencing bank immediately after the paint shop that basically partitions the overall problem into two more or less independent subproblems. Another factor is that some cars may need two passes through the paint shop. The grouping and spacing heuristics that have been implemented in industry are, therefore, significantly more complicated than the simple procedure described above.

Paced assembly sequencing problems have also been tackled with constraint programming techniques. Appendix D describes a program for a paced assembly system that is encoded in the constraint programming language OPL.

6.4 Scheduling of Flexible Flow Systems with Bypass

Consider an assembly system with a number of stages in series and, at each stage, a number of machines in parallel. A job, which in this case often is equivalent to a batch of identical items such as Printed Circuit Boards (PCB's), needs processing at every stage, but only on one machine. Usually, any of the machines will do, but it may be the case that at a given stage not all machines are identical and that a given job has to be processed on a certain machine. If a job does not need processing at a stage, the material handling system will allow that job to bypass that stage and all the jobs residing there, see Figure 6.2. A buffer at a stage may have a limited capacity and when a buffer is full, then either the material handling system must come to a standstill, or, if there is an option to recirculate, the jobs must bypass that stage and recirculate. The manufacturing process is repetitive and therefore it is of interest to find a good cyclic schedule (similar to the one in Section 6.2).

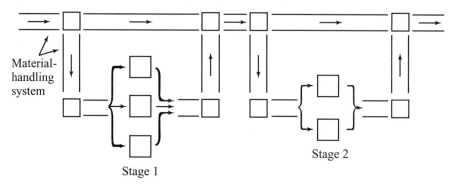

Fig. 6.2. Flexible flow line with bypass

The *Flexible Flow Line Loading (FFLL)* algorithm was designed at IBM for the machine environment described above. It was originally conceived for an assembly system used for the insertion of components in PCB's. The two main objectives of the algorithm are the maximization of throughput and the minimization of WIP. With the goal of maximizing the throughput an attempt is made to minimize the makespan of a whole day's mix. The FFLL algorithm actually attempts to minimize the cycle time of a Minimum Part Set (MPS). (For the definition of the MPS, see Section 6.2.) As buffer spaces are limited, it tries to minimize the WIP to reduce blocking probabilities. The FFLL algorithm consists of three phases:

(i) The machine allocation phase.
(ii) The sequencing phase.
(iii) The release timing phase.

The *machine allocation phase* assigns each job to a particular machine in each bank of machines. Machine allocation is done before sequencing and timing, because, in order to perform the last two phases, the workload assigned to each machine must be known. The lowest conceivable maximum workload for a bank would be obtained if all the machines in a bank were given an equal workload. In order to obtain nearly balanced workloads for the machines at a bank, the Longest Processing Time first (LPT) heuristic is used (see Section 5.2). In this heuristic all the jobs are, for the time being, assumed to be available at the same time and are allocated one at a time to the next available machine in decreasing order of their processing times. After the allocation is determined in this way, the items assigned to a machine may be resequenced, which does not alter the workload balance over the machines at a given bank. The output of this phase is merely the allocation of jobs to machines and not the sequencing of the jobs or the timing of their processing.

The *sequencing phase* determines the order in which the jobs of the MPS are released into the system. This has a significant impact on the MPS cycle

6.4 Scheduling of Flexible Flow Systems with Bypass

time. The FFLL algorithm uses a so-called *Dynamic Balancing* heuristic for sequencing an MPS. This heuristic is based on the intuition that jobs tend to queue up in the buffer of a machine when a large workload is sent to that machine in a short period of time. This occurs when there is an interval in the loading sequence that contains many jobs with long processing times allocated to one machine. Let n be the number of jobs in an MPS and m the number of machines in the entire system. Let p_{ij} denote the processing time of job j on machine i. Note that $p_{ij} = 0$ for all but one machine in a bank. Let

$$\mathcal{W}_i = \sum_{j=1}^{n} p_{ij}$$

denote the workload in an MPS that is assigned to machine i, and let

$$\mathcal{W} = \sum_{i=1}^{m} \mathcal{W}_i$$

denote the total workload of an MPS. Assuming a fixed sequence, let J_k denote the set of jobs loaded into the system up to and including job k, and let

$$\alpha_{ik} = \sum_{j \in J_k} \frac{p_{ij}}{\mathcal{W}_i}.$$

The α_{ik} represent the fraction of the total workload of machine i that has entered the system by the time job k is loaded. Clearly, $0 \leq \alpha_{ik} \leq 1$. The Dynamic Balancing procedure attempts to keep the $\alpha_{1k}, \alpha_{2k}, \ldots, \alpha_{mk}$ as close to one another as possible, i.e., as close to an ideal target α_k^*, that is defined as

$$\alpha_k^* = \sum_{j \in J_k} \sum_{i=1}^{m} p_{ij} \bigg/ \sum_{j=1}^{n} \sum_{i=1}^{m} p_{ij}$$
$$= \sum_{j \in J_k} p_j / \mathcal{W},$$

where

$$p_j = \sum_{i=1}^{m} p_{ij}$$

is the workload on the entire system due to job j. Hence α_k^* is the fraction of the total system workload that is released into the system by the time job k is loaded. The cumulative workload on machine i, i.e., $\sum_{j \in J_k} p_{ij}$, should be close to the target $\alpha_k^* \mathcal{W}_i$. Now, let o_{ik} denote a measure of *overload* at machine i due to job k entering the system, defined as

$$o_{ik} = p_{ik} - p_k \mathcal{W}_i / \mathcal{W}.$$

Clearly, o_{ik} may be negative, which implies an underload. Let

$$O_{ik} = \sum_{j \in J_k} o_{ij} = \left(\sum_{j \in J_k} p_{ij} \right) - \alpha_k^* \mathcal{W}_i$$

denote the cumulative overload (or underload) on machine i due to the jobs in the sequence up to and including job k. To be exactly on target means that machine i is neither overloaded nor underloaded when job k enters the system, i.e., $O_{ik} = 0$. The Dynamic Balancing heuristic attempts to minimize

$$\sum_{k=1}^{n} \sum_{i=1}^{m} \max(O_{ik}, 0).$$

The procedure is basically a greedy heuristic, which selects from among the remaining items in the MPS the one that minimizes the objective at that point in the sequence.

The *release timing phase* works as follows. From the allocation phase the MPS workload at each machine is known. The machine with the greatest MPS workload is the bottleneck since the MPS cycle time cannot be smaller than the MPS workload at the bottleneck machine. We wish to determine a timing mechanism that yields a schedule with a minimum MPS cycle time. First, let all jobs in the MPS enter the system as rapidly as possible. Consider the machines one at a time. At each machine the jobs are processed in the order in which they arrive and processing starts as soon as the job is available. The release times are now modified as follows. Assume that the starting and completion times at the bottleneck machine are fixed. First, consider the machines that are upstream of the bottleneck machine and delay the processing of all jobs on each of these machines, as much as possible, without altering the job sequences. The delays are thus determined by the starting times at the bottleneck machine. Second, consider the machines that are positioned downstream from the bottleneck machine. Process all jobs on these machines as early as possible, again without altering job sequences. These modifications in release times tend to reduce the number of jobs waiting for processing, thus reducing required buffer space.

This three phase procedure attempts to find the cyclic schedule with minimum MPS cycle time in a steady state. If the system starts out empty at some point in time, it may take a few MPS's to reach a steady state. Usually, this transient period is very short. The algorithm tends to achieve short cycle times during the transient period as well.

Extensive experiments with the FFLL algorithm indicates that the method is a valuable tool for the scheduling of flexible flow lines.

Example 6.4.1 (Application of FFLL Algorithm). Consider a flexible flow shop with three stages. At stages 1 and 3 there are two machines in parallel. At stage 2, there is a single machine. There are five jobs in an MPS. Let p'_{hj} denote the processing time of job j at stage h, $h = 1, 2, 3$.

6.4 Scheduling of Flexible Flow Systems with Bypass

Jobs	1 2 3 4 5
p'_{1j}	6 3 1 3 5
p'_{2j}	3 2 1 3 2
p'_{3j}	4 5 6 3 4

The first phase of the FFLL algorithm performs an allocation procedure for stages 1 and 3. Applying the LPT heuristic to the five jobs on the two machines in stage 1 results in an allocation of jobs 1 and 4 to one machine and jobs 5, 2 and 3 to the other machine. Both machines have to perform a total of 9 time units of processing. Applying LPT to the five jobs on the two machines in stage 3 results in an allocation of jobs 3 and 5 to one machine and jobs 2, 1 and 4 to the other machine (in this case LPT does *not* yield an optimal balance). Note that machines 1 and 2 are at stage 1, machine 3 is at stage 2 and machines 4 and 5 are at stage 3. If p_{ij} denotes the processing time of job j on machine i, we have

Jobs	1 2 3 4 5
p_{1j}	6 0 0 3 0
p_{2j}	0 3 1 0 5
--	-----
p_{3j}	3 2 1 3 2
--	-----
p_{4j}	4 5 0 3 0
p_{5j}	0 0 6 0 4

The workload $\mathcal{W}_i$ of machine i due to a single MPS can now be computed. The workload vector $\mathcal{W}_i$ is $(9, 9, 11, 12, 10)$ and the entire workload $\mathcal{W}$ is 51. The workload imposed on the entire system due to job j, p_j, can also be computed. The p_j vector is $(13, 10, 8, 9, 11)$. Based on these numbers, all values of o_{ik} can be determined, e.g.,

$$o_{11} = 6 - 13 \times 9/51 = +3.71$$
$$o_{21} = 0 - 13 \times 9/51 = -2.29$$
$$o_{31} = 3 - 13 \times 11/51 = +0.20$$
$$o_{41} = 4 - 13 \times 12/51 = +0.94$$
$$o_{51} = 0 - 13 \times 10/51 = -2.55$$

Computing the entire o_{ik} matrix yields:

$$\begin{Vmatrix} +3.71 & -1.76 & -1.41 & +1.41 & -1.94 \\ -2.29 & +1.23 & -0.41 & -1.59 & +3.06 \\ +0.20 & -0.16 & -0.73 & +1.06 & -0.37 \\ +0.94 & +2.64 & -1.88 & +0.88 & -2.59 \\ -2.55 & -1.96 & +4.43 & -1.76 & +1.84 \end{Vmatrix}$$

Of course, the solution also depends on the initial job in the MPS. If the initial job is chosen according to the Dynamic Balancing heuristic, then job 4 goes first and
$$O_{i4} = (+1.41, -1.59, +1.06, +0.88, -1.76).$$
There are four jobs that qualify to go second, namely jobs 1, 2, 3 and 5. If job k goes second, the respective O_{ik} vectors are:
$$O_{i1} = (+5.11, -3.88, +1.26, +1.82, -4.31)$$
$$O_{i2} = (-0.35, -0.36, +0.90, +3.52, -3.72)$$
$$O_{i3} = (+0.00, -2.00, +0.33, -1.00, +2.67)$$
$$O_{i5} = (-0.53, +1.47, +0.69, -1.71, +0.08)$$
The dynamic balancing heuristic then selects job 5 to go second. Proceeding in the same manner the dynamic balancing heuristic selects job 1 to go third and
$$O_{i1} = (+3.18, -0.82, +0.89, -0.77, -2.47).$$
Proceeding, we see that job 3 goes fourth. Then
$$O_{i3} = (+1.76, -1.23, +0.16, -2.64, +1.96).$$
The final cycle is thus $4, 5, 1, 3, 2$.

Applying the release timing phase to this cycle results initially in the schedule presented in Figure 6.3. The MPS cycle time of 12 is actually determined by machine 4 (the bottleneck machine) and there is therefore no idle time allowed between the processing of jobs on this machine. It is clear that the processing of jobs 3 and 2 on machine 5 can be postponed by three time units.

The model discussed in this section can be compared to a flexible flow shop (which is a special case of a flexible job shop). There are similarities as well as differences between the model considered in this section and a flexible flow shop that fits within the framework of Chapter 5. The machine environments in the two models are similar, since both settings are flexible flow shops that allow bypass. In the model described in this section there is the material handling system that imposes additional constraints that have an impact on the objectives. The limited buffers in the flexible assembly system require a minimization of the Work-In-Process. The objectives in the two models are also different from another point of view. The models described in Chapter 5 tend to be more Make-To-Order; an important objective in such models may be the minimization of the total weighted tardiness. The model considered in this section is basically a Make-To-Stock model. The main objective is the maximization of throughput.

6.5 Mixed Model Assembly Sequencing at Toyota

Among the major car manufacturers Toyota has been always one of the more innovative ones as far as manufacturing and assembly is concerned. Toyota

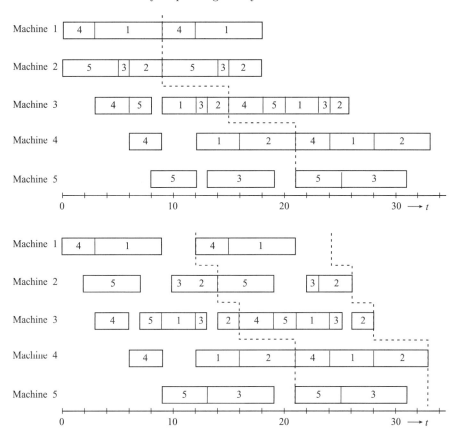

Fig. 6.3. Gantt charts for the FFLL algorithm

operates according to the Just-In-Time (JIT) principle. To make the JIT operation run smoothly, it is important that the consumption rates of all the parts at each station are kept as regular as possible. Toyota's most important goal in the operation of its mixed model assembly lines is to keep the rates of consumption of all parts more or less constant. In other words, the quantity of each part used per hour must be maintained as regular as possible.

As described in the section on paced assembly systems, balancing the workload at each one of the workstations over time is often an important objective in car assembly. Some cars may need more than the average amount of processing at a particular workstation, while others may need less. However, if Toyota manages to keep the consumption of all the parts at each one of the workstations more or less constant, then the workloads are balanced as well.

In order to formalize the Toyota objective, let $\mathcal{N}$ denote the total number of cars to be sequenced. This number is a measure of the planning horizon. Let ℓ denote the number of different models and $\mathcal{N}_j$, $j = 1, \ldots, \ell$, the number of units of model j that have to be produced over the period under consideration. Thus

$$\mathcal{N} = \sum_{j=1}^{\ell} \mathcal{N}_j.$$

In practice the planned production quantity may be around 500 and the number of different models around 180. Let ν denote the number of different types of parts needed by all workstations for the assembly of the $\mathcal{N}$ cars and let b_{ij}, $i = 1, \ldots, \nu$, $j = 1, \ldots, \ell$, denote the number of parts of type i needed for the assembly of one unit of model j. Let R_i denote the total number of parts of type i required for the assembly of all $\mathcal{N}$ cars and x_{ik} the total number of parts of type i needed in the assembly of the first k units in the sequence. So $R_i/\mathcal{N}$ is the average number of part i required for the assembly of a car and $kR_i/\mathcal{N}$ the average number of part i required for the assembly of k cars.

To keep the rate of consumption of part i as regular as possible, the variable x_{ik} should be as close as possible to the value $kR_i/\mathcal{N}$. Consider now all ν parts and the two vectors

$$\left(\frac{kR_1}{\mathcal{N}}, \ldots, \frac{kR_\nu}{\mathcal{N}}\right)$$

and

$$(x_{1k}, \ldots, x_{\nu k}).$$

The first vector plays the role of the target or goal, while the second vector represents the actual numbers of parts needed for the assembly of the first k units in the given sequence. Let

$$\Delta_k = \sqrt{\sum_{i=1}^{\nu} \left(\frac{kR_i}{\mathcal{N}} - x_{ik}\right)^2}$$

denote a measure of the difference between the two vectors. Let $\Delta_k(j)$ denote the value of this difference if model j has been put in the k-th position, i.e.,

$$\Delta_k(j) = \sqrt{\sum_{i=1}^{\nu} \left(\frac{kR_i}{\mathcal{N}} - x_{i,k-1} - b_{ij}\right)^2}.$$

In order to minimize this objective, Toyota developed the Goal Chasing method which can be described as follows.

Algorithm 6.5.1 (Goal Chasing Method).
Step 1.
   Set $x_{i0} = 0$, $S_0 = \{1, 2, \ldots, \ell\}$, and $k = 1$.
Step 2.
   Select for the k-the position in the sequence model j^* that minimizes the measure $\Delta_k(j)$, i.e.,

$$\Delta_k(j^*) = \min_{j \in S_{k-1}} \left\{ \sqrt{\sum_{i=1}^{\nu} \left(\frac{kR_i}{\mathcal{N}} - x_{i,k-1} - b_{ij}\right)^2} \right\},$$

Step 3.

If more units of model j^* remain to be sequenced, set $S_k = S_{k-1}$

If all units of model j^* have now been sequenced, set $S_k = S_{k-1} - \{j^*\}$.

Step 4.

If $S_k = \emptyset$, then STOP.

If $S_k \neq \emptyset$, set $x_{ik} = x_{i,k-1} + b_{ij^*}$, $\quad i = 1, \ldots, \nu$.

Set $k = k + 1$ and go to Step 2.

Since the number of parts in a car is around 20,000, it is in practice difficult to apply the Goal Chasing method to all parts. Therefore, the parts are represented only by their respective subassembly. The number of subassemblies is around 20 and Toyota gives the important subassemblies additional weight. The subassemblies include:

(i) engines,
(ii) transmissions,
(iii) frames,
(iv) front axles,
(v) rear axles,
(vi) bumpers,
(vii) steering assemblies,
(viii) wheels,
(ix) doors, and
(x) air conditioners.

The fact that the consumption rates of the parts are the main objective gives an indication of how important JIT is for Toyota.

Note the similarity between the Goal Chasing method and the Dynamic Balancing routine described in Section 6.4. The physical environment at Toyota is, of course, more similar to the environment of the paced assembly systems described in Section 6.3.

6.6 Discussion

Scheduling problems in flexible assembly systems usually are not as easy to formulate as the more conventional job shop scheduling problems. The material handling systems or conveyor systems impose constraints that tend to be application-specific and difficult to formulate mathematically. Since these problems are hard to formulate as mathematical programs the solution procedures are usually not based on branch-and-bound, but rather based on heuristics. The framework of a heuristic is typically modular and application-specific. Other approaches include constraint programming techniques. Appendix D presents a constraint programming application of a mixed model assembly line.

Another class of systems that share many of the scheduling complexities of flexible assembly systems are the flexible manufacturing systems. In general, a flexible manufacturing system may be defined as a production system that is capable of producing a variety of part types; it consists of (numerically controlled) machines that are connected to one another by an automated material handling system. The entire system typically is under computer control. With the appropriate tool setups, each machine in the system can perform different operations. An operation may therefore be performed at any one of a number of machines. The routing of a job through the system is therefore flexible and is one of the types of decisions that have to be made in the scheduling process. Two other types of decisions are the sequencing of the operations on the machines and the setups of the tools on the machines. These scheduling problems are harder than the classical job shop scheduling problems and the scheduling problems that occur in flexible assembly systems. As in flexible assembly systems, material handling systems and limited buffers increase the complexity. Just as a flexible assembly system often can be viewed as a (flexible) flow shop with a number of additional constraints, a flexible manufacturing system can be viewed as a (flexible) job shop with a number of additional constraints. While a limited buffer in a flexible assembly system may cause blocking, a limited buffer in a flexible manufacturing system may not only cause blocking but also deadlock.

The scheduling problems in flexible manufacturing systems are not easy to formulate as mathematical programs when all decisions and constraints are included in the formulation. Even when they can be formulated as mathematical programs, they are very hard to solve. Many formulations therefore incorporate only one or two of the three types of decisions listed above and ignore some of the resource constraints.

Most of the research in flexible assembly systems and flexible manufacturing systems have focused on specific problems arising in industry. While a classification scheme has been available for job shop scheduling problems for quite some time, it is only recently that such a scheme has been emerging for scheduling problems in flexible manufacturing systems.

Exercises

6.1. Consider in the model of Section 6.2 an MPS of 5 jobs.

(a) Show that when all jobs are different the number of different cyclic schedules is 4!.

(b) Compute the number of different cyclic schedules when two jobs are the same (i.e., there are four different job types among the 5 jobs).

6.2. Consider the model discussed in Section 6.2 with 4 machines and an MPS of 4 jobs.

Exercises

Jobs	1	2	3	4
p_{1j}	6	4	6	8
p_{2j}	2	10	4	6
p_{3j}	4	8	0	2
p_{4j}	8	2	6	6

(a) Apply the unweighted PF heuristic to find a cyclic schedule. Choose job 1 as the initial job and compute the MPS cycle time.

(b) Apply again the unweighted PF heuristic. Choose job 2 as the initial job and compute the MPS cycle time.

(c) Find the optimal schedule.

6.3. Consider the same problem as in the previous exercise.

(a) Apply a weighted PF heuristic to find a cyclic schedule. Choose the weights associated with machines $1, 2, 3, 4$ as $2, 2, 1, 2$, respectively. Select job 1 as the initial job.

(b) Apply again a weighted PF heuristic but now with weights $3, 3, 1, 3$. Select again job 1 as the initial job.

(c) Repeat again (a) and (b) but select job 2 as the initial job.

(d) Compare the impact of the weights on the heuristic's performance with the impact of the selection of the first job on the performance.

6.4. Consider the model discussed in Section 6.2. Assume that a system is in a steady state if each machine is in a steady state. That is, at each machine the departure of job j in one MPS occurs exactly the cycle time before the departure of job j in the next MPS. Construct an example with 3 machines and an MPS of a single job that takes more than 100 MPS's to reach a steady state, assuming the system starts out empty.

6.5. In order to model a paced assembly line consider a single machine and n jobs. The processing times of each one of the jobs is 1. Job j has a due date d_j and a weight w_j. If job j is completed after its due date, then a penalty $w_j T_j$ is incurred. There are sequence dependent setup costs c_{jk} but no setup times. The sequence dependent setup costs are determined as follows: job j has an attribute a_j that can be either 0 or 1. Most jobs have an attribute value a_j equal to 0. If there are two jobs with a_j equal to 1 sequenced in such a way that there are less than 3 jobs with a_j equal to 0 in between, then a penalty cost $c_1 = 3$ is incurred. (This implies that this attribute a_j corresponds to a capacity constrained operation).

(a) Design an algorithm for finding a sequence with minimum cost. (*Hint:* Consider, for example, a composite dispatching rule followed by a local search; see Appendix C.)

(b) Apply the algorithm developed to the following instance.

Jobs	1 2	3 4	5 6 7	8 9 10
a_j	0 1	1 0	0 0 1	0 0 0
d_j	∞ 2	∞ 1	∞ 5 6	∞ 2 ∞
w_j	0 4	0 1	0 3 4	0 3 0

Compute the total cost of the sequence.

6.6. Apply the GS heuristic to the instance in Exercise 6.5. Compare the result with that obtained in the previous exercise.

6.7. Consider the model in Exercise 6.5. Now job j has two attributes a_j and b_j. The a_j attribute is the same as in Exercise 6.5. The b_j values also can be either 0 or 1. If two jobs with b_j value equal to 1 are positioned in such a way that there are 4 or less jobs with b_j value equal to 0 in between, then a penalty cost $c_2 = 3$ is incurred. (This situation corresponds to a line with two capacity constrained operations).

(a) Describe how the algorithm of Exercise 6.5 has to be modified to take this generalization into account.

(b) Apply your algorithm to the instance below.

Jobs	1 2	3 4	5 6 7	8 9 10
a_j	0 1	1 0	0 0 1	0 0 0
b_j	0 1	0 1	1 0 1	0 0 0
d_j	∞ 2	∞ 1	∞ 5 6	∞ 2 ∞
w_j	0 4	0 1	0 3 4	0 3 0

Compute the total cost of the sequence.

6.8. Consider the application of the FFLL algorithm in Example 6.4.1. Instead of letting the dynamic balancing heuristic minimize

$$\sum_{i=1}^{m}\sum_{k=1}^{n}\max(O_{ik},0),$$

let it minimize

$$\sum_{i=1}^{m}\sum_{k=1}^{n}|O_{ik}|.$$

Redo Example 6.4.1 and compare the performances of the two dynamic balancing heuristics.

6.9. Consider the application of the FFLL algorithm to the instance in Example 6.4.1. Instead of applying LPT in the first phase of the algorithm, find the optimal allocation of jobs to machines (which leads to a perfect balance of machines 4 and 5). Proceed with the sequencing phase and release timing phase based on this new allocation.

6.10. Consider the instance in Example 6.4.1 again.

(a) Compute in Example 6.4.1 the number of jobs waiting for processing at each stage as a function of time and determine the required buffer size at each stage.

(b) Consider the application of the FFLL algorithm to the instance in Example 6.4.1 with the machine allocation as prescribed in Exercise 6.9. Compute the number of jobs waiting for processing at each stage as a function of time and determine the required buffer size.

(Note that with regard to the machines before the bottleneck, the release timing phase in a sense postpones the release of each job as much as possible and tends to reduce the number of jobs waiting for processing at each stage.)

Comments and References

Flexible assembly systems constitute a subcategory of flexible manufacturing systems (FMS). (Another subcategory of FMS are the general flexible machining systems (GFMS).) A significant amount of research has been done on scheduling in FMS in general. For an overview of such research on scheduling in FMS, see MacCarthy and Liu (1993), Rachamadugu and Stecke (1994), and Basnet and Mize (1994). Liu and MacCarthy (1996) clarify the basic concepts of scheduling in FMS and classify these scheduling problems according to a classification scheme that is based on the configurations of these systems. For a comprehensive mixed integer linear programming model for the basic scheduling problem that occurs in an FMS, see Liu and MacCarthy (1997). The scheduling of a single flexible machine is studied by Tang and Denardo (1988).

A certain amount of research has been done on the scheduling of flexible assembly systems. The unpaced assembly system with limited buffers is analyzed by Pinedo, Wolf and McCormick (1987), McCormick, Pinedo, Shenker and Wolf (1989), and Abadie, Hall and Sriskandarajah (2000). Its transient analysis is discussed in McCormick, Pinedo, Shenker and Wolf (1990). For more results on cyclic scheduling, see Matsuo (1990) and Roundy (1992), and for more results on flow shops with limited buffers, see Wismer (1972) and Pinedo (1982). Hall and Sriskandarajah (1996) present a survey that includes many of these cyclic scheduling models with limited buffers. For the scheduling of a robotic assembly system, see Hall, Kamoun and Sriskandarajah (1997).

The book by Scholl (1998) focuses on the balancing and sequencing of assembly lines (paced assembly systems). For scheduling problems and solutions in the automotive industry, see, for example, Burns and Daganzo (1987), Yano and Bolat (1989), Bean, Birge, Mittenthal and Noon (1991), Garcia-Sabater (2001) and Drexl and Kimms (2001). Important aspects of scheduling problems in the automotive industry are the sequence dependent setup costs and setup times. The CD-ROM that is attached to this book contains several mini-cases that focus on assembly line sequencing in the automotive industry (projects at Nissan and Peugeot).

A considerable amount of research has focused specifically on sequence dependent setups; see, for example, Gilmore and Gomory (1964), and Bianco, Ricciardelli, Rinaldi and Sassano (1988). Crama, Kolen and Oerlemans (1990) develop a comprehensive hierarchical decision model for planning a multiple machine flow line with sequence dependent setups.

The scheduling problem concerning the flexible flow line with limited buffers and bypass is based on a setting found at IBM and analyzed by Wittrock (1985, 1988, 1990).

The way Toyota schedules its assembly lines is discussed in Monden (1983); Section 6.5 is based on Monden's Appendix 2.

Chapter 7

Economic Lot Scheduling

7.1 Introduction 141
7.2 One Type of Item and the Economic Lot Size ... 142
7.3 Different Types of Items and Rotation Schedules 146
7.4 Different Types of Items and Arbitrary Schedules 150
7.5 More General ELSP Models 159
7.6 Multiproduct Planning and Scheduling at
    Owens-Corning Fiberglas...................... 162
7.7 Discussion 164

7.1 Introduction

In a job shop, each job has its own identity and its own set of processing requirements. In a flexible assembly system, there are a number of different types of jobs and jobs of the same type have identical processing requirements; in such a system, setup times and setup costs are often not important and a schedule may alternate many times between jobs of different types. In a flexible assembly system an alternating schedule is often more efficient than a schedule with long runs of identical jobs.

In the models considered in this chapter, a set of identical jobs may be large and setup times and setup costs between jobs of two different types may be significant. A setup typically depends on the characteristics of the job about to be started and the one just completed. If a job's processing on a machine requires a major setup then it may be advantageous to let this job be followed by a number of jobs of the same type.

In this chapter we refer to jobs as items and we call the uninterrupted processing of a series of identical items a run. If a facility or machine is geared to produce identical items in long runs, then the production tends to be Make-To-Stock, which inevitably involves inventory holding costs. This form of production is, at times, also referred to as continuous manufacturing (in contrast to the forms of discrete manufacturing considered in the previous chapters). The time horizon in continuous manufacturing is often in the order of months or even years. The objective is to minimize the total cost, which includes inventory holding cost as well as setup cost. The optimal schedule is typically a trade-off between inventory holding costs and setup costs and is often repetitive or cyclic.

The associated scheduling problem has several aspects. First, the lengths of the runs have to be determined and, second, the order of the different runs has to be established. The run lengths are typically referred to as the lot sizes and they are the result of trade-offs between setup costs and inventory holding costs. The lots have to be sequenced in such a way that the setup times and setup costs are minimized. This scheduling problem is referred to as the *Economic Lot Scheduling Problem (ELSP)*.

In the standard ELSP a single facility or machine has to produce n different items. The machine can produce items of type j at a rate of Q_j per unit time. If an item of type j is regarded as a job with processing time p_j, then $Q_j = 1/p_j$. We assume that the demand rate for type j is constant at D_j items per unit time. The inventory holding cost for one item of type j is h_j dollars per unit time. If an item of type j is followed by an item of type k a setup cost c_{jk} is incurred; moreover, a setup time s_{jk} may be required. In some models we assume that a setup involves a cost but no machine time and in other, more general, models we assume that a setup involves a cost as well as machine time. The setup cost and time may be either sequence dependent or independent. If the setup cost (time) is sequence independent, then $c_{jk} = c_k$ ($s_{jk} = s_k$). The problem can be viewed as one of deciding a cycle length x and a sequence of runs or cycle $j_1, j_2, \ldots, j_\nu$. This sequence may contain repetitions, so $\nu \geq n$. The associated run times are $\tau_{j_1}, \tau_{j_2}, \ldots, \tau_{j_\nu}$ and there may be idle time between two consecutive runs.

In practice, there are many applications of economic lot scheduling. In the process industries (e.g., the chemical, paper, pharmaceutical, aluminum and steel industries) setup costs and inventory holding costs are significant. When minimizing the total costs, a scheduling problem often reduces to an economic lot scheduling problem (see Example 1.1.4). There are applications of lot scheduling in the service industries as well. In the retail industry (e.g., Sears, Wal-Mart) the procurement of each item has to be controlled carefully. Placing an order for additional supplies entails an ordering cost and keeping a supply in inventory entails a holding cost. The retailer has to determine the trade-off between the inventory holding costs and the ordering costs.

7.2 One Type of Item and the Economic Lot Size

In this section we consider the simplest case, namely a single machine and one type of item. Since there is only one type of item the subscript j can

7.2 One Type of Item and the Economic Lot Size

be dropped, i.e., the production rate is Q and the demand rate is D items per unit time. We assume that the machine capacity is sufficient to meet the demand, i.e., $Q > D$. The problem is to determine the length of a production run. After a run has been terminated and sufficient inventory has been built up, the machine remains idle until the inventory has been depleted and a new run is about to start. Clearly, the length of a production run is determined by the trade-off between inventory holding costs and setup costs. In order to minimize the total cost per unit time we have to find an expression for the total cost over a cycle.

Let x denote the cycle time that has to be determined. If D denotes the demand rate, then the demand over a cycle is Dx and the length of a production run to meet the demand over a cycle is Dx/Q. If the inventory level at the beginning of the production run is zero, then the inventory level goes up during the run at a rate $Q - D$ until it reaches

$$(Q - D)\frac{Dx}{Q}.$$

During the idle period the inventory level goes down at a rate D until it reaches zero and the next production run starts. So the average inventory level is

$$\frac{1}{2}\left(Dx - \frac{D^2 x}{Q}\right).$$

Each production run incurs a setup cost c. The average cost per unit time due to setups is therefore c/x. Let h denote the inventory holding cost per item per unit time. The total average cost per unit time due to inventory holding costs and setups is therefore

$$\frac{1}{2}h\left(Dx - \frac{D^2 x}{Q}\right) + \frac{c}{x}.$$

To determine the optimal cycle length we take the derivative of this expression with respect to x and set it equal to zero, yielding

$$\frac{1}{2}hD\left(1 - \frac{D}{Q}\right) - \frac{c}{x^2} = 0.$$

Straightforward algebra gives the optimal cycle length

$$x = \sqrt{\frac{2Qc}{hD(Q-D)}}.$$

The total amount to be produced during a cycle, i.e., the lot size, is

$$Dx = \sqrt{\frac{2DQc}{h(Q-D)}}.$$

The lot size Dx is not necessarily an integer number. (This is one of the differences between continuous models and discrete models; this difference is examined more closely in Example 7.2.2.)

The idle time of the machine during a cycle turns out to be

$$x\left(1 - \frac{D}{Q}\right).$$

The ratio D/Q is at times denoted by ρ, and may be regarded as the utilization of the machine, i.e., the proportion of time that the machine is busy.

Now consider the limiting case when the production rate Q is arbitrarily high, i.e., $Q \to \infty$. Then,

$$x = \lim_{Q \to \infty} \sqrt{\frac{2Qc}{hD(Q-D)}} = \sqrt{\frac{2c}{hD}}.$$

In this case the lot size is equal to

$$Dx = \sqrt{\frac{2Dc}{h}},$$

which is often called the *Economic Lot Size (ELS)* or *Economic Order Quantity (EOQ)*.

All the expressions above are based on the assumption that there is a setup cost but not a setup time. If, in addition to the setup cost, there is also a setup time s and $s \leq x(1-\rho)$, then the solution presented above is still feasible and optimal. If

$$s > x(1-\rho),$$

then the lot size computed above is infeasible. The optimal solution then is the solution where the machine alternates between setups and production runs with a cycle length

$$x = \frac{s}{1-\rho}.$$

That is, the machine is either producing or being set up for the next run. The machine is never idle.

The first example illustrates the use of these formulae.

Example 7.2.1 (The ELSP with and without Setup Times). Consider a facility with a production rate $Q = 90$ items per week, a demand rate $D = 50$ items per week, a setup cost $c = \$2000$, a holding cost $h = \$20$ per item per week, and no setup times. From the analysis above it follows that the cycle time x is 3 weeks and the quantity produced in a cycle is 150. Figure 7.1.a depicts the inventory level over the cycle. The idle time during a cycle is $3(1 - 5/9) = 1.33$ weeks, which is approximately 9 days.

Now suppose that there are setup times. If the setup time is less than 9 days (the length of the idle period), then the 3 week cycle remains optimal.

7.2 One Type of Item and the Economic Lot Size 145

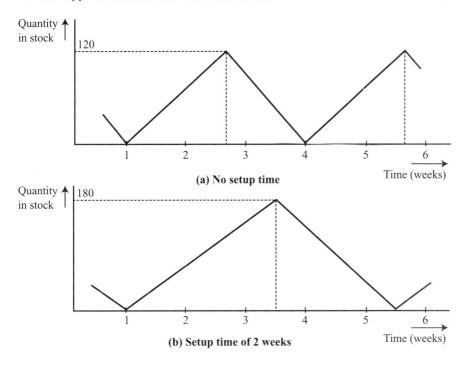

Fig. 7.1. Inventory levels in Example 7.2.1

If the setup time is longer than 9 days, then the cycle time has to be longer. For example, if a setup lasts 2 weeks (because of maintenance and cleaning), then the cycle time is 4.5 weeks. Figure 7.1.b depicts the inventory level over a cycle.

The next example highlights the differences between continuous and discrete settings.

Example 7.2.2 (Continuous Setting vs. Discrete Setting). Consider a production rate Q of 0.3333 items per day, a holding cost h of \$5.00 per item per day and a setup cost c of \$90.00. The demand rate D is 0.10 items per day. Applying the cycle length formula gives

$$x = \sqrt{\frac{60}{0.5(0.3333 - 0.1)}} = 22.678$$

and the number of items in a lot is $Dx = 2.2678$.

In a discrete setting such a number is not feasible. Consider the following discrete counterpart of this instance. The time to produce one item (or job) is $p = 1/Q = 3$ days. The demand rate is 1 item every 10 days. A lot of size k, k integer, has to be produced every $10k$ days. (The solution in the continuous setting suggests that the optimal solution in the discrete setting is either a lot of size 2 every 20 days or a lot of size 3 every 30 days.) The total cost per day with a lot of size 1 every 10 days is 90/10 = $9.00. The total cost per day with a lot of size 2 every 20 days is

$$(90 + 7 \times 5)/20 = \$6.25$$

and the total cost per day with a lot of size 3 every 30 days is

$$(90 + 7 \times 5 + 14 \times 5)/30 = \$6.50.$$

So in a discrete setting it is optimal to produce every 20 days a lot of size 2.

7.3 Different Types of Items and Rotation Schedules

Consider again a single machine, but now with n different items. The demand rate for item j is D_j and the machine is capable of producing item j at a rate Q_j. In order to start a production run for item j, a setup cost c_j is incurred. We assume, for the time being, that this setup cost is sequence independent. In this section we determine the best production cycle that contains a single run of each item. Thus, the cycle lengths of the n items have to be identical. Such a schedule is referred to as a *rotation* schedule. The length of the cycle determines the length of each of the production runs. Hence there is only a single decision variable, the cycle length x. In order to determine the optimal cycle length, it is again necessary to find an expression for the total cost per unit time as a function of the cycle length x.

If setups require machine time, then it may not be possible to make the cycle length arbitrarily small since frequent setups may take up too much machine time.

The length of the production run of item j in a cycle is $D_j x / Q_j$. Assume that the inventory level at the beginning of the production run of item j is zero. During the production run, the level increases at rate $Q_j - D_j$ until it reaches level $(Q_j - D_j) D_j x / Q_j$. During the idle period, the inventory decreases at a rate D_j until it reaches zero and the next production run starts. So the average inventory level of item j is

$$\frac{1}{2}\left(D_j x - \frac{D_j^2 x}{Q_j}\right).$$

The facility incurs a setup cost c_j for each production run of item j. The average cost per unit time due to setups for item j is therefore c_j/x. The total average cost per unit time due to inventory holding costs and setup costs is therefore

$$\sum_{j=1}^{n}\left(\frac{1}{2}h_j\left(D_j x - \frac{D_j^2 x}{Q_j}\right) + \frac{c_j}{x}\right).$$

7.3 Different Types of Items and Rotation Schedules

To find the optimal cycle length we take the derivative with respect to x and set it equal to zero, obtaining

$$\sum_{j=1}^{n}\left(\frac{1}{2}h_j D_j\left(1-\frac{D_j}{Q_j}\right)\right) - \frac{\sum_{j=1}^{n} c_j}{x^2} = 0,$$

Straightforward algebra yields the optimal cycle length

$$x = \sqrt{\left(\sum_{j=1}^{n} \frac{h_j D_j (Q_j - D_j)}{2 Q_j}\right)^{-1} \sum_{j=1}^{n} c_j}.$$

The machine idle time during a cycle can be computed in a manner similar to the single item case. This idle time is equal to

$$x\left(1 - \sum_{j=1}^{n} \frac{D_j}{Q_j}\right).$$

The ratio $\rho_j = D_j/Q_j$ can be regarded as the utilization factor of the machine due to item j.

Consider the limiting case where the production rates are arbitrarily fast, i.e., $Q_j = \infty$ for $j = 1, \ldots, n$. In this special case the optimal cycle length is

$$x = \sqrt{\left(\sum_{j=1}^{n} \frac{h_j D_j}{2}\right)^{-1} \sum_{j=1}^{n} c_j}.$$

Example 7.3.1 (Rotation Schedules without Setup Times). Consider four different items with the following production rates, demand rates, holding costs and setup costs.

items	1	2	3	4
D_j	50	50	60	60
Q_j	400	400	500	400
h_j	20	20	30	70
c_j	2000	2500	800	0

The optimal cycle length x is 1.24 months and the total idle time is $0.48x = 0.595$ months. Figure 7.2 displays the optimal rotation schedule. The total average cost per unit time can be computed easily and is $2155 + 2559 + 1627 + 2213 = 8554$.

As the setup cost of item 4 is zero, it is clear that a rotation schedule does not make sense here. It makes more sense to spread the production of item 4 uniformly over the cycle to reduce inventory holding costs. In the next section we consider this example again and allow for more general schedules.

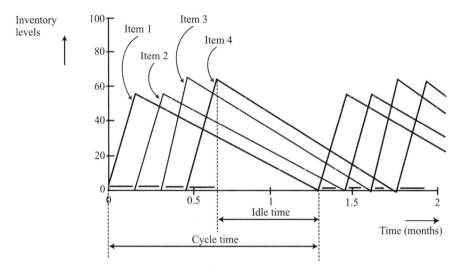

Fig. 7.2. Rotation schedule in Example 7.3.1

In the analysis above the order in which the different runs are sequenced does not matter. We assumed that there were no setup times and that setup costs were sequence independent. So, up to now, there was not any scheduling problem, only a lot sizing problem.

If there are setup times that are sequence independent, i.e., $s_{jk} = s_k$ for all j and k, then the problem still does not have a sequencing component, since the sum of the setup times does not depend on the sequence. If the sum of the setup times is less than the idle time in the rotation schedule computed above, the length of the rotation schedule remains optimal. If the sum of the setup times exceeds the idle time computed above, then the actual optimal cycle length has to be larger than the optimal cycle length obtained before. Actually, the optimal cycle length again turns out to be the cycle length that corresponds to a schedule in which the machine is never idle, i.e.,

$$x = \Big(\sum_{j=1}^{n} s_j\Big) \Big/ \Big(1 - \sum_{j=1}^{n} \rho_j\Big).$$

If the setup times are sequence dependent, then there is a sequencing problem and a sequence that minimizes the sum of the setup times has to be found. Minimizing the sum of the setup times in a rotation schedule is equivalent to the so-called *Travelling Salesman Problem (TSP)*, which can be described as follows. A salesman has to visit n cities and the distance from city j to city k is d_{jk}. His objective is to find a tour with the minimum total travel distance. That this TSP is equivalent to our sequencing problem can be shown easily. City j corresponds to item j and the distance from city j to

7.3 Different Types of Items and Rotation Schedules

city k, d_{jk}, is equivalent to the setup time needed when item k follows item j, i.e., s_{jk}. The TSP is known to be NP-hard.

If, in the case of sequence dependent setup times, a sequence can be found that minimizes the sum of the setup times and this sum is less than the idle time in the rotation schedule, then the lot sizes computed above, as well as the sequence, are optimal.

However, if the optimal sequence results in a total setup time that is larger than the machine idle time obtained before, then the optimal cycle length has to be larger than the cycle length given in the formula above. The optimal cycle length then again will be such that the machine is always either producing or being setup for the next production run. In any case, the lot sizing problem and the scheduling problem can still be analyzed separately.

The scheduling problem with arbitrary setup times is known to be extremely hard. However, when the setup times have a special structure, an easy solution may exist. Consider, for example, the following setup times:

$$s_{jk} = 0, \qquad j \leq k$$

and

$$s_{jk} = (j-k)s, \qquad j > k.$$

An optimal sequence can be obtained by starting out with the item with the lowest index, continuing with the item with the second lowest index, and so on. At the end of the run of the item with the highest index, a changeover is made to the item with the lowest index in order to start a new cycle. This sequence is obtained by applying the *Shortest Setup Time first (SST)* rule, which is often used as a heuristic in cases with arbitrary setup times (see Appendix C).

Example 7.3.2 (Rotation Schedules with Setup Times). Consider the same four items as in Example 7.3.1. However, there are now sequence dependent setup times. There are $3! = 6$ possible sequences. The setup times are given in the table below.

k	1	2	3	4
s_{1k}	-	0.064	0.405	0.075
s_{2k}	0.448	-	0.319	0.529
s_{3k}	0.043	0.234	-	0.107
s_{4k}	0.145	0.148	0.255	-

This setup matrix is asymmetric, i.e., s_{jk} is not necessarily equal to s_{kj}.

Recall that in the case without setup times the cycle time is 1.24 months and the total idle time is 0.595 months. Since there are only six sequences, all sequences can be enumerated and the best one can be selected. The sequence $1, 4, 2, 3$ requires a total setup of 0.585 months, which is feasible and therefore optimal. However, if SST is used starting with item 1, then the sequence $1, 2, 3, 4$ is selected. This sequence requires a total setup of 0.635 months which exceeds the idle time under the optimal cycle.

7.4 Different Types of Items and Arbitrary Schedules

We now generalize the model described in the previous section to allow for schedules that are more general than rotation schedules. Within a cycle there may be multiple runs of any given item. For example, if there are three different items 1, 2 and 3, then the cycle 1, 2, 1, 3 is allowed. There may be setup costs as well as setup times.

If ρ_j denotes the utilization factor D_j/Q_j of item j, then a feasible solution exists if and only if

$$\rho = \sum_{j=1}^{n} \rho_j < 1.$$

This necessary and sufficient condition is the same as the condition for the model in Section 7.3. It is intuitive that the setup times do not have any impact on this feasibility condition. The setup times do take up machine time; but, if a machine is operating close to capacity, then the cycle time and individual production runs just have to be made long enough in order to minimize the impact of the setup times.

In contrast to the model in the previous section for which there exists a closed form solution (at least, when the setup times are sequence independent), the problem considered in this section is very hard. There does not exist an efficient algorithm for this problem. However, there are good heuristics that usually lead to satisfactory solutions. In what follows, we describe one such heuristic for the case with sequence independent setup times, i.e., $s_{jk} = s_k$. Let $\mathcal{S}$ denote the set of all possible sequences of arbitrary length and j_l the index of the item produced in position l of the sequence. So $j_1, \ldots, j_\nu$, where $\nu \geq n$, denotes the production sequence of a given cycle. The sequence may contain repetitions. Consider the item that is produced in the l-th position of the sequence. If $j_l = k$, then item k is produced in the l-th position of the sequence. In the remainder of this section, the superscript l is used to refer to data related to the item produced in the l-th position of the sequence, e.g., $Q^l = Q_{j_l}$ and if the item in the l-th position is item k, then $Q^l = Q_{j_l} = Q_k$.

The production of the item in position l involves a setup cost c^l, a setup time s^l, a production time τ^l, and a subsequent idle time u^l which may be zero. If item k is produced in the l-th position, item k may be produced again within the same cycle. Let x denote the cycle time and v the time from the start of the production of item k in the l-th position till the start of the next production of item k (this may be in the same cycle or in the next cycle). So

$$v = \frac{Q^l \tau^l}{D^l}$$

and if $j_l = k$, then

$$v = \frac{Q_k \tau^l}{D_k}.$$

7.4 Different Types of Items and Arbitrary Schedules

The highest inventory level is $(Q^l - D^l)\tau^l$. The total inventory cost for the production run of item k in position l is

$$\frac{1}{2}h^l(Q^l - D^l)\left(\frac{Q^l}{D^l}\right)(\tau^l)^2.$$

Let I_k denote the set of all positions in the sequence in which item k is produced and L_l all the positions in the sequence starting with position l (when item k is produced) up to, but not including, the position in the sequence where item k is produced next. The definition of L_l assumes that the sequence $j_1, \ldots, j_\nu$ repeats itself. Let $\mathcal{S}$ denote the set of all possible cyclic schedules. The ELSP can now be written as

$$\min_{\mathcal{S}} \min_{x,\tau^l,u^l} \frac{1}{x}\left(\sum_{l=1}^{\nu} \frac{1}{2}h^l(Q^l - D^l)\left(\frac{Q^l}{D^l}\right)(\tau^l)^2 + \sum_{l=1}^{\nu} c^l\right)$$

subject to

$$\sum_{j \in I_k} Q_k \tau^j = D_k x \qquad \text{for } k = 1, \ldots, n,$$

$$\sum_{j \in L_l} (\tau^j + s^j + u^j) = \left(\frac{Q^l}{D^l}\right)\tau^l \qquad \text{for } l = 1, \ldots, \nu,$$

$$\sum_{j=1}^{\nu} (\tau^j + s^j + u^j) = x$$

The first set of constraints ensures that enough time is allocated to the production of item k to meet its demand over the cycle. The second set ensures that enough of the item in position l is produced to meet the demand till the next time that item is produced.

The problem described above may be viewed as being composed of a master problem and a subproblem. The master problem focuses on the search for the best sequence $j_1, \ldots, j_\nu$ (an element of $\mathcal{S}$), and the subproblem must determine the optimal production times, idle times, and cycle length (τ^l, u^l, x) given the sequence.

That the subproblem is relatively simple can be argued as follows. If the sequence $j_1, \ldots, j_\nu$ is fixed, then the first set of constraints in the nonlinear programming formulation is redundant, since substitution of the third set into the first set yields

$$\sum_{j \in I_k}\left(\frac{Q^j}{D^j}\right)\tau^j = \sum_{j=1}^{\nu}(\tau^j + s^j + u^j),$$

which is the sum of the second set over all positions in I_k. So, given a fixed sequence, the nonlinear programming problem that determines the optimal production times and idle times can be formulated as follows:

$$\min_{x,\tau^l,u^l} \frac{1}{x}\Big(\sum_{l=1}^{\nu}\frac{1}{2}h^l(Q^l-D^l)\Big(\frac{Q^l}{D^l}\Big)(\tau^l)^2 + \sum_{l=1}^{\nu}c^l\Big)$$

subject to

$$\sum_{j\in L_l}(\tau^j + s^j + u^j) = \Big(\frac{Q^l}{D^l}\Big)\tau^l \qquad \text{for } l=1,\ldots,\nu,$$

$$\sum_{j=1}^{\nu}(\tau^j + s^j + u^j) = x$$

The master problem, i.e., finding the best sequence $j_1,\ldots,j_\nu$, is more complicated. One particular heuristic yields good sequences in practice. This heuristic is in what follows referred to as the *Frequency Fixing and Sequencing (FFS)* heuristic. This FFS heuristic consists of three phases:

(i) The computation of relative frequencies phase.
(ii) The adjustment of relative frequencies phase.
(iii) The sequencing phase.

The first phase determines the relative frequencies with which the various items have to be produced. The number of times item k is produced during a cycle is denoted by y_k. In the second phase, these production frequencies are adjusted so they can be spaced out evenly over the cycle; the adjusted frequency of item k is denoted by y'_k. In the third and last phase these adjusted frequencies are used to produce an actual sequence.

The first phase determines, besides the relative frequencies y_k, also the corresponding run times τ_k. If the runs of item k are of equal length and evenly spaced, then the frequency y_k and the cycle time x determine the run time τ_k, i.e.,

$$\tau_k = \frac{\rho_k x}{y_k}.$$

To compute the y_k we relax the original nonlinear programming formulation by dropping the second set of the three sets of constraints. Without these interlinking constraints the actual sequence is no longer important. Optimizing over sequences now becomes optimizing over the cycle time x and run times τ_k or, equivalently, over production frequencies y_k.

Substitutions lead to the following modifications in the objective function of the original nonlinear programming formulation of the ELSP problem:

$$\frac{1}{x}\Big(\sum_{l=1}^{\nu}\frac{1}{2}h^l(Q^l-D^l)\Big(\frac{Q^l}{D^l}\Big)(\tau^l)^2 + \sum_{l=1}^{\nu}c^l\Big)$$
$$= \frac{1}{x}\Big(\sum_{k=1}^{n}\frac{1}{2}y_k h_k(Q_k-D_k)\Big(\frac{Q_k}{D_k}\Big)(\tau_k)^2 + \sum_{k=1}^{n}y_k c_k\Big)$$

7.4 Different Types of Items and Arbitrary Schedules

$$= \sum_{k=1}^{n} \frac{1}{2} h_k (Q_k - D_k) \left(\frac{Q_k}{D_k}\right) \rho_k \tau_k + \sum_{k=1}^{n} \frac{c_k \rho_k}{\tau_k}$$

$$= \sum_{k=1}^{n} \frac{1}{2} h_k (Q_k - D_k) \tau_k + \sum_{k=1}^{n} \frac{c_k \rho_k}{\tau_k}$$

$$= \sum_{k=1}^{n} \frac{a_k \tau_k}{\rho_k} + \sum_{k=1}^{n} \frac{c_k \rho_k}{\tau_k}$$

$$= \sum_{k=1}^{n} \frac{a_k x}{y_k} + \sum_{k=1}^{n} \frac{c_k y_k}{x},$$

where

$$a_k = \frac{1}{2} h_k (Q_k - D_k) \rho_k = \frac{1}{2} h_k (1 - \rho_k) D_k.$$

Of course, in any feasible schedule the relative frequencies y_k have to be integers. This implies that the run times τ_k cannot assume just any values, since they are determined by the relative frequencies. However, in order to make the problem easier, we delete the integrality constraints on the y_k and thus relax the constraints on the τ_k as well (basically deleting the first set of constraints in the original nonlinear programming formulation). Disregarding the integrality constraints on the y_k results in a relatively easy nonlinear programming problem.

$$\min_{y_k, x} \sum_{k=1}^{n} \frac{a_k x}{y_k} + \sum_{k=1}^{n} \frac{c_k y_k}{x},$$

subject to

$$\sum_{k=1}^{n} \frac{s_k y_k}{x} \leq 1 - \rho.$$

The constraint in this problem is equivalent to the last constraint in the original formulation. If the left hand side of this constraint is strictly smaller than the right hand side, then the sum of the setup times is less than the time the machine is not producing, implying there is still some idle time remaining.

Before presenting a solution for this simplified nonlinear programming problem some observations have to be made. First, a solution that is feasible for the simplified nonlinear programming problem may not be feasible for the original nonlinear programming problem; the solution to the simplified problem may require some tweaking. Second, it is clear that the simplified nonlinear programming problem has an infinite number of optimal solutions. If the solution $x^*, y_1^*, \ldots, y_n^*$, is optimal and

$$\sum_{k=1}^{n} \frac{s_k y_k}{x} < 1 - \rho$$

(i.e., the inequality is strict), then any solution $kx^*, ky_1^*, \ldots, ky_n^*$, with k integer, is also optimal. Basically, multiplying the first cyclic schedule by an integer k results in an identical cyclic schedule. A third observation is the following: if the inequality above is strict, then the solution $x^*, y_1^*, \ldots, y_n^*$ would also be optimal if all setup times are zero. However, if in the optimal solution

$$\sum_{k=1}^{n} \frac{s_k y_k}{x} = 1 - \rho,$$

then the setup times play an important role. The fact that there is no idle time implies that the optimal solution requires relatively high production frequencies with relatively short run times (possibly due to high holding costs or low setup costs). These high frequencies require that the maximum proportion of machine time, i.e., $(1 - \rho)$, is dedicated to setup times.

The nonlinear programming problem can be dealt with as follows. Incorporating the constraint in the objective function using a Lagrangean multiplier λ ($\lambda \geq 0$), results in an unconstrained nonlinear optimization problem with objective function

$$\min_{y_k, x} \sum_{k=1}^{n} \frac{a_k x}{y_k} + \sum_{k=1}^{n} \frac{c_k y_k}{x} + \lambda \left(\sum_{k=1}^{n} \frac{s_k y_k}{x} - (1 - \rho) \right).$$

Taking the partial derivative of this function with respect to y_k and setting it equal to zero yields

$$y_k = x \sqrt{\frac{a_k}{c_k + \lambda s_k}}.$$

The cycle length x can be adjusted so that the production frequencies take appropriate values (for example, one may choose the cycle length x so that the smallest frequency value is approximately equal to 1). If there are idle times, then λ is set equal to zero. If there are no idle times, then the λ has to satisfy the equation

$$\sum_{k=1}^{n} \left(s_k \sqrt{\frac{a_k}{c_k + \lambda s_k}} \right) = 1 - \rho,$$

since

$$\sum_{k=1}^{n} s_k y_k = (1 - \rho) x.$$

The solution y_k is unlikely to be integer. To find an integer solution that is close to the values obtained for y_k may require the construction of a long sequence with high frequencies.

The second phase of the FFS heuristic makes adjustments in the frequencies y_k. It has been shown in the literature that it is possible to find a new set of frequencies y_k' that are integers and powers of 2 with the cost of this

7.4 Different Types of Items and Arbitrary Schedules

new solution being within 6% of the cost of the original solution. Of course, the run times of item k have to change then as well. The new run times, τ'_k, can be computed by assuming that the total idle time remains the same and the runs of item k are of equal length and equally spaced.

The third phase of the FFS heuristic generates the actual sequence. The heuristic used here has its roots in the heuristic used for scheduling n different jobs on a number of parallel machines to minimize the makespan. (Recall from the second section of Chapter 5 that the most popular heuristic for this problem is the LPT rule). Let

$$y'_{\max} = \max(y'_1, \ldots, y'_n).$$

For each item k, there are y'_k jobs with the same estimated processing time τ'_k (assuming that the lots will be equally spaced). Now consider a scheduling problem with $y'_{\max}$ machines in parallel and y'_k jobs of length τ'_k, $k = 1, \ldots, n$, (implying a total of $\sum_{k=1}^{n} y'_k$ jobs). There is an additional restriction in that item k with frequency y'_k must have the y'_k lots (jobs) placed on machines that are equally spaced. For example, if $y'_{\max} = 6$ and $y'_k = 3$, then there are two choices: the three jobs are assigned either to machines 1, 3 and 5 or to machines 2, 4 and 6. Now the following variation of the LPT heuristic can be used: the pairs (y'_k, τ'_k) are listed in decreasing order of y'_k. Pairs with identical frequencies y'_k are listed in decreasing order of the estimated processing time τ'_k. The pairs are taken from the list one by one starting at the top. When the pair (y'_k, τ'_k) is taken from the list, the corresponding y'_k jobs of length τ'_k are put on the machines (satisfying the spacing restriction) so that the maximum of the total processing assigned so far to the selected y'_k machines is minimized. After all pairs in the list have been assigned, the resulting sequences on the $y'_{\max}$ machines are concatenated, i.e., machine 1, followed by machine 2, and so on, to obtain a single sequence. This idea is based on the fact that after all jobs have been scheduled and the total processing is more or less equally partitioned over all the machines, the concatenated sequence will maintain the equal spacing of the various runs of any given item.

The following example illustrates the application of the FFS heuristic to an instance without setup times.

Example 7.4.1 (Application of the FFS Heuristic without Setup Times). Consider again the situation described in Example 7.3.1. However, now the schedule does not necessarily have to be a rotation schedule. There are setup costs but no setup times. Since item 4 has no setup cost and a fairly high holding cost, its production should be spread out as uniformly as possible in between the production of the other three items.

In order to find the frequencies y_k, we have to first solve the unconstrained optimization problem. Note that

$$(1 - \rho)x = 0.48x,$$

and
$$y_k = \frac{\rho_k x}{\tau_k},$$

$$a_k = \frac{1}{2} h_k (Q_k - D_k) \rho_k.$$

The following values can be computed easily.

items	1	2	3	4
D_j	50	50	60	60
Q_j	400	400	500	400
h_j	20	20	30	70
c_j	2000	2500	800	0
ρ_j	0.125	0.125	0.12	0.15
a_j	437.5	437.5	792	1785

It immediately follows that

$$y_1 = 0.46x$$
$$y_2 = 0.42x$$
$$y_3 = 0.99x$$
$$y_4 = \infty$$

Suppose that the cycle time x is set equal to 2 months. This cycle time corresponds to the following approximate values for $y_1, \ldots, y_4$: $y_1 = y_2 = 1$, $y_3 = 2$ and $y_4 = 16$. The choice of y_4 is somewhat arbitrary but it has to be made high. The higher y_4, the more uniform the production of item 4 can be made in the final solution. Given a cycle time of 2 months and the production frequencies above, the runtimes τ_k of the four items are $\tau_1 = \tau_2 = 0.25$, $\tau_3 = 0.12$, $\tau_4 = 0.3/16$.

Now we apply the LPT-like heuristic. The number of machines in parallel is $y_{\max} = 16$. Item 4, the first to be assigned, is assigned to all 16 machines with all 16 processing times equal to $0.3/16$. Item 3 is assigned next and is assigned to machines 1 and 9 with the two processing times equal to 0.12. Item 1 is then put on machine 5 and item 2 on machine 13. Concatenating the sequences of the 16 parallel machines results in the cyclic schedule

| 4 , 3 | 4 | 4 | 4 | 4 , 1 | 4 | 4 | 4 | 4 , 3 | 4 | 4 | 4 | 4 , 2 | 4 | 4 | 4 |.

Item 4 goes first for $0.3/16$ months, followed by item 3 for 0.12 months. Item 4 goes next four times in a row, each time for $0.3/16$ months (these four runs are separated in the final solution by idle times). Item 1 follows for 0.25 months. Item 4 goes again four times, and so on.

A feasibility check has to be done. It is clear that, in an ideal solution, the runs of each item are spaced evenly over the cycle. If the runs of an item are evenly spaced, we are assured that there are no stockouts. Attempting to

7.4 Different Types of Items and Arbitrary Schedules

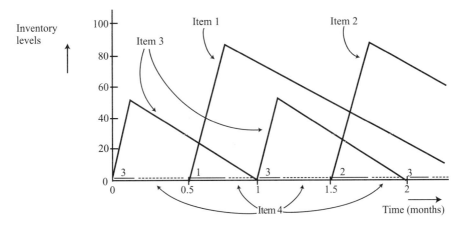

Fig. 7.3. Schedule in Example 7.4.1

uniformize the production of item 4 over the cycle results in many short runs that are either separated by idle times or by production runs of other items. We need to check whether, whenever items 1, 2, or 3 are produced, the stock of item 4 is sufficient to cover the demand in the periods that the other items are in production.

This solution could have been obtained in another way as well. As mentioned above, the y_4 was selected somewhat arbitrarily. The reason for choosing a high value is that it enables us to uniformize the item's production over the cycle. It is clear that the y_4 has to be chosen at least as large as the sum of all the other y's, i.e., at least 4. If $y_4 = 4$, then the algorithm yields the sequence

$$4, 3, 4, 1, 4, 3, 4, 2.$$

A schedule can now be constructed as follows. First the production runs of items 1, 2, and 3 are spaced evenly over the cycle and fixed. Then the production of item 4 is scheduled separately in between the remaining idle times. This production of item 4 is scheduled evenly over time (in *many* short runs). However, before any of the items 1, 2, or 3 has to go into production, a given amount of inventory of item 4 has to be built up. The inventory level of item 4 depends on the length of the runs of items 1, 2, and 3. It is only during these buildups of inventory that holding costs are incurred for item 4.

The entire schedule is depicted in Figure 7.3. The average total cost per unit time can be computed and is equal to

$$1875 + 2125 + 1592 + 190 = 5782.$$

Recall that the cost of the rotation schedule in Example 7.3.1 is 8554.

The next example illustrates the application of the FFS heuristic on an instance with setup times. In contrast to the previous example there is now, under the optimal sequence, no idle time on the machine.

Example 7.4.2 (Application of the FFS Heuristic with Setup Times).
Consider the instance in the previous example but now with setup times. The setup times are sequence independent.

items	1	2	3	4
D_j	50	50	60	60
Q_j	400	400	500	400
h_j	20	20	30	70
c_j	2000	2500	800	0
s_j	0.5	0.2	0.1	0.2
ρ_j	0.125	0.125	0.12	0.15
a_j	437.5	437.5	792	1785

Note that, because of the nonzero setup time of item 4 the frequency of item 4 cannot be made arbitrarily high. In order to find the frequencies y_k, we first have to find a λ that satisfies the equation

$$\sum_{k=1}^{n} \left(s_k \sqrt{\frac{a_k}{c_k + \lambda s_k}} \right) = 1 - \rho.$$

It can be verified easily that $\lambda \approx 8000$ satisfies this equation. With this value of λ the y_k frequencies can be computed as a function of the cycle time x, i.e.,

$$y_k = x \sqrt{\frac{a_k}{c_k + \lambda s_k}}.$$

$$y_1 = 0.27x$$
$$y_2 = 0.33x$$
$$y_3 = 0.70x$$
$$y_4 = 1.05x$$

If the cycle time x is fixed at 3 months, then the approximate values $y'_1, \ldots, y'_4$ can be either $(1, 1, 2, 2)$ or $(1, 1, 2, 4)$. Both solutions are power of two solutions. Compare these solutions with the frequency values in the previous example, i.e., $(1, 1, 2, 16)$. It is clear that, because of the setup times, the frequency of item 4 cannot be as high as in the previous example. There is simply not enough idle time for that many setups.

Consider the solution $x = 3$ with frequencies $(1, 1, 2, 2)$. The item sequence is $1, 3, 4, 2, 3, 4$. It has to be verified whether this solution is feasible. The idle time before taking setups into account is $0.48 \times 3 = 1.44$ months and the total amount of setup time required is 1.3, which implies that the schedule is feasible. Actually, this means that the cycle time x can be made slightly smaller than 3, and a slightly smaller cycle time will give a better solution. (In the previous example, without setup times, the cycle length was 2 months; here the cycle time was made 3 months assuming that there would not be any idle time.) The average total cost per unit time can be computed as in the previous examples.

Consider the solution $x = 3$ with frequencies $(1, 1, 2, 4)$. The order of the items is
$$1, 4, 3, 4, 2, 4, 3, 4.$$
The total setup time required during a cycle is 1.7 months. This implies that a cycle length of 3 months is not feasible with these frequencies. In order to have these frequencies the cycle length has to be made larger (see Exercise 7.10).

7.5 More General ELSP Models

All models considered in the previous sections are single machine models. Some of these models can be extended fairly easily. For example, consider the model with multiple products on m identical machines in parallel. There are setup costs but no setup times. Any particular item has to be processed on one and only one of the m machines. The utilization factor of item j is again defined as $\rho_j = D_j/Q_j$ and in order for a feasible solution to exist we must have
$$\sum_{j=1}^{n} \rho_j \leq m.$$
Suppose that the schedule for each of the machines has to be a rotation schedule. If the cycle times of the m rotation schedules have to be equal, the problem is relatively easy and not much different from the one described in Section 7.3. The only additional issue that needs to be resolved is the assignment of the items to the different machines. Assuming that each item has to be produced on one and only one machine, the loads have to be balanced and the sum of the ρ_j's of the items assigned to any one machine has to be less than one. To find a good balance or, equivalently, a good partition of the n different items over the m machines, we can use the LPT heuristic with the ρ_j values playing the role of processing times. The LPT heuristic (used for minimizing the makespan in a parallel machine environment) will result in a reasonably good assignment of the items to the m machines.

Example 7.5.1 (Rotation Schedules with Machines in Parallel). Consider the situation in Example 7.3.1. Instead of a single machine, we now have 2 machines in parallel. The production rate of each of the two machines is half the production rate of the machine in Example 7.3.1. The data are presented below.

items	1	2	3	4
D_j	50	50	60	60
Q_j	200	200	250	200
h_j	20	20	30	70
c_j	2000	2500	800	0

Because the two machines have the same cycle length, the formula in Example 7.3.1 can be used here as well. The optimal cycle x in this case turns out to be 1.35 months (the optimal cycle length with a single machine at twice the speed is 1.24 months).

However, if the cycle times of the m rotation schedules are allowed to be different, then the problem becomes more difficult. We can reduce total cost by taking advantage of different cycle times. Again, an assignment of the items to the machines has to be found while maintaining a proper machine load balance. There is now an additional difficulty. If an item with a cost structure that favors a long cycle time is assigned to the same machine as an item with a cost structure that favors a short cycle time, then the solution is not likely to be a good one. One can deal with this difficulty as follows. Consider each item as a single product model (as in Section 7.2) and compute its cycle time. Rank the items in decreasing order of their cycle times. Start taking the items from the top of the list and put them on one machine. Keep assigning items to this machine until its capacity is exhausted, i.e., the allocation of an item makes the sum of the ρ_j's larger than one. This last item is then reallocated to the second machine, and so on. This procedure may not lead to a good load balance, and may even lead to an infeasible solution (i.e., the sum of the ρ_j's on the last machine may be larger than one). If that is the case, then items on adjacent machines have to be swapped in order to obtain a better balance.

The parallel machine model with rotation schedules and sequence dependent setups is of course harder. The assignment of items to the m machines now has to consider machine balance, preferred cycle times, as well as setup times on all machines. The setup time structure becomes especially important when there are large differences between setup times. Very little research has been done on this problem.

When the schedules on the different machines do not have to be rotation schedules, i.e., the parallel machines generalization of the model considered in Section 7.4, the problem is even harder. However, now it is no longer necessary to assign items with similar cycle times to the same machine. It is clear that in the parallel machine environment there are still many unresolved issues that require more research.

Another important extension of the single machine setting is the environment with machines in series, i.e., the flow shop. Consider a single machine feeding another single machine with the production rates and setup costs of the two machines being identical. Hence the cost structures of an item on the upstream machine and on the downstream machine are the same. The machines can be scheduled and synchronized so that the items, when they leave the upstream machine, can immediately start their processing on the downstream machine without having to wait. Because of this synchronization, the results in Sections 7.3 and 7.4 can be extended to the case of similar machines with identical costs in series.

Example 7.5.2 (Rotation Schedules with Machines in Series). Consider the same product mix as in Example 7.3.1. However, now instead of a single machine, we have two machines in series. After an item has completed its processing on the upstream machine, it has to go to the downstream machine and complete its processing there. There are setup costs but no setup times. The setup costs of an item on the two machines are the same. A rotation schedule is needed for the two machines (i.e., the same rotation schedule must be used for both machines).

If the rotation schedule is such that an item with a long processing time (i.e., with a low production rate) is followed by an item with a short processing time (i.e., with a high production rate), then the item with the short processing time may have to wait between the two machines. Assume that at the beginning of the rotation schedule machine 1 starts with the production of the item with the shortest processing time (i.e., with the highest production rate), it then continues, without stopping, with the item with the second shortest processing time, and so on. After it completes the run of the item with the longest processing time machine 1 remains idle until it is necessary to start the new cycle. In this way any item that comes out of machine 1 can start immediately on machine 2 without having to wait, i.e., there is no Work-In-Process in between the two machines.

The inventory costs of the finished goods are exactly the same as in the single machine case, so this system can be analyzed as a single machine. However, there are now two setup costs, instead of only one. This implies that the optimal cycle length is $\sqrt{2} = 1.4142$ times longer than the optimal cycle length for a single machine. This result can be extended easily to m identical machines in series.

The results in the previous example can be further generalized. Consider m machines in series with identical production rates for each product type, but different setup costs. This problem can still be reduced to a single machine problem with production rates identical to those of one of the machines in the original problem. However, now the setup costs have to be set equal to the sum of the setup costs of the original m machines, i.e., the setup cost for item j in the new problem is

$$c_j = \sum_{i=1}^{m} c_{ij}.$$

When the machines do not have identical production rates for each product type the problem is not that easy. Consider first the case with two machines that have identical setup costs but different speeds. But the speed structure is uniform over the items, i.e., the production rate of item j on machine i is $Q_{ij} = v_i Q_j$, where v_i is a speed factor of machine i. One approach is to first analyze the slow machine as a single machine in isolation and then adapt the fast machine accordingly (since the high speed machine provides more flexibility). The schedule of the fast machine can be adapted in such a way that there is no WIP in between the two machines. This model can then be analyzed as a single machine with the production rates of the slow machine and setup costs that are the sum of the setup costs of the two machines.

When the production rates are not uniform, it may become necessary to schedule Work-In-Process (WIP) between the machines. The carrying cost of the WIP in between the machines may be different from the holding cost of the finished goods. This makes the model more complicated. Very little research has been done on this problem.

An even more general machine environment is the flexible flow shop, i.e., a number of stages in series with at each stage a number of machines in parallel. Under very special conditions optimal rotation schedules can be determined for such a machine environment. For example, consider two stages in series with at each stage two machines in parallel. For any given product type the production rates of the four machines are the same and so are the setup costs. Under these circumstances there is no need for any WIP in between two stages. This makes it possible to determine the optimal rotation schedules relatively easily when the cycle times of all four machines have to be the same.

7.6 Multiproduct Planning and Scheduling at Owens-Corning Fiberglas

Owens-Corning Fiberglas is a leading manufacturer of fiberglass products and has a large manufacturing facility in Anderson, South Carolina. In its manufacturing process, molten fiberglass is formed and the glass is spun onto spools of various sizes. This material is used to weave fabric and to produce chopped strand mat. Fiberglass mat is sold in rolls of various widths and weights, treated with one of three process binders, and trimmed at one or both edges or not at all. The demand for the products comes mainly from the marine industry for the manufacture of boat hulls, and from the residential construction industry for bath tubs and shower booths.

At the time when a production planning and scheduling system was developed for this facility, the product line consisted of over 200 distinct mat items. Twenty-eight of these represented over 80% of the total annual demand and were treated as high volume standard (stock) products. The remaining items were made to order. The manufacturing facility had two main production lines referred to as Mat Lines 1 and 2. Line 1 had approximately three times the capacity of Line 2 and could produce mat 76 inches wide, whereas Line 2 was limited to 60 inches. The product came off these lines in the form of 175- to 230-pound cylinders. The average cost of down time on Line 1 was approximately $300 per hour; the average cost of down time on Line 2 was less. Maintenance costs were related to the frequency of job changeovers. In addition, each time a product change was made on a line, there was a sequence dependent setup cost partly due to material waste. The monthly costs due to setups ranged from $15,000 to $50,000 (with approximately 75 job changes and 50 hours of down time).

The production planning and scheduling system developed for the mat lines focused on three issues, namely aggregate planning (focusing on inventory costs and workforce scheduling), production run quantities and lot sizing (taking line assignments and inventory levels into account), and detailed line sequencing of Make-to-Stock and Make-to-Order products (taking setup costs into account). The system developed for the mat lines consisted therefore of three main modules, namely,

(i) the aggregate planning module,
(ii) the lot sizing module, and
(iii) the sequencing module.

The aggregate planning module used as input the aggregate demand forecast for the next twelve months. Its objective was to minimize the sum of direct payroll costs, overtime costs and hiring and firing costs. The time horizon ranged from three to twelve months. The optimization method in this module was based on a production switching heuristic; this rule considered inventory levels and forecasts of future demand and, based on these data, determined the appropriate production rates. The output of this model included targets with respect to aggregate inventory levels, production rates and levels of employment.

The output of the aggregate planning module served as input to the lot sizing module. The lot sizing module also required a detailed short term demand forecast for each stock item. The time horizon considered in this module was up to three months. The output from this module were the line assignments and the lot sizes. The optimization in this module was based on a linear program formulation. The objective was to minimize all the relevant setup costs and production costs subject to several sets of constraints. The inventory level constraints ensured that aggregate inventory levels and safety stock levels were met and the production balance constraints guaranteed demand satisfaction as well as inventory conservation. The linear program was solved using the MPSX package and the program was run on a monthly basis providing the plant with specific inventory levels, lot sizes and line assignments for the coming months.

The output of the lot sizing module served as input for the sequencing module. The time horizon of this module was one month, and its main objective was the minimization of the sequence dependent setup costs. The dominant components of setup costs were direct downtime and mat waste. Changeovers were classified as fiber changes, width changes, weight changes, and slitter changes. A distinction could be made within each family of changeovers based upon the direction of the change. For example, it was easier to decrease weight and width than to increase them. The sequencing heuristic was based on sim-

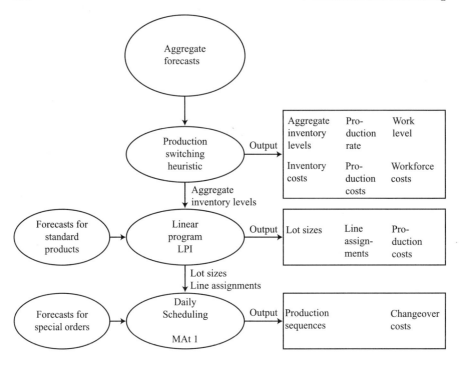

Fig. 7.4. Overview of system at Owens-Corning Fiberglass

ple dispatching rules and the sequencing module was run on a weekly basis. The entire system is depicted in Figure 7.4.

Implementation of the system led to major improvements in the operation of the plant. The average number of changeovers went down from an average of 70 before the system's implementation to an average of 40 after the system's implementation.

The Owens-Corning Fiberglas environment is somewhat similar to the setting described in Section 7.4. However, Owens-Corning had two machines in parallel instead of the single machine in Section 7.4 (a parallel machine environment was considered in Section 7.5). Note that the FFS heuristic described in Section 7.4 is based on a nonlinear programming formulation, whereas the lot sizing module in the Owens-Corning Fiberglas system was based on a linear programming formulation.

7.7 Discussion

The models discussed in this chapter have similarities as well as differences with the models for the the flexible assembly systems described in Chapter 6. In both chapters the planning horizons are basically unbounded. (This is in

7.7 Discussion

contrast to the models described in Chapters 4 and 5, which all have a finite number of jobs.) However, the objectives considered in Chapter 6 are fundamentally different from the objectives considered in this chapter. In Chapter 6, the typical objective is to maximize the throughput or, equivalently, to minimize the cycle time. In this chapter, the throughput is basically given, since the demand levels are known. The objective is to minimize the sum of the inventory carrying costs and the setup costs. Nonetheless, the objectives in this chapter display some similarities with the objectives in paced assembly systems.

The models considered in this chapter are very important for industries that produce Make-To-Stock and for environments with setup times and costs. Examples of these industries include the paper industry, the aluminum industry and the steel industry. If one compares the models described in this chapter with the problems that have to be solved in those industries, then a number of issues arise. The problems in practice are, of course, more complicated than the models considered in this chapter. Often, if there are multiple machines in parallel, the production rates of any given item on the various machines may vary. Run lengths have in practice, besides an impact on the inventory costs and the total change-over costs, also an effect on various other factors, including

(i) the quality of the finished product,
(ii) the production yield or the amount of waste incurred,
(iii) the productivity of the facility and its total production capacity.

These three factors, which are not independent, are seldom included in medium term or long term planning models. The three factors are somewhat related to one another. The quality of the products in process industries (which typically is a continuous measure rather than a discrete measure) depends strongly on the length of the run. The longer the length of the run, the higher the average quality of production. In the process industries there is usually also a yield or waste problem (cutting stock or trim related issues). If the run length is very small, then the average waste tends to increase. So the more changeovers there are, the lower the average quality of the product, and the lower the productivity and the capacity. The cost of machine capacity is basically determined by opportunity costs (i.e., the shadow prices or dual prices of the resources involved).

Practical problems are often a combination of Make-To-Stock and Make-To-Order. The Make-To-Stock aspects involve problems such as those described in this chapter whereas the Make-To-Order aspects are related to job shop scheduling problems. Researchers have been analyzing inventory problems in which the facilities are assumed to be set up in series. This area of research, often referred to as multi-echelon inventory theory or supply chain management, is considered in the next chapter.

Exercises

7.1. Consider four different products with the following demand rates, production rates, holding costs and setup costs.

items	1	2	3	4
D_j	50	50	60	60
Q_j	400	500	500	400
h_j	20	20	30	70
c_j	2000	1000	1000	100

(a) Find the optimal rotation schedule. Determine its cycle length, and the total idle time.

(b) Suppose now that item 4 can be produced many times during a cycle. Items 1, 2 and 3 still can be produced only once during a cycle. Find the optimal production schedule. How does the optimal cycle length compare with the original optimal cycle length?

7.2. Consider two identical machines in parallel. Four items have to be produced.

items	1	2	3	4
D_j	50	50	60	60
Q_j	200	200	300	300
h_j	20	20	30	70
c_j	2000	2500	800	0

(a) Find the optimal rotation schedule assuming that the cycle lengths of the two machines have to be the same. Compute the total average cost per unit time.

(b) Find the optimal rotation schedules of the two machines assuming the cycle lengths of the two machines do not have to be the same (determine which items have to be combined with one another on the same machine to obtain the best result). Compute the average cost per unit time and compare the result with the result found under (a).

7.3. Consider the following two stage production process in a paper mill with a downstream converting operation. At the first stage there is a single paper machine. The output of this operation consists of large rolls of paper. The second stage is a single machine cutting operation that produces cutsize paper. To simplify the problem assume that only two items have to be produced. Also, each item that comes out of the second stage corresponds to one of the items that comes out of the first stage. The production rates, setup costs and holding costs are different at the two stages. In the table below Q_{ij} denotes the production rate of item j at stage i, c_{ij} the setup cost of item j at stage i, and h_{ij} the holding cost of item j after processing at stage i (so h_{1j} denotes

Exercises

the holding cost of keeping item j in inventory in between the two stages, while h_{2j} denotes the holding cost of item j as a finished good).

items	1	2
D_j	100	50
Q_{1j}	400	400
Q_{2j}	600	1000
h_{1j}	20	20
h_{2j}	60	80
c_{1j}	3000	2500
c_{2j}	1000	1250

The schedules at both stages have to be rotation schedules (i.e., it is, for example, not allowed to produce item 1 at the first stage for a while, leave the machine idle for some time, produce item 1 again, and then item 2).

(a) Assuming that the cycle length x of the two stages have to be the same, what is the cycle length with the minimum total cost?

(b) Assume that the cycle lengths at the two stages are allowed to be different. Determine the optimal cycle lengths of the two stages.

7.4. Connsider the environment with two machines in parallel in Example 7.5.1. Suppose now that the production rate of one machine is .7 times the production rate of the machine in Example 7.3.1 and the production rate of the second machine .3 times.

a) Determine the optimal rotation schedules when both machines must have the same cycle time.

b) Determine the optimal rotation schedules when the two machines do not have to have the same cycle times.

7.5. Consider the data in Example 7.3.1. Consider now m identical machines in parallel. The production rate of item j on any one of the machines is Q_j/m. (This implies that the total production capacity does not depend on the number of machines in parallel.) Assume that all the machines have to be scheduled according to rotation schedules with the same cycle time x. Compute the optimal x and the total cost for $m = 2, 3, 4$. Plot the total cost against m.

7.6. Consider m identical facilities with each facility having a production rate Q_j/m. There are no setup times. Assume that all the facilities have to follow rotation schedules with the same cycle length x. Derive an expression for the optimal cycle length x. How does x depend on m? Discuss the monotonicity and the convexity of the function.

7.7. Consider a paper mill with two paper machines. There are 5 different types of paper that have to be produced. Items 1 and 2 have to be produced on machine 1 and item 3 has to be produced on machine 2. Items 4 and 5 can be produced on either one of the two machines.

items	1	2	3	4	5
D_j	60	60	80	80	100
Q_j	200	200	300	300	400
h_j	20	30	40	20	20
c_j	3000	2000	800	4000	1500

(a) Determine the optimal rotation schedule assuming that the cycle lengths have to be the same.

(b) Determine the optimal rotation schedules assuming the cycle lengths do not have to be the same.

(c) Determine the optimal schedule if the schedule on machine 1 has to be a rotation schedule and the schedule on machine 2 may be an arbitrary schedule.

7.8. Consider the following generalization of the paper making facility of Exercise 7.3. Again, there are two stages. The entire facility produces three different items. However, the paper machine at the first stage produces only two different intermediate products. One of the intermediate products that comes out of stage 1 is used at stage 2 to produce items 1 and 2. (This implies that the data correponding to items 1 and 2 regarding stage 1 are the same.) The other intermediate product that comes out of stage 1 is converted at stage 2 into item 3.

items	1	2	3
D_j	50	70	100
Q_{1j}	250	250	300
Q_{2j}	200	400	300
h_{1j}	30	30	40
h_{2j}	40	20	30
c_{1j}	2000	2000	900
c_{2j}	1000	3000	2000

Determine the optimal rotation schedule and its cycle length.

7.9. Consider the setting in Exercise 7.2. Instead of a single stage with two machines in parallel, we have now two stages in series with two machines in parallel at each stage. All four machines are identical with regard to production rates and setup costs (the data being the same as in Exercise 7.2). Determine the optimal rotation schedules of the four machines assuming that the cycle times of the four machines are the same.

7.10. Consider the setting in Example 7.4.2.

(a) What is the minimum cycle time when the frequency values are $y_1' = y_2' = 1$ and $y_3' = y_4' = 2$? Compute the total average cost of this solution.

(b) What is the minimum cycle time when the frequency values are $y_1' = y_2' = 1$, $y_3' = 2$ and $y_4' = 4$? Compute the total average cost of this solution.

(c) Compare the results obtained under (a) and (b) with the total average cost obtained in Example 7.4.1. Explain your results.

Comments and References

Various books and monographs focus on lot sizing and scheduling; see, for example, Haase (1994), Brüggemann (1995), Kimms (1997), and Zipkin (2000). Several excellent survey papers cover this topic also in depth, see Graves (1981) and Drexl and Kimms (1997).

The material in Section 2 is very basic. The EOQ formula was first mentioned by Harris (1915) and studied in detail by Wilson (1934). This material is covered in every elementary textbook on production planning and operations management.

Maxwell (1964) did an exhaustive study of rotation schedules. Gallego (1988) and Gallego and Roundy (1992) generalized these results allowing for backorder costs. Jones and Inman (1989, 1996) made an in depth study of the worst case behavior of rotation schedules and compared rotation schedules with other types of schedules. Gallego and Queyranne (1995) extended some of these results.

The FFS heuristic, described in Section 7.4, for generating arbitrary schedules is due to Dobson (1987, 1992). Gallego and Shaw (1997) established the NP-hardness of the ELSP with arbitrary cyclic schedules.

A fair amount of work has been done on lot scheduling in more complicated machine environments; see Crowston, Wagner and Williams (1973), Carreño (1990), Brüggemann (1995), Jones and Inman (1996), Chao and Pinedo (1996), and Pinedo and Chao (1999).

The production planning and scheduling system developed for Owens-Corning Fiberglas is discussed in Oliff and Burch (1985).

Chapter 8

Planning and Scheduling in Supply Chains

8.1 Introduction 171
8.2 Supply Chain Settings and Configurations 173
8.3 Frameworks for Planning and Scheduling in
    Supply Chains................................ 178
8.4 A Medium Term Planning Model for a Supply
    Chain... 184
8.5 A Short Term Scheduling Model for a Supply
    Chain... 190
8.6 Carlsberg Denmark: An Example of a System
    Implementation 193
8.7 Discussion 197

8.1 Introduction

This chapter focuses on models and solution approaches for planning and scheduling in supply chains. It describes several classes of planning and scheduling models that are currently being used in systems that optimize supply chains. It also discusses the architecture of the decision support systems that have been implemented in industry and the problems that have come up in the implementation and integration of systems for supply chains. In the implementations considered the total cost in the supply chain has to be minimized, i.e., the stages in the supply chain do not compete with one another in any form, but collaborate in order to minimize total cost.

This chapter basically embeds medium term planning models, such as the lot sizing models described in Chapter 7, and detailed scheduling models, such as the job shop scheduling models described in Chapter 5, into a single framework. The models in this chapter are quite general. There is a network of interconnected facilities and the demands for the various end-products may not be stationary. The planning and scheduling may be done at the same point in time but with different horizons and with different levels of detail.

A medium term production planning model typically optimizes several consecutive stages in a supply chain (i.e., a multi-echelon model), with each stage having one or more facilities. Such a model is designed to allocate the production of the different products to the various facilities in each time period, while taking into account inventory holding costs and transportation costs. A planning model may make a distinction between different product families, but often does not make a distinction between different products within a family. It may determine the optimal run length (or, equivalently, batch size or lot size) of a given product family when a decision has been made to produce that family in a given facility. If there are multiple families being produced in the same facility, then there may be setup costs and setup times. The optimal run length of a product family is a function of the trade-off between the setup cost and/or setup time and the inventory carrying cost. The main objectives in medium term planning involve inventory carrying costs, transportation costs, tardiness costs and the major setup costs. However, in a medium term planning model it is typically not customary to take the sequence dependency of setup times and setup costs into account. The sequence dependency of setups is difficult to incorporate in such an integer programming formulation and it can increase the complexity of the formulation significantly.

A short term detailed scheduling model is typically only concerned with a single facility, or, at most, with a single stage. Such a model usually takes more detailed information into account than a planning model. It is typically assumed that there are a given number of jobs and each one has its own parameters (including sequence dependent setup times and sequence dependent setup costs). The jobs have to be scheduled in such a way that one or more objectives are minimized, e.g., the number of jobs that are shipped late, the total setup time, and so on. The related models have been discussed in Chapters 5 and 6.

Clearly, planning models differ from scheduling models in a number of ways. First, planning models often cover multiple stages and optimize over a medium term horizon, whereas scheduling models are usually designed for a single stage (or facility) and optimize over a short term horizon. Second, planning models use more aggregate information, whereas scheduling models use more detailed information. Third, the objective to be minimized in a planning model is typically a cost objective and the unit in which it is measured is a monetary unit; the objective to be minimized in a scheduling model is typically a function of the completion times of the jobs and the unit in which it is measured is often a time unit. Nevertheless, even though there are fundamental differences between these two types of models, they often have to be incorporated into a single framework, share information, and interact extensively with one another.

Planning and scheduling models may also interact with other types of models, such as long term strategic models, facility location models, demand management models, and forecasting models; these models are not discussed in this chapter. The interactions with these other types of models tend to be less intensive and less interactive. In this chapter, we assume that the physical setting of the supply chain has already been established; the configuration of the chain is given, and the number of facilities at each stage is known.

Supply chains in the various industries are often not very similar and may actually give rise to different issues and problems. This chapter considers applications of planning and scheduling models in supply chains in various industry sectors. A distinction can be made between two types of industries, namely the continuous manufacturing industries (which include the process industries) and the discrete manufacturing industries (which include, for example, automotive and consumer electronics). Each one of these two main categories is subdivided into several subcategories. This categorization is used because of the fact that the planning and scheduling procedures in the two main categories tend to be different. We focus on the frameworks in which the planning and scheduling models have to be embedded; we describe the type of information that has to be transferred back and forth between the modules and the kind of optimization that is done within the modules.

This chapter is organized as follows. The second section describes and categorizes some of the typical industrial settings. The third section discusses the overall frameworks in which planning models and scheduling models have to be embedded. The fourth section describes a standard mixed integer programming formulation of a planning model for a supply chain. The fifth section covers a typical formulation of a scheduling problem in a facility within a supply chain. The sixth section describes an actual implementation of a planning and scheduling software system at the Danish beerbrewer Carlsberg A/S. The last section presents the conclusions and discusses the current trends in the design and development of decision support systems for supply chains.

8.2 Supply Chain Settings and Configurations

This section gives a concise overview of the various types of supply chains. It describes the differences in the characteristics and parameters of the various categories. It first describes the various different industry groups and their supply chain characteristics and then discusses how the different planning and scheduling models analyzed in the literature can be used in the management of these chains. One can make a distinction between two types of manufacturing industries, namely:

(I) Continuous manufacturing industries (e.g., the process industries),
(II) Discrete manufacturing industries (e.g., cars, semiconductors).

These two industry sectors are not all encompassing; the borderlines are somewhat blurry and may overlap. However, planning and scheduling in continuous manufacturing (the process industries) often have to deal with issues that are quite different from those in discrete manufacturing.

Continuous Manufacturing: Continuous manufacturing (process) industries often have various types of operations. The most common types of operations can be categorized as follows:

(I-a) Main processing operations
(I-b) Finishing or converting operations,

Main Processing Operations in Continuous Manufacturing (I-a): The main production facilities in the process industries are, for example, paper mills, steel mills, aluminum mills, chemical plants, and refineries. In paper, steel, and aluminum mills the machines take in the raw material (e.g., wood, iron ore, alumina) and produce rolls of paper, steel or aluminum, which afterwards are handled and transported with specialized material handling equipment. Machines that do the main processing operations typically have very high startup and shutdown costs and usually work around the clock. A machine in the process industries also incurs a high changeover cost when it has to switch over from one product to another. Various methodologies can be used for analyzing and solving the models for such operations, including cyclic scheduling procedures and Mixed Integer Programming approaches.

Finishing Operations in Continuous Manufacturing (I-b): Many process industries have some form of finishing operations that do some converting of the output of the main production facilities. This converting usually involves cutting of the material, bending, folding and possibly painting or printing. These operations often (but not always) produce commodity type items, for which the producer has many clients. For example, a finishing operation in the paper industry may produce cut size paper from the rolls that are supplied by a paper mill. The paper finishing business is often a mixture of Make-To-Stock (MTS) and Make-To-Order (MTO). If it operates according to MTO, then the scheduling is based on customer due dates and sequence dependent setup times. This leads often to single machine and parallel machine scheduling models. If it operates according to MTS, then it may follow a so-called s-S or Q-R inventory control policy. If it is a mixture of MTO and MTS, then the scheduling policies become a mixture of inventory control and detailed scheduling rules.

Discrete Manufacturing: The discrete manufacturing industry sector is quite diverse and includes the automotive industry, the appliances industry, and the PC industry. From the perspective of planning and scheduling a distinction can be made between three different types of operations in this sector. The reason for making such a distinction is based on the fact that planning and scheduling in these three segments are quite different.

(II-a) Primary converting operations (e.g., cutting and shaping of sheet metal),

8.2 Supply Chain Settings and Configurations

(II-b) Main production operations (e.g., production of engines, PCBs, wafers), and

(II-c) Assembly operations (e.g., cars, PCs).

Primary Converting Operations in Discrete Manufacturing (II-a): Primary converting operations are somewhat similar to the finishing operations in the process industries. These operations may typically include stamping, cutting, and bending. The output of such an operation is often a particular part that is cut and bent into a given shape. There are usually few operations done on such an item and the routing in such a facility is relatively simple. The final product coming out of a primary converting facility is usually not a finished good, but rather a part or a piece made of a single material (boxes, containers, frames, body parts of cars, and so on). Examples of the types of operations in this category are stamping facilities that produce body parts for cars and facilities that produce epoxy boards of various sizes for the plants that produce Printed Circuit Boards. The planning and scheduling procedures under *II-a* may be similar to those under *I-b*. However, they may be here more integrated with the operations downstream.

Main Production Operations in Discrete Manufacturing (II-b): The main production operations are those operations that require multiple different operations by different machine tools and the product (as well as its parts) may have to follow a certain route through the facility going through various work centers. Capital investments have to be made in various types of machine tools (lathes, mills, chip fabrication equipment). For example, in the semiconductor industry wafers typically have to undergo hundreds of steps. These operations include oxidation, deposition, and metallization, lithography, etching, ion implantation, photoresist stripping, and inspection and measurements. It is often the case that certain operations have to be performed repeatedly and that some orders have to visit certain workcenters in the facility several times, i.e., they have to recirculate through the facility. In semiconductor and Printed Circuit Board manufacturing the operations are often organized in a job shop fashion. Each order has its own route through the system, its own quantity (and processing times) and its own committed shipping date. An order typically represents a batch of identical items that requires sequence dependent setup times at many operations.

Assembly Operations in Discrete Manufacturing (II-c:) The main purpose of an assembly facility is to put different parts together. An assembly facility typically does not alter the shape or form of any one of the individual parts (with the possible exception of the painting of parts). Assembly operations usually do not require major investments in machine tools, but do require investments in material handling systems (and possibly robotic assembly equipment). An assembly operation may be organized in workcells, in assembly lines, or according to a mixture of workcells and assembly lines. For example, PCs are assembled in workcells, whereas cars and TVs are typically put together in assembly lines. Workcells typically do not require any sequencing, but they may be subject to learning curves. In assembly operations that are set up in a line, the sequencing is based on grouping and spacing heuristics combined with committed shipping dates. The schedules that are generated by the grouping and spacing heuristics typically affect not only the throughput of the line, but also the quality of the items produced.

Supply chains in both continuous and discrete manufacturing may have, in addition to the stages described above, additional stages. In a supply chain in a process industry there may be a stage preceding Stage I-a in which the raw material is being gathered at its point of origination (which may be a forest or a mine) and taken to the main processing operations. There may also be a distribution stage following stage I-b. A company may have its own distribution centers in different geographical locations, where it keeps certain SKUs in stock for immediate delivery. The company may also ship directly from its manufacturing operations to customers. A supply chain in a discrete manufacturing industry also may have other types of stages. There may be a stage preceding the first stage II-a in which raw material is being collected at a supplier (which may be an operation of the type I-b) and brought to a primary converting operation. There may also be a stage following stage II-c that would consist of distribution operations (e.g., dealerships).

Supply chains in both continuous and discrete manufacturing may have several facilities at each one of the stages, each one feeding into several facilities at stages downstream. The configuration of an entire chain may be quite complicated: For example, there may be assembly operations that produce subassemblies that have to be fed into a main assembly operation.

There are some basic differences between the parameters and operating characteristics of the facilities in the two main categories described above. Several of these differences have an impact on the planning and scheduling processes, including the differences in (i) the planning horizon, (ii) the clock-speed, (iii) the level of product differentiation.

(i) The planning horizon in continuous manufacturing facilities tends to be longer than the planning horizon in discrete manufacturing facilities. In continuous as well as in discrete manufacturing the planning horizons tend to be shorter the more downstream in the supply chain.

(ii) The so-called "clock-speed" tends to be higher in a discrete manufacturing facility than in a continuous manufacturing facility. A high clock-speed implies that existing plans and schedules often have to be changed or adjusted; that is, planning and scheduling is more reactive. In continuous as well as in discrete manufacturing the clockspeed increases the more downstream in the supply chain.

(iii) In discrete manufacturing there may be a significant amount of mass customization and product differentiation. In continuous manufacturing mass-customization does not play a very important role. The number of SKUs in discrete manufacturing tends to be significantly larger than the number of SKUs in continuous manufacturing. The number of SKUs tends to increase more downstream in the supply chain.

8.2 Supply Chain Settings and Configurations

These operating characteristics are summarized in table 8.1.

Sector Processes	Time Horizon	Clock-Speed	Product Differentiation
(I-a) planning	long-medium	low	very low
(I-b) planning/scheduling	medium/short	medium/high	medium/low
(II-a) planning/scheduling	medium/short	medium	very low
(II-b) planning/scheduling	medium/short	medium	medium/low
(II-c) scheduling	short	high	high

Table 8.1. Operating charcateristics

Because of these differences, the planning and scheduling issues in each one of the sectors tend to be quite different. Table 8.2 presents a summary of the model types that can be used in the different categories as well as the corresponding solution techniques.

Sector Models	Solution Techniques
(I-a) Lot sizing models (multi-stage); cyclic scheduling models	Mixed Integer Programming formulations
(I-b) Single machine scheduling models; parallel machine scheduling models	Batch scheduling; mixtures of inventory control rules and dispatching rules
(II-a) Single machine scheduling models; Parallel machine scheduling models	Batch scheduling and dispatching rules
(II-b) Flow Shop and Job Shop Scheduling Models with specific routing patterns	Integer Programming formulations; shifting bottleneck heuristics; dispatching rules
(II-c) Assembly Line Models; Workcell Models	Grouping and Spacing Heuristics; Make-to-Order/Just-In-Time

Table 8.2. Model types and the corresponding solution techniques.

Note that problems that have continuous variables may result in Mixed Integer Programming (MIP) formulations, whereas problems that have only discrete variables may result in pure Integer Programming (IP) formulations (or Disjunctive Programming formulations). However, a discrete problem in which certain variables assume large values (i.e., the number of units to be produced) may be replaced by a continuous problem, resulting in a Mixed Integer Programming formulation rather than a pure Integer Programming formulation. Planning models typically result in Mixed Integer Programming formulations with a mix of continuous and discrete variables. Scheduling models usually do not have any continuous variables (they may have continuous variables when preemptions and job splitting are allowed). When there are few discrete variables, it makes a lot of sense to solve the Linear Programming relaxation of the MIP. The solution may provide a useful lower bound and may give an indication with regard to the structure of the optimal solutions of the MIP. If the formulation of the problem is a pure Integer Program (which is often the case with a scheduling problem), then solving the linear relaxation typically does not provide much of a benefit.

8.3 Frameworks for Planning and Scheduling in Supply Chains

The main objective in a supply chain or production distribution network is to produce and deliver finished products to end consumers in the most cost effective and timely manner. Of course, this overall objective forces each one of the individual stages to formulate its own objectives.

Since planning and scheduling in a global supply chain requires the coordination of operations in all stages of the chain, the models and solution techniques described in the previous section have to be integrated within one framework. Different models that represent successive stages have to exchange information and interact with one another in various ways. A continuous model for one stage may have to interact with a discrete model for the next stage.

Planning and scheduling procedures in supply chains are typically used in several phases: a first phase involves a multi-stage medium term planning process (using aggregate data) and a subsequent phase performs a short term detailed scheduling at each one of those stages separately. Typically, whenever a planning procedure has been applied and the results have become available, each facility can apply its scheduling procedures. However, scheduling procedures are usually applied more frequently than planning procedures. Each facility in every one of these stages has its own detailed scheduling issues to deal with, see Figure 8.1.

If successive stages in a supply chain belong to the same company, then it is usually the case that these stages are incorporated into a single planning model. The medium term planning process attempts to minimize the total cost over all the stages. The costs that have to be minimized in this optimization process include production costs, holding or storage costs, transportation costs, tardiness costs, non-delivery costs, handling costs, costs for increases

8.3 Frameworks for Planning and Scheduling in Supply Chains

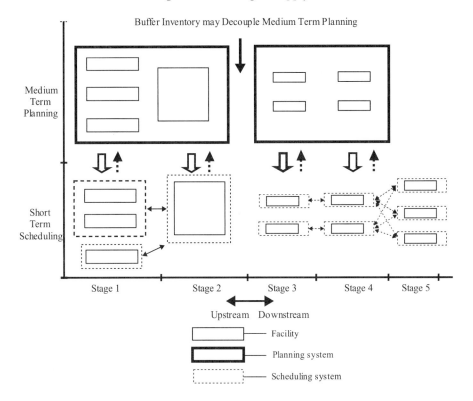

Fig. 8.1. Planning and Scheduling in Supply Chains

in resource capacities (e.g., scheduling third shifts), and costs for increases in storage capacities.

In this medium term optimization process many input data are only considered in an aggregate form. For example, time is often measured in weeks or months rather than days. Distinctions are usually only made between major product families, and no distinctions are made between different products within one family. A setup cost may be taken into account, but it may only be considered as a function of the product itself and not as a function of the sequence.

The results of this optimization process are daily or weekly production quantities for all product families at each location or facility as well as the amounts scheduled for transport every week between the locations. The production of the orders require a certain amount of the capacities of the resources at the various facilities, but no detailed scheduling takes place in the medium term optimization. The output consists of the allocations of resources to the various product families, the assignment of products to the various facilities in each time period, and the inventory levels of the finished goods at the various locations. As stated before, in this phase of the optimization process a distinction may be made between different product families, but not between different products within the same family. The model is typically formulated as a Mixed Integer Program. Variables that represent production quantities are often continuous. Integer (discrete) variables are often 0-1 variables; they are, for example, needed in the formulation when a decision has to be made whether or not a particular product family will be produced at a certain facility during a given time period.

The output of the medium term planning process serves as an input to the detailed (short term) scheduling process. The detailed scheduling problems typically attempt to optimize each stage and each facility separately. So in the scheduling phase of the optimization process, the process is partitioned according to

(i) the different stages and facilities and

(ii) the different time periods.

So in each detailed scheduling problem the scope is considerably narrower (with regard to time as well as space), but the level of detail taken into consideration is considerably higher, see Figure 8.2. This level of detail is increased in the following dimensions:

(i) the time is measured in a smaller unit (e.g., days or hours); the process may be even time continuous,

(ii) the horizon is shorter,

(iii) the product demand is more precisely defined, and

(iv) the facility is not a single entity, but a collection of resources or machines.

The product demand now does not consist, as in the medium term planning process, of aggregate demands for entire product families. In the detailed scheduling process the demand for each individual product within a family is taken into account. The minor setup times and setup costs in between different products from the same family are taken into account as well as the sequence dependency.

The facility is now not a single entity; each product has to undergo a number of operations on different machines. Each product has a given route and given processing requirements on the various machines. The detailed scheduling problem can be analyzed as a job shop problem and various techniques can be used, including

(i) dispatching rules,

(ii) shifting bottleneck techniques,

(iii) local search procedures (e.g., genetic algorithms), or

(iv) integer programming techniques.

The objective takes into account the individual due dates of the orders, sequence dependent setup times, sequence dependent setup costs, lead times, as well as the costs of the resources. However, if two successive facilities (or stages) are tightly coupled with one another (i.e., the two facilities operate according to the JIT principle), then the short term scheduling process may optimize the two facilities jointly. It actually may consider them as a sin-

8.3 Frameworks for Planning and Scheduling in Supply Chains

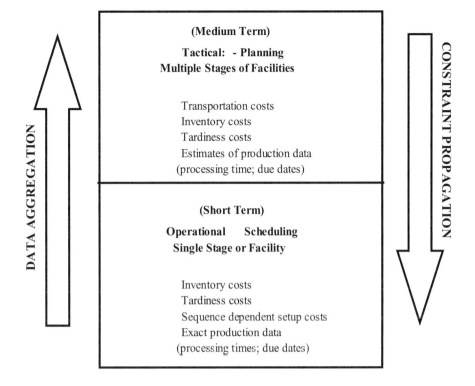

Fig. 8.2. Data aggregation and constraint propogation

gle facility with the transportation in between the two facilities as another operation.

The interaction between a planning module and a scheduling module may be intricate. A scheduling module may cover only a relatively short time horizon (e.g., one month), whereas the planning module may cover a longer time horizon (e.g., six months). After the schedule has been fixed for the first month (fixing the schedule for this month requires some input from the planning module), the planning module does not consider this first month any more; it assumes the schedule for the first month to be fixed. However, the planning module still tries to optimize the second up to the sixth month. Doing so, it considers the output of the scheduling module as a boundary condition (see Figure 8.3). However, it also may be the case that the time periods covered by the detailed scheduling process and the medium term planning process overlap.

A planning and scheduling framework for a supply chain typically must have a mechanism that allows feedback from a scheduling module to the planning module, see Figure 8.4. This feedback mechanism enables the optimization process to go through several iterations. It may be used under various

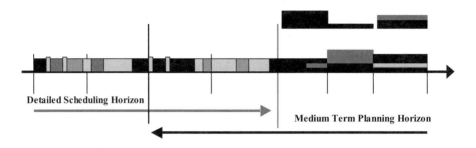

Fig. 8.3. Scheduling and planning horizons

circumstances: First, the results of the detailed short term optimization process may indicate that the estimates used as input data for the medium term planning process were not accurate. (The average production times in the planning processes do not take the sequence dependency of the setup times into account; setup times are estimated and embedded in the total production times. The total setup times in the detailed schedule may actually be higher than the setup times anticipated in the planning procedure). If the results of the detailed scheduling process indicate that the input to the planning process has to be modified, then new input data for the planning process have to be generated and the planning process has to be redone.

Second, there may be an exogenous reason necessitating a feedback from the detailed scheduling process to the medium term planning process. A major disruption may occur on the shop floor level, e.g., an important machine goes down for an extended period of time. A disruption may be of such magnitude that it not only affects the facility where it occurs, but other facilities as well. The entire planning process may be affected. A framework with a feedback mechanism allows the overall optimization process to iterate, see Figure 8.4.

The individual modules within the planning and scheduling framework for a given chain may have other interesting features. Two types of features that are often incorporated are decomposition features and so-called discretization features. Each feature can be activated and deactivated by the user of the system.

8.3 Frameworks for Planning and Scheduling in Supply Chains

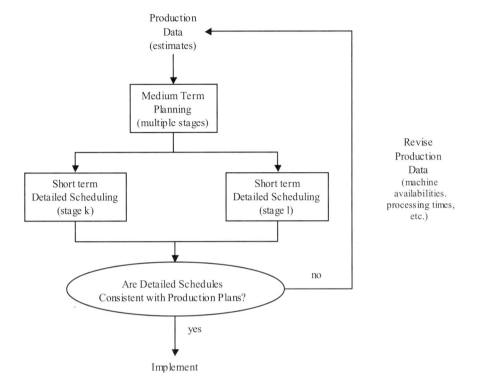

Fig. 8.4. Information flows between planning and scheduling systems

Decomposition is often used when the optimization problem is simply too large to be dealt with effectively by the routines available. A decomposition process partitions the overall problem in a number of subproblems and solves the (smaller) subproblems separately. At the end of the process the partial solutions are stitched together into one overall solution. Decomposition can be done according to

(i) time;
(ii) available resources (facilities or machines);
(iii) product families;
(iv) geographical areas.

Some of the decompositions may be designed in such a way that they are activated automatically by the system itself and other decompositions may be designed in such a way that they have to be activated by the user of the system. Decomposition is used in medium term modules as well as in detailed scheduling modules. In medium term planning the decomposition is often based on time and/or on product family (these may be internal decompositions activated by the system itself). The user may specify in a medium term planning process a geographical decomposition. In the detailed scheduling process the decomposition is often machine based (such a decomposition may be done internally by the system or imposed by the user).

One type of discretization feature may be used when the continuous version of a problem (for example, a linear programming relaxation of a more realistic integer programming formulation) does not yield results that are sufficiently accurate. In order to obtain more accurate results certain constraints may have to be imposed on given variables. For example, production quantities are often not allowed to assume just any values, but only values that are multiples of given fixed amounts or lot sizes (e.g., the capacity of a tank in the brewing of beer). The quantities that have to be transported between two facilities also have to be multiples of a fixed amount (e.g., the size of a container). This type of discretization feature may transform a continuous optimization problem (i.e., a linear program) into a discrete optimization problem.

Another type of discretization can be done with respect to time. It allows the user of the system to determine the size of the time unit. If the user is only interested in a rough plan, he may set the time unit to be equal to a week. That is, the results of the optimization then only specify what is going to be produced that week, but will not specify what is going to be produced within each day of that week. If the user sets the time unit equal to one day, the result will be significantly more precise. Besides specifying the sizes of the time units a system may use different time units in different periods. The discretization feature is often implemented in the medium term planning modules. The first week of a three month planning period may be specified on a daily basis, the next three weeks maybe determined on a weekly basis, and all activities beyond the first month are planned based on a continous model. This type of discretization does not change the nature of the problem; if the original problem is a linear program, then it remains a linear program.

The APO system of SAP Germany enables the modeler to activate and deactivate the discretization of various types of constraints in order to improve the performance of the optimization process. For example, discretization may be used for daily and weekly time buckets, but not for monthly time buckets in which Linear Programming is used without discretization.

8.4 A Medium Term Planning Model for a Supply Chain

This section considers a typical medium term planning model for a supply chain. It does not present the model in its full generality; the notation needed for a more general model is simply too cumbersome. In order to simplify the notation a description of the model is given with many of the relevant parameters having fixed values. It also does not incorporate all of the features described in the previous section (e.g., all the time units are of the same size).

Consider three stages in series. The first and most upstream stage (Stage 1) has two facilities (or factories) in parallel. They both feed into Stage 2 which is a distribution center (DC). Both Stages 1 and 2 can deliver to a

8.4 A Medium Term Planning Model for a Supply Chain

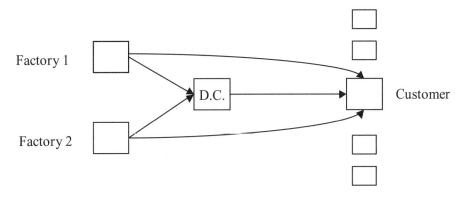

Fig. 8.5. A system with three stages

customer which is part of Stage 3, see Figure 8.5. The factories do not have any room to store finished goods and the customer does not want to receive any early deliveries.

The problem has the following parameters and input data. The two factories work around the clock; so the weekly production capacity available is $24 \times 7 = 168$ hours. There are two major product families F_1 and F_2. As stated before, in the medium term planning process all the products within a family are considered identical. The demand forecasts for the next 4 weeks are known (the time unit being 1 week). In this section, the subscripts and superscripts have the following meaning.

The subscript t ($t = 1, \ldots, 4$) refers to week t.
The subscript i ($i = 1, 2$), refers to factory i.
The subscript j ($j = 1, 2$), refers to product family j.
The subscript l ($l = 1, 2, 3$) refers to stage l;
  $l = 1$ refers to the two factories,
  $l = 2$ refers to the distribution center, and
  $l = 3$ refers to the customer.
The superscript p refers to a production parameter.
The superscript m refers to a transportation (i.e., moving) parameter.

The demand for product family j, $j = 1, 2$, at the DC level (Stage 2) by the end of week t, $t = 1, \ldots, 4$, is denoted by D_{2jt}. The demand for product family j, $j = 1, 2$, at the customer level (Stage 3) by the end of week t, $t = 1, \ldots, 4$, is denoted by D_{3jt}. Production times and costs are given:

c^p_{ij} = the cost to produce one unit of family j in factory i.
$\hat{p}_{ij}$ = the time (in hours) to produce one unit of family j in factory i.

The $\hat{p}_{ij}$ is just an estimate of the average time required to produce one unit, since it combines processing times with setup times. Because the run lengths have not been determined yet, it is not clear yet what the average production time will be. The $\hat{p}_{ij}$ is the reciprocal of the rate of production, see Chapter 7. Holding costs and transportation data include:

$h =$ the weekly holding (storage) cost in the DC for one unit of any type.
$c^m_{i2o} =$ the cost of moving one unit of any type from factory i to the DC.
$c^m_{io3} =$ the cost of moving one unit of any type from factory i to the customer.
$c^m_{o23} =$ the cost of moving one unit of any type from the DC to the customer.
$\tau =$ the transportation time from any one of the two factories to the DC, from any one of the two factories to the customer, and from the DC to the customer; all transportation times are assumed to be identical and equal to 1 week.

The following weights and penalty costs are given:

$w''_j =$ the tardiness cost per unit per week for an order of family j products that arrive late at the DC.
$w'''_j =$ the tardiness cost per unit per week for an order of family j products that arrive late at the customer.
$\psi =$ the penalty for never delivering one unit of product.

The objective is to minimize the total of the production costs, holding or storage costs, transportation costs, tardiness costs, and penalty costs for non-delivery over a horizon of four weeks. In order to formulate this problem as a Mixed Integer Program the following decision variables have to be defined:

$x_{ijt} =$ number of units of family j produced at factory i during week t.
$y_{i2jt} =$ number of units of family j transported from factory i to the DC in week t
$y_{i3jt} =$ number of units of family j transported from factory i to customer in week t
$z_{jt} =$ number of units of family j transported from the DC to customer in week t.
$q_{2j0} =$ number of units of family j in storage (being held) at the DC at time 0.
$q_{2jt} =$ number of units of family j in storage (being held) at the DC at the end of week t.
$v_{2jt} =$ number of units of family j that are tardy (have not yet arrived) at the DC in week t.
$v_{2j4} =$ number of units of family j that have not been delivered to the DC by the end of the planning horizon (the end of week 4).
$v_{3j0} =$ number of units of family j that are tardy (have not yet arrived) at the customer at time 0.
$v_{3jt} =$ number of units of family j that are tardy (have not yet arrived) at the customer by the end of week t.
$v_{3j4} =$ the number of units of family j that have not been delivered to the customer by the end of the planning horizon (the end of week 4).

8.4 A Medium Term Planning Model for a Supply Chain

Note that if y_{i2jt} (y_{i3jt}) are transported in week t from factory i to the DC (customer), then they leave the factory in week t and arrive at their destination in week $t+1$. The same is true with regard to the variable z_{jt}. There are various constraints in the form of upper bounds UB_{ilj} and lower bounds LB_{ilj} on the quantities of family j to be shipped from factory i to stage l.

The integer program can now be formulated as follows:

minimize

$$\sum_{t=1}^{4}\sum_{j=1}^{2}\sum_{i=1}^{2} c_{ij}^{p} x_{ijt} + \sum_{t=1}^{4}\sum_{j=1}^{2}\sum_{i=1}^{2} c_{i2o}^{m} y_{i2jt} + \sum_{t=1}^{4}\sum_{j=1}^{2}\sum_{i=1}^{2} c_{io3}^{m} y_{i3jt}$$

$$+ \sum_{t=1}^{4}\sum_{j=1}^{2} c_{o23}^{m} z_{jt} + \sum_{t=1}^{4}\sum_{j=1}^{2} h\, q_{2jt} + \sum_{t=1}^{3}\sum_{j=1}^{2} w_{j}'' v_{2jt}$$

$$+ \sum_{t=1}^{3}\sum_{j=1}^{2} w_{j}''' v_{3jt} + \sum_{j=1}^{2} \psi\, v_{2j4} + \sum_{j=1}^{2} \psi\, v_{3j4}$$

subject to the following weekly production capacity constraints:

$$\sum_{j=1}^{2} \hat{p}_{1j} x_{1jt} \leq 168 \qquad t=1,\ldots,4;$$

$$\sum_{j=1}^{2} \hat{p}_{2j} x_{2jt} \leq 168 \qquad t=1,\ldots,4;$$

subject to the following transportation constraints:

$$y_{1ljt} \leq UB_{1lj} \qquad\qquad t=1,\ldots,4;$$
$$y_{1ljt} \geq LB_{1lj} \text{ or } y_{1ljt}=0 \qquad t=1,\ldots,4;$$
$$y_{2ljt} \leq UB_{2lj} \qquad\qquad t=1,\ldots,4;$$
$$y_{2ljt} \geq LB_{2lj} \text{ or } y_{2ljt}=0 \qquad t=1,\ldots,4;$$

$$\sum_{l=2}^{3} y_{iljt} = x_{ijt} \qquad t=1,\ldots,4;\ j=1,2;\ i=1,2;$$

$$\sum_{i=1}^{2} y_{i3jt} + z_{jt} \leq D_{3,j,t+1} + v_{3jt} \qquad t=1,\ldots,3;\ j=1,2;$$

$$z_{j1} \leq \max(0, q_{2j0}) \qquad j=1,2;$$
$$z_{jt} \leq q_{2,j,t-1} + y_{1,2,j,t-1} + y_{2,2,j,t-1} \qquad t=2,3,4;\ j=1,2;$$

subject to the following storage constraints:

$$q_{2j1} = \max(0, q_{2j0} - D_{2j1} - z_{j1}) \qquad j = 1, 2;$$
$$q_{2jt} = \max(0, q_{2,j,t-1} + y_{1,2,j,t-1} + y_{2,2,j,t-1} - v_{2,j,t-1} - D_{2jt} - z_{jt})$$
$$j = 1, 2;\ t = 2, 3, 4;$$

subject to the following constraints regarding number of jobs tardy and number of jobs not delivered:

$$v_{2j1} = \max(0, D_{2j1} - q_{2j0}) \qquad j = 1, 2;$$
$$v_{2jt} = \max(0, D_{2jt} + v_{2,j,t-1} + z_{jt} - q_{2,j,t-1} - y_{1,2,j,t-1} - y_{2,2,j,t-1})$$
$$j = 1, 2;\ t = 2, 3, 4;$$
$$v_{3j1} = \max(0, D_{3j1}) \qquad j = 1, 2;$$
$$v_{3jt} = \max(0, D_{3jt} + v_{3,j,t-1} - z_{j,t-1} - y_{1,3,j,t-1} - y_{2,3,j,t-1})$$
$$j = 1, 2;\ t = 2, 3, 4.$$

It is clear that most variables in this mixed integer programming formulation are continuous variables. However, the transportation variables y_{iljt} are subject to disjunctive constraints. A different formulation of the problem could have some integer $(0-1)$ variables to make sure that the continuous (transportation) variables y_{iljt} are either 0 or larger than the lower bound LB_{ilj}. Moreover, note that those constraints in which a variable is equal to the max of an expression and 0 are nonlinear. In order to ensure that the given variable remains nonnegative, an additional binary variable has to be introduced. Note that a relaxation of the formulation described above without the disjunctive constraints provides a valid lower bound on the total cost.

The following numerical example illustrates an application of the model described above.

Example 8.4.1 (Medium Term Planning). Consider the following instance of the problem described above. The production times and costs in factory 1 are: $\hat{p}_{11} = 0.001$ hours (i.e., 3.6 sec.) and $\hat{p}_{12} = 0.002$ hours (i.e., 7.2 sec.); $c^p_{11} = \$1.00$ and $c^p_{12} = \$0.50$. The production times and costs concerning factory 2 are: $\hat{p}_{21} = 0.002$ hours (i.e., 7.2 sec.) and $\hat{p}_{22} = 0.003$ hours (i.e., 10.8 sec.); $c^p_{21} = \$0.50$ and $c^p_{22} = \$0.25$.

The holding cost for one unit of any type of product at the DC (h) is $0.10 per unit per week. The transportation costs are $c^m_{12o} = \$0.10$ per unit; $c^m_{22o} = \$0.30$ per unit; $c^m_{io3} = \$0.05$ for $i = 1, 2$, $c^m_{o23} = \$0.50$ per unit.

The forecast demand at the DC and from the customer for the two different product families are presented in the table below.

	week 1	week 2	week 3	week 4
D_{21t}	20,000	30,000	15,000	40,000
D_{22t}	0	50,000	30,000	50,000
D_{31t}	10,000	5,000	15,000	40,000
D_{32t}	0	10,000	0	5,000

8.4 A Medium Term Planning Model for a Supply Chain

The weekly shipments of product family 1 from factory 1 to the DC have to contain at least 10,000 units or otherwise there is no shipment, i.e., $LB_{121} = 10,000$. From factory 2 to the DC there has to be each week at most a shipment of 10,000 units of product family 2, i.e., $UB_{222} = 10,000$. The transportation time τ is 1 week.

The tardiness cost w_1'' (w_2'') is \$70.00 (\$35.00) per unit per week. The tardiness cost w_1''' (w_2''') is \$140.00 (\$105.00) per unit per week. The penalty cost ψ for not delivering at all is \$1000.00 per unit.

Running these data through a Mixed Integer Programming solver (assuming that the boundary conditions v_{3j0} and q_{2j0} are zero) yields the following production and transportation decisions.

	week 1	week 2	week 3	week 4
x_{11t}	0	0	0	0
x_{12t}	47,333	20,000	50,000	0
x_{21t}	65,000	33,500	76,500	0
x_{22t}	12,667	10,000	5,000	0

	week 1	week 2	week 3
y_{121t}	0	0	0
y_{131t}	0	0	0
y_{221t}	50,000	18,500	36,500
y_{231t}	15,000	15,000	40,000
y_{122t}	47,333	20,000	50,000
y_{132t}	0	0	0
y_{222t}	2,667	10,000	0
y_{232t}	10,000	0	5,000

The total cost of this solution is approximately \$3,004,750.00. (The exact total cost may depend on whether or not integrality assumptions concerning the quantities produced and transported are in effect as well as on other settings in the program.)

If an additional constraint is added to this problem requiring the production lot sizes to be multiples of 10,000, then we obtain the following solution.

	week 1	week 2	week 3	week 4
x_{11t}	0	0	0	0
x_{12t}	60,000	20,000	60,000	0
x_{21t}	70,000	30,000	80,000	0
x_{22t}	0	10,000	0	0

	week 1	week 2	week 3
y_{121t}	0	0	0
y_{131t}	0	0	0
y_{221t}	55,000	15,000	40,000
y_{231t}	15,000	15,000	40,000
y_{122t}	50,000	20,000	55,000
y_{132t}	10,000	0	5,000
y_{222t}	0	10,000	0
y_{232t}	0	0	0

The total cost in this case is indeed higher than the total cost without the production constraint that requires items to be produced in lots of 10,000. The total cost is approximately $3,017,000.00. The increase is less than 0.5%. The increased costs are mainly due to excess production (the total production quantities now exceed the total demand quantities) and, as a consequence, additional transportation and holding costs.

It is clear that the formulation of this medium term planning problem can be extended very easily to more time periods, more factories at the first stage and more product families. An extension to more stages may be a little bit more involved when there is an increase in the complexity of routing patterns.

8.5 A Short Term Scheduling Model for a Supply Chain

The short term scheduling problem for a facility in a supply chain can be described as follows: The output of the medium term planning problem specifies that over the short term n_j items of family j have to be produced. The scheduling problem can either be modeled as a job shop (or flexible flow shop) that takes all the production steps in the facility into account, or as a somewhat simpler single (or parallel) machine scheduling problem that focuses only on the bottleneck operation. If the operations in a facility are well balanced and the location of the bottleneck depends on the types of orders that are in the system, then the entire facility may have to be modeled as a job shop. If the bottleneck in the facility is a permanent bottleneck (that never moves), then a focus on the bottleneck may be justified. If the bottleneck stage is modeled as a parallel machine scheduling model, then the parallel machines may not be identical. They may also be subject to different maintenance and repair schedules.

There is, of course, a close relationship between the time $\hat{p}_{ij}$ in the medium term planning process and the processing time of an order in the short term detailed scheduling problem. The $\hat{p}_{ij}$ in the medium term planning process is an estimate and may be a value anywhere in between the average processing

8.5 A Short Term Scheduling Model for a Supply Chain

time of an order at the bottleneck operation and the total (estimated) throughput time of an order through the facility. The $\hat{p}_{ij}$ is a function of the exact processing times, the sequence dependent setup times, and the run lenghts.

In the remaining part of this section the subscript i will refer to a "machine" instead of a factory and the subscript j will refer to an order or a job rather than a product family. An order cannot be released before all the required raw material has arrived (these dates are typically stored in a Material Requirements Planning (MRP) system). That is, order j has an earliest possible starting time that is typically referred to as a release date r_j, a committed shipping date d_j and a priority factor or weight w_j. Dependent upon the manufacturing environment, preemptions may or may not be allowed. Every time a machine switches over from one type of item to another type of item a setup cost may be incurred and a setup time may be needed. If a schedule calls for a large number of preemptions, a large number of setups may be incurred.

The objective to be minimized may include the total setup times on the machines at the bottleneck as well as the total weighted tardiness, which is denoted by $\sum w_j T_j$ (see Chapter 5). The objective may be formulated as

$$\alpha_1 \sum w_j T_j + \alpha_2 \sum I_{ijk} s_{ijk},$$

where the α_1 and the α_2 denote the weights of the two parts of the objective function. The first part of the objective function is the total weighted tardiness and the second part of the objective represents the total of all setups; the indicator variable I_{ijk} is 1 if job k follows job j on machine i, the indicator variable is 0 otherwise.

This scheduling problem may be tackled via a number of different techniques, including a combination of dispatching rules, such as the Shortest Setup Time (SST) first rule, the Earliest Due Date first (EDD) rule, and the Weighted Shortest Processing Time first (WSPT) rule, see Chapter 5 and Appendix C. Other techniques may include genetic algorithms or integer programming approaches. In this phase, however, integer programming approaches are not often used because they are computationally quite intensive.

Example 8.5.1 (Short Term Scheduling). Consider the two factories described in the medium term planning process in the previous section. In the detailed scheduling process the two factories may be scheduled independently from one another and the scheduling is done one week at the time. Consider factory 1 with the two product families. The production process in this factory consists of various steps, but one of these steps is the bottleneck. This bottleneck consists of a number of resources in parallel. Consider in the same way the operations of factory 2 in its first week of operations. The solution of the integer program yields $x_{21t} = 65,000$ and $x_{22t} = 12,667$. Of the 65,000 of product family 1 a total of 50,000 has to be shipped to the DC and the remainder has to go to the customer. Of the 12,667 of product family 2 a total of 2667 has to be shipped to the DC and the remaining 10,000 has to go to the customer.

Assume that for this process the following more detailed information is available (which was not taken into account into the medium term planning process). The time unit in the scheduling process is 1 hour; this is in contrast to the 1 week in the medium term planning process (in actual implementations the time unit in the scheduling process can be made arbitrarily small). The scheduling horizon is 1 week.

Recall that 2 hours of the bottleneck resource are required to produce 1000 units of family 1 in factory 2, whereas 3 hours of the bottleneck resource are required for 1000 units of family 2. This implies that, based on these estimated production times, the planned production takes the full capacity of the bottleneck resource (in hours)

$$65 \times 2 + 12.667 \times 3 = 168.$$

However, the 2 and 3 hours requirement of the bottleneck resource are only estimates. They are estimates that are being used in the medium term planning process in order not to have to make a distinction between sequence dependent setup times and run times. The actual run times (or processing times), excluding any setup times are the following: To produce in factory 2 1000 units of family 1, 1.75 hours of the bottleneck resource is required, whereas 1000 units of family 2 requires 2.5 hours of the bottleneck resource. To start producing units of family 1 a setup of 16 hours is required. To start producing units of family 2 a setup of 6 hours is required. If each one of the products were to be produced in a single run in that week, then the entire production could be done within 168 hours, since

$$16 + 65 \times 1.75 + 6 + 12.66 \times 2.5 = 167.4$$

So, if there are not too many setup times the original assumptions for the medium term planning model are appropriate.

However, the shipment to the customer is supposed to go on a truck at time 120 (after 5 days), whereas the shipment to the DC takes place at the end of the week at time 168. All the raw material required to produce family 1 products are available at time 0, whereas the material necessary to produce family 2 products are only available after 2 days, i.e., at time 48.

This problem can be modeled as a single machine scheduling problem with jobs having different release dates and being subject to sequence dependent setups. The objective is

$$\alpha_1 C_{\max} + \alpha_2 \sum w_j T_j.$$

There are 4 different jobs, with the following processing times, release dates, and sequence dependent setup times. Each job is characterized by its family type and its destination. Jobs 1 and 2 are from the same family, so there is a zero setup time if one job is followed by the other. If either job 1 or job 2 follows job 3 or 4, then a setup of 16 hours is required. If job 3 or 4 follows job 1 or 2 a setup of 6 hours is required.

job	1	2	3	4
p_j	87.5	26.25	6.67	25
r_j	0	0	36	36
d_j	168	120	168	120

Two scheduling approaches may be appealing. One would schedule the jobs according to the Shortest Setup Time first rule (with ties broken according to the Earliest Due Date rule). This approach would yield the schedule 1, 2, 4, 3; after job 1 has been completed job 2 has to start because it has a zero setup time. All jobs are completed by time 168. However, job 4 is completed late. It had to be shipped by time 120 and it is shipped by time 161.

The second approach follows the Earliest Due Date first rule (with ties broken according to the Shortest Setup Time first rule). This approach yields schedule 2, 4, 3, 1. Since there is an additional setup time the makespan is 167.4 + 16 = 183.4. The shipment to the customer leaves on time, but the shipment to the DC leaves late.

The weights α_1 and α_2 in the objective function determine which schedule is more preferable.

The results from the detailed scheduling analysis may, for various reasons, not be useful. When trying to minimize the makespan (in order to ensure that the production of the required quantities are completed within that week), it may turn out that there does not exist a schedule that can complete the requested production within a week. The reason may be the following: the production times $\hat{p}_{ij}$ that were entered in the medium term planning problem were mere estimates based on factory data, including average processing times on bottleneck machines, expected throughput times, expected setup times, and so on. However, the value $\hat{p}_{ij}$ did not represent an accurate cycle time, since the average production time may depend on the run length of the batches at the bottleneck. It may be that the schedule generated in the detailed scheduling process has batch sizes that are very short with an average production time that is larger than the estimates used in the medium term planning process. If there is a major discrepancy (i.e., the frequency of the setups is considerably higher than usual), then a new estimate may have to be developed for the $\hat{p}_{ij}$ in the medium term planning process and the integer programming problem has to be solved again.

8.6 Carlsberg Denmark: An Example of a System Implementation

There are many software vendors that sell custom made solutions for supply chain planning and scheduling. One of the largest companies in this field is SAP, which is based in Walldorf (Germany). SAP has a division that develops its so-called Advanced Planner and Optimizer (APO) system; this supply

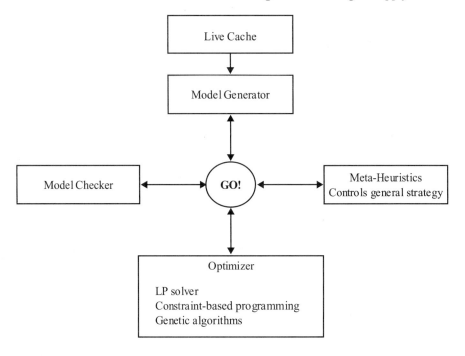

Fig. 8.6. The SAP-APO optimizer architecture

chain planning and scheduling system has functionalities at various levels, including the tactical level and the operational level.

On the tactical level, medium term planning scenarios can be monitored for a global chain from distribution centers to plants and suppliers. The optimizer automatically processes bills of materials while taking capacities into account, and it minimizes transportation costs, production costs, and holding or storage costs. The sheer complexity of this global view is handled through a rough-cut model that aggregates time units in buckets (e.g., day or week) and products and resources in families.

On the operational level, APO relies on a detailed scheduling model. At this level the short term, day to day operations are monitored, taking into account the idiosyncrasies of all the operations in the supply chain. The optimizer schedules the orders taking into account all the rules and constraints that are prevalent in a complex manufacturing environment (which may be a multi-stage production process with primary resources, secondary resources, and alternative routings).

Figure 8.6 shows the architecture of the optimizer in APO. For long term and medium term planning APO uses its LP solvers (CPLEX). APO has various approaches for short term planning and detailed scheduling, including Constraint Programming, Genetic Algorithms, and Repair Algorithms.

8.6 Carlsberg Denmark: An Example of a System Implementation

This section describes an implementation of the SAP-APO system at the beerbrewer Carlsberg A/S in Denmark. The modeling that forms the basis for this case is somewhat similar to the models described in the fourth and fifth section of this chapter. Carlsberg Denmark A/S, the largest beerbrewer in Scandinavia, started in 2001 a supply chain project with the objective to decrease inventory costs, to optimize sourcing decisions, to increase customer service level and in general to change the business to a more demand driven process. Carlsberg selected APO. The project is an example of how a real-life implementation takes planning and scheduling issues in various stages of a supply chain into account. The system has been operational since the end of 2002.

The supply chain considered in the project consists of three stages. The first stage is the production process of the beer at two breweries with 2 and 4 filling lines, respectively. Each filling line has a different capacity. The second stage consists of a distribution center (DC) and the third stage consists of the local warehouses, see Figure 8.5. In the first stage there are three production steps, namely brewing (and fermentation), filtering and filling of the beer. All three steps have a limited capacity, but the bottleneck is usually the filling step. The resources for the filling operations at the two plants have different costs and processing times. When creating the production orders for brewing and filling, different lot size constraints have to be taken into account. Production orders for the brewing have always a fixed lot size because the brewing tank has to be filled. If the demand quantity is higher than the fixed lot size, then additional production orders have to be created for the brewing process (each with the fixed lot size as the production quantity). Orders below the minimal lot size are increased to the minimal lot size and orders above the minimal lot size are either rounded up or down to the closest integer value. The filling resources have to be filled up to 100%. There is further a split in the business processes according to the sales volumes of the various products. There are three categories: A, B, and C. Category A are the fast movers and include the well-known brands Carlsberg Pils and Tuborg Green. Category C are the (more expensive) slow movers.

Once the beer is bottled, it has to be transported either to the distribution center (DC) or to a local warehouse. Depending on the different products and the quantities to be transported, either a direct delivery from the plant to a local warehouse or a transport via the DC is better. Again, lot size constraints have to be taken into consideration when creating transport orders. The transport durations depend, of course, on the origin and the destination.

One of the main objectives of Carlsberg is to provide a given level of service to its customers. A typical way to achieve a given service level is to keep safety stocks at the warehouses. The higher the safety stock levels, the higher the service levels, but also the higher the inventory costs. One function of a supply chain management system is the computation of the lowest levels of safety stocks that achieve the desired service levels. Carlsberg uses advanced safety stock methods to compute safety stock values for all its products at its

DC as well as at its local warehouses. These safety stock levels depend on the given service level, the demand forecast, the uncertainty in the forecast, the replenishment lead time and the typical lot sizes.

The medium term planning module plans ahead for 12 weeks, with the first 4 weeks in days and the remaining 8 weeks in weekly periods. Assuming a given demand pattern (sales orders and forecasts), APO creates a Mixed Integer Program, along the lines described in the previous section, and tries to find a solution with minimum cost. The total costs include production costs, storage costs, transportation costs, late delivery (tardiness) costs, non-delivery costs, and violation of the safety stock levels computed in the first step. Some of the costs mentioned above can be specified in an exact way, such as the production and transportation costs. Other costs, such as storage, violation of safety stock, and late and non-delivery costs, merely represent the priorities of Carlsberg. If, for example, Carlsberg considers the safety stock in the local warehouses more critical than the safety stock in the DC, then the cost assigned to the violation of safety stock for a product at the DC is less than the costs of violating safety stocks of the same products at the local warehouses. If neither safety stock can be maintained, then the system will create a transport from the DC to the local warehouse (provided the difference between the costs of safety stock violations at the DC and at the local warehouse is higher than the transportation cost from the DC to the local warehouse). Clearly, all cost types are strongly related with one another and modifying one type of cost can have many unforeseen consequences in the solution generated. Carlsberg developed its own cost model for storage costs; this model, for example, takes into account the location occupied by a pallet, the maximum number of levels pallets can be stacked, the number of products per pallet, and the warehouse itself. Based on these parameters for each product at each location, storage costs can be computed.

The following constraints have to be taken into consideration, namely the production times in the three production steps, the capacity of the bottling resources on a daily or weekly level, the transportation times between locations, the lot size constraints, the existing stock and the resource consumptions.

The medium term plan is the result of various costs trade-offs and material consumption. The system generates for the next 12 weeks the planned production quantities for the three production steps in detail (including the quantity of each product to be bottled on each filling resource as well as the quantities to be transported from one location to another.

The short term scheduling starts its computations using results obtained from the medium term plan. The planned production orders for the first week that come out of the medium term planning system are transformed into short term production orders on which a detailed scheduling procedure has to be applied. These production orders are then scheduled on the filling resources by applying a genetic algorithm with as objective the minimization of the sum of the sequence dependent setup times and the sum of the tardinesses. The due dates are specified by the medium term planning problem and are equal to the starting times of the transportation orders. It is possible that the results of the medium term plan are changed by the short term scheduling procedure (i.e., a different filling resource may be selected in the same plant). After the detailed scheduling has been completed, the transportation planning and scheduling has to be done. In this step the trucks that provide the transportation between the different locations are loaded based on the results that come out of the medium term plan and the results from the detailed scheduling procedures. In order to maximize the utilization, a truck may transport on any given trip various different products.

A new medium term plan is generated every day. The daily run takes into account the most up-to-date capacity situation of all the available resources, the results of the previous day detailed schedule, and the most current demand forecast. Afterwards, a new detailed schedule and transportation plan are generated.

The generation of the medium term plan is split into three Mixed Integer Programs, which are solved in consecutive runs. Each MIP has between 100,000 and 500,000 variables and between 50,000 and 150,000 constraints. Total running time is about 10-12 hours. Each MIP uses product decomposition methods, which creates 5 to 10 subproblems each. The generation of the subproblems has to take into account different priorities for the finished beer products and the fact that the same brewed and filtered beer type may end up in different end products. The quality of the solution is measured in business terms as well as in technical terms. Only by considering all dimensions one can speak of a "good" or a "bad" solution. The technical data and measures tend to be easy to collect and understand; the more important business measures are harder to understand and verify. The two most important technical measures are (i) the difference between the costs of the MIP solution and the LP relaxation solution, and (ii) the difference between the overall delivery percentages of the MIP solution and the LP relaxation solution. The difference between the two costs is on average between 0.2 and 10%, but sometimes it reaches 400%. A huge cost difference between the MIP and the LP relaxation can occur when the LP can fulfill all demands while the MIP cannot (because of the lot size constraints). As unfilled demand brings about a very high penalty cost, the cost difference between the MIP and the LP relaxation may then be very high.

The user interfaces of the systems are, of course, quite elaborate and include the typical Gantt charts, see Figures 8.7 and 8.8.

8.7 Discussion

The purpose of this chapter is to provide insights into the use of planning and scheduling models in supply chain management, as well as into the information sharing and interactions that occur between the different types of models that form the basis for a system. In the literature, planning models have often been

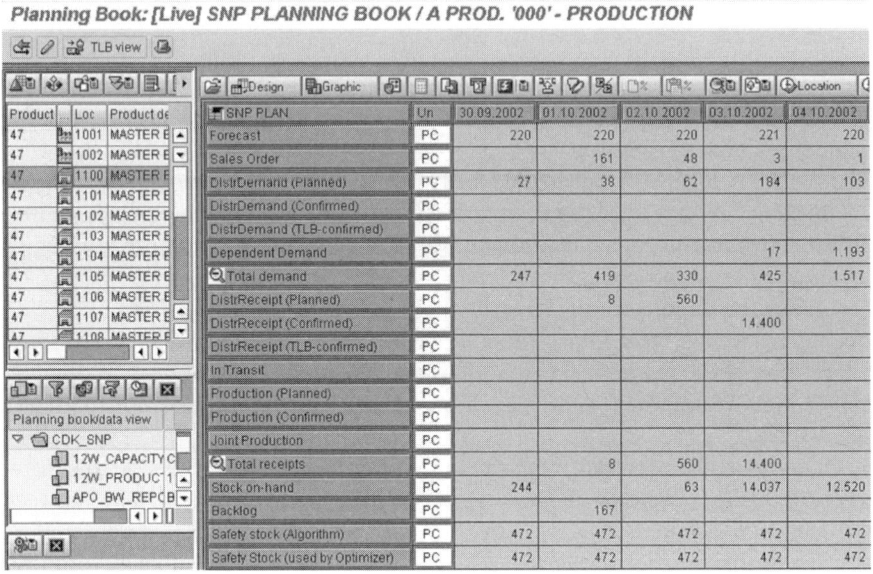

Fig. 8.7. Carlsberg-Denmark planning system user interface

analyzed in detail; scheduling models, on the other hand, have been studied less often within a supply chain management framework. The interactions and information sharing between the planning models and the scheduling models also deserve more attention.

There are several reasons why it is not that easy to incorporate planning systems and scheduling systems in one framework. One reason is that in the planning stage the objective is measured in dollar terms (e.g., minimization of total cost), whereas in the schedule stage the objective is typically measured in time units (e.g., minimization of total tardiness). A second reason is that the time periods over which the planning module and the scheduling module optimize may overlap only partially, see Figure 8.3. The horizon over which the scheduling module optimizes typically starts at the current time and covers a relatively short term. The planning module optimizes over a period that starts at some time in the future (since it may assume that all schedules before this point in time already have been fixed) and covers a long term. The units of time used in the two modules may be different as well. In the scheduling module the unit may be an hour or a day; in the planning module it may be a week or a month.

Comparing the modeling that is being done in practice for medium term planning processes with the models that have been studied in the research literature, it becomes clear that there are differences in emphasis. When multi-stage models are considered in the planning and scheduling research literature, there is more of an emphasis on setup costs (typically sequence independent)

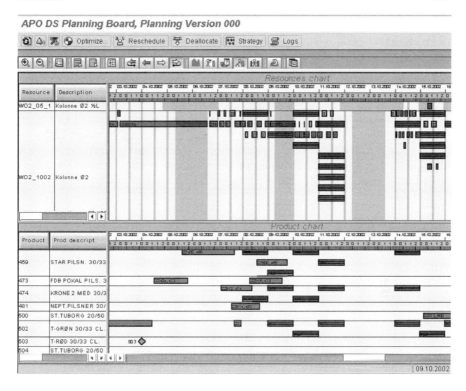

Fig. 8.8. Carlsberg-Denmark scheduling system user interface

and less of an emphasis on transportation costs; in the modeling that is done in practice, there is a very strong emphasis on transportation costs and less of an emphasis on setup costs. Incorporating both setup costs and transportation costs in a multi-stage planning model may cause the number of variables to become prohibitively large.

Exercises

8.1. Consider the model in Section 8.4. How do the number of variables and constraints depend on the number of customers?

8.2. Consider the model in Section 8.4.
(a) How do the number of variables and the number of constraints increase when the number of factories in stage 1 is increased from 2 to 3.
(b) How do the number of variables and constraints increase if the number of distribution centers is increased from 1 to 2 (assume that both distribution centers can receive from both factories and can deliver to the same customer).

8.3. Consider Example 8.4.1.

(a) Solve the same instance but assume now that $LB_{112} = 0$. (Use either Ilog's OPL code or Dash Optimization's Mosel code for this problem on the accompanying CD-ROM.) Compare the new solution with the solution in Example 8.4.1.

(b) If this lower bound is zero, does the problem reduce to a linear programming problem? Explain your answer.

8.4. Consider Example 8.4.1. Solve the same instance again, but assume now that $LB_{222} = LB_{223} = 5000$. Compare the solution obtained with the solution of Example 8.4.1.

8.5. Consider Example 8.4.1. Solve the same instance (again use either the Dash code or the OPL code on the CD), but now assume that the production lot size is:

(a) 20,000, and
(b) 30,000.
(c) Does the total cost increase as a function of the production lot size in a linear manner, concavely, or convexly? Give an explanation for your conclusion.

8.6. Modify either the Dash code or the OPL code on the CD for Example 8.4.1 in such a way that the transportation quantities are always multiples of a fixed value K (i.e., they do not have just a lower bound K).

8.7. Use the code developed in Exercise 8.6 to solve the instances with transportation quantities that are multiples of

(a) 10,000;
(b) 20,000;
(c) 30,000.
(d) Does the total cost increase as a function of the transportation quantity in a linear manner, concavely, or convexly? Give an explanation for your conclusion.

8.8. Compare the sensitivity of the total cost with respect to the production lot size to the sensitivity with respect to the transportation quantity.

8.9. Formulate an integer program for a model with two distribution centers in the second stage. Each distribution center has its own customer (i.e., each one of the two customers can only be supplied by one of the distribution centers). Compare the number of variables and the number of constraints in this Mixed Integer Program with the number of variables and constraints in the original formulation.

8.10. Formulate an integer program for a model with two factories, three products, two distribution centers in the second stage and one customers in the third stage. Each distribution center can supply the customer. Compare the number of variables and the number of constraints in this Mixed Integer Program with the number of variables and constraints in the original formulation.

Comments and References

There is an extensive literature on supply chain management. Many papers and books focus on supply chain coordination; however, a significant amount of this work has an emphasis on inventory control, pricing issues, and the value of information, see Hax and Meal (1975), Bowersox and Closs (1996), Simchi-Levi, Kaminsky and Simchi-Levi (2000), Chopra and Meindl (2001), and Stadtler and Kilger (2002). A fair amount of research has been done on the solution methods applicable to planning and scheduling models for Supply Chain Management, see Muckstadt and Roundy (1993), Shapiro (2001), and Miller (2002). Some of the planning and scheduling models have been studied in a rather restricted manner in order to obtain elegant theoretical results; see, for example, Hall and Potts (2003). Some research has been done on more integrated models in the form of hierarchical planning systems; this research has resulted in frameworks that incorporate planning as well as scheduling models, see Barbarosoglu and Ozgur (1999), Dhaenens-Flipo and Finke (2001).

Only a limited amount of the research has focused on the details of actual applications and implementations of the more integrated systems, since not many of these installations have been successful, see Hadavi and Voigt (1987), Hadavi (1998), Shepherd and Lapide (1998), and Sadeh, Hildum, and Kjenstad (2003). Examples of planning and scheduling applications in continuous manufacturing can be found in Haq (1991), Akkiraju, Keskinocak, Murthy and Wu (1998), Murthy, Akkiraju, Goodwin, Keskinocak, Rachlin, Wu, Kumaran, Yeh, Fuhrer, Aggarwal, Sturzenbecker, Jayaraman and Daigle (1999), Rachlin, Goodwin, Murthy, Akkiraju, Wu, Kumaran, and Das (2001), and Keskinocak, Wu, Goodwin, Murthy, Akkiraju, Kumaran and Derebail (2002). Examples of planning and scheduling in discrete manufacturing are described in Arntzen, Brown, Harrison and Trafton (1995), De Bontridder (2001) and Vandaele and Lambrecht (2001).

This chapter is mainly based on the paper by Kreipl and Pinedo (2004). The third, fourth, fifth, and sixth section of this chapter have been inspired primarily by the architecture of systems developed by SAP Germany AG, see Braun (2001), Braun and Groenewald (2000), and Strobel (2001). The model described in the fourth section is a simplified example of a class of standard planning models that has been the basis for a number of actual implementations by SAP Germany AG in various different industries, from semiconductor (DRAM) manufacturing companies to beerbrewers with elaborate distribution systems.

The OPL code for Example 8.4.1 on the CD-ROM is due to Irvin Lustig from ILOG. The Mosel code for Example 8.4.1 on the CD-ROM is due to Alkis Vazacopoulos and Nitin Verma from Dash Optimization.

Part III

Planning and Scheduling in Services

9   Interval Scheduling, Reservations, and Timetabling 205

10 Scheduling and Timetabling in Sports and Entertainment. 229

11 Planning, Scheduling, and Timetabling in Transportation . 253

12 Workforce Scheduling 289

Chapter 9

Interval Scheduling, Reservations, and Timetabling

9.1 Introduction 205
9.2 Reservations without Slack 207
9.3 Reservations with Slack 210
9.4 Timetabling with Workforce Constraints 213
9.5 Timetabling with Operator or Tooling
    Constraints 216
9.6 Assigning Classes to Rooms at U.C. Berkeley ... 221
9.7 Discussion 224

9.1 Introduction

Scheduling activities in an environment with resources in parallel may require at times a reservation system. Each activity (i.e., reservation) is supposed to occupy one of the resources for a given time period. Activity j, $j = 1, \ldots, n$, has a duration p_j and has to fit within a time window that is specified by an earliest starting time r_j and a latest termination time d_j. There may or may not be any slack, i.e., either

$$p_j < d_j - r_j$$

or

$$p_j = d_j - r_j.$$

Such a setting is equivalent to the environment described in Chapter 5 which has m machines in parallel and n jobs; job j has a processing time p_j, a release date r_j, a due date d_j, and a weight w_j.

It may not be possible to schedule all n activities and the decision-maker may have to decide which activities to schedule and which ones not. Several objectives may be of interest, e.g., maximizing the number of activities scheduled or maximizing the total utilization of the resources.

We first consider models with no slack in the time windows for the activities, i.e., the starting time and completion time of an activity have to be equal to its release date and due date. These models are at times referred to as fixed interval, or simply interval scheduling models. We consider also the more general case where there may be a slack in any given time window, i.e., the difference between the due date and the release date may be greater than the duration of the activity.

Reservation systems are ubiquitous. Consider, for example, a system in which customers attempt to reserve hotel rooms for specific time periods. The hotel has to decide which reservations to confirm and which ones not. This problem occurs in many other settings as well (e.g., in a manufacturing setting a customer may want to reserve one or more machines for specific time periods).

A different class of models that is also considered in this chapter is the class of timetabling models. In these models there are typically n activities to be scheduled with an *unlimited* number of identical resources in parallel. However, timetabling problems have an additional dimension: an activity can be done by one of the resources only if specific operators are available at the time of its execution. So an activity can be scheduled on any one of the resources at any time as long as the necessary operators are available during that period. The availability of the operators may be subject to certain constraints; these constraints may have as an effect that certain combinations of activities cannot be done at the same time, even though the resources are available. A typical objective may be to complete all activities in minimum time, i.e., to minimize the makespan. In a more general timetabling problem the timing of activity j may also be constrained by an earliest starting time r_j and a latest completion time d_j.

This chapter considers two types of timetabling problems. The first type of timetabling problem assumes that all operators are identical, i.e., the operators constitute a single homogeneous workforce. The total number of operators available is W and in order to do activity j on one of the resources W_j operators have to be present. If the sum of the people required by activities j and k is larger than W (i.e., $W_j + W_k > W$), then activities j and k may not overlap in time. This type of timetabling is in what follows referred to as timetabling with workforce or personnel constraints.

In the second type of timetabling problem each operator has its own identity and is unique. (An operator may now be equivalent to a specific tool or fixture that is required in order to perform certain activities.) Each activity now requires a specific subset of the operators and/or tools. In order for an activity to be scheduled all the operators or tools in its subset have to be available. Two activities that need the same operator can therefore not be processed at the same time.

This second type of timetabling can occur in many different settings. Consider, for example, a large repair shop for aircraft engines. In order to do certain types of repairs it is necessary to have specific tools, equipment, and operators. A given tool or piece of equipment may be required for certain types of repairs; timetabling may therefore be necessary. A second example of this type of timetabling occurs when meetings have to be scheduled. The operators are now the people who have to attend the meetings and each meeting has to be assigned to a time period in which those who have to attend are able to do so. The meeting rooms correspond to the resources. A third example of this type of timetabling occurs when exams have to be scheduled. Each operator represents a student (or a group of students). Two exams that have to be taken by the same student (or group of students) cannot be scheduled at the same time. The objective is to schedule all the exams within a given time period.

One reason for covering reservation problems and timetabling problems in the same chapter is that both types of problems lead to well-known graph coloring problems. The reservation problem with zero slack and the timetabling problem with operator or tooling constraints are both closely related to a well-known node coloring problem in graph theory. The reservation model with slack is, of course, a generalization of the reservation model without slack. The timetabling problem with workforce constraints cannot be compared that easily to the timetabling problem with operator constraints. In the timetabling problem with workforce constraints there is only one type of operator, but there are a number of them and they are interchangeable. In the timetabling problem with operator or tooling constraints there are several different types of operators, but of each type there is only one. In any case, both the timetabling problem with operator or tooling constraints and the timetabling problem with workforce constraints are special cases of the project scheduling problem with workforce constraints described in Section 4.6.

In this chapter we often make a distinction between the feasibility version of a problem and its optimization version. In the feasibility version we need to determine whether or not a feasible schedule exists; in the optimization version an objective has to be minimized. If no efficient algorithm exists for the feasibility version, then no efficient algorithm exists for the optimization version either.

Throughout this chapter we assume that all data are integer and that preemptions are not allowed.

9.2 Reservations without Slack

In this section we consider the following reservation model. There are m resources in parallel and n activities. Activity j has a release date r_j, a due date d_j, and a weight w_j. As stated before, all data are integer. The fact that there is no slack between release date and due date implies that

$$p_j = d_j - r_j.$$

If we decide to do activity j, then it has to be done within the specified time frame. However, it may be the case that activity j cannot be done by just any one of the m resources; it may have to be done by a resource that belongs to a specific subset M_j of the m resources. When all activities have equal weights, the objective is to maximize the number of activities done. In contrast, when the activities have different weights, the objective is to maximize the weighted number of activities scheduled. A weight is often equivalent to a profit that is made by doing the activity. In a more general model the weight of activity j may also depend on the resource to which it is assigned, i.e., the weight is w_{ij} (i.e., the profit depends on the activity as well as on the resource to which the activity is assigned).

Example 9.2.1 (A Car Rental Agency). Consider a car rental agency with four types of cars: subcompact, midsize, full size and sport-utility. Of each type there are a fixed number available. When customer j calls to make a reservation for p_j days, he may, for example, request a car of either one of two types and will accept the price quoted by the agency for either type. The set M_j for such a customer includes all cars belonging to the two types. The profit made by the agency for a car of type i is π_i dollars per day. So, the weight of this particular reservation is $w_{ij} = \pi_i p_j$.

However, if customer j specifically requests a subcompact and all subcompacts have been rented out, the agency may decide to give him a midsize for the price of a subcompact in order not to lose him as a customer. The set M_j includes subcompacts as well as midsizes (even though customer j requested a subcompact), but the agency's daily profit is a function of the car as well as of the customer, i.e., π_{ij} dollars per day, since the agency gives him a larger car at a lower price. The weight is $w_{ij} = \pi_{ij} p_j$.

Most reservation problems can be formulated as integer programs. Time is divided in periods or slots of unit length. If the number of slots is fixed, say H, then the problem is referred to as an H-slot problem. Let x_{ij} denote a binary variable that assumes the value 1 if activity j is assigned to resource i and 0 otherwise, and let J_t denote the set of activities that need a resource in slot t, i.e., during the period $[t-1, t]$. The following constraints have to be satisfied:

$$\sum_{i=1}^{m} x_{ij} \leq 1 \quad j = 1, \ldots, n$$

$$\sum_{j \in J_t} x_{ij} \leq 1 \quad i = 1, \ldots, n, \ t = 1, \ldots, H.$$

The first set of constraints ensures that every activity is assigned to at most one resource and the second set ensures that a resource is not assigned to more than one activity in any given slot.

The easiest version of the reservation problem is a feasibility problem: does there exist an assignment of activities to resources with every activity

9.2 Reservations without Slack

being assigned? A more general and harder version is the following feasibility problem. Does there exist an assignment of activities to resources with activity j being done by a resource that belongs to a given subset M_j? In the optimization problem the objective is to maximize the total profit

$$\sum_{i=1}^{m}\sum_{j=1}^{n} w_{ij}x_{ij},$$

where the weight w_{ij} is equivalent to a profit associated with assigning activity j to resource i.

Some special cases of this optimization problem can be solved in polynomial time. For example, consider the case where all n activities have a duration equal to one, i.e., $p_j = 1$ for all j. Each time slot can be considered as a separate subproblem and each subproblem can be solved as an independent assignment problem (see Appendix A). This decomposition can be applied even with arbitrary weights w_{ij} and with arbitrary resource subsets M_j.

Another version of the reservation model that allows for an efficient solution assumes arbitrary durations, identical weights (i.e., $w_{ij} = 1$ for all i and j), and each set M_j consisting of all m resources (i.e., the m resources are identical). The durations, earliest starting times (release dates) and latest completion times (due dates) are arbitrary integers and the objective is to maximize the number of activities assigned. This problem cannot be decomposed into a number of independent subproblems (one for each time slot), since the durations of the different activities may overlap. However, it can be shown that the following relatively simple algorithm maximizes the total number of activities. In this algorithm the activities are ordered in increasing order of their release dates, i.e.,

$$r_1 \leq r_2 \leq \cdots \leq r_n.$$

Set J denotes the set of activities already selected.

Algorithm 9.2.2 (Maximizing Number of Activities Assigned).
Step 1.
   Set $J = \emptyset$ and $j = 1$.
Step 2.
   If a resource is available at time r_j, then assign activity j to that resource; include activity j in J, and go to Step 4.
   Otherwise go to Step 3.
Step 3.
   Let j^* be such that
   $$C_{j^*} = \max_{k \in J}(C_k) = \max_{k \in J}(r_k + p_k).$$

If $C_j = r_j + p_j > C_{j^*}$, do not include activity j in J and go to Step 4.

Otherwise, delete activity j^* from J, assign activity j to the resource freed and include activity j in J.

Step 4.

If $j = n$, STOP,

otherwise set $j = j + 1$ and return to Step 2.

Another version of this reservation model with zero slack, arbitrary durations, equal weights, and identical resources is also of interest. Assume there are an unlimited number of identical resources in parallel and all activities have to be assigned. However, the assignment must be done in such a way that a minimum number of resources is used. This problem is, in a sense, a dual of the problem discussed before. It turns out that minimizing the number of resources when all activities have to be done is also an easy problem.

It can be solved as follows. Again, the activities are ordered in increasing order of their release dates, i.e., $r_1 \leq r_2 \leq \cdots \leq r_n$. First, activity 1 is assigned to resource 1. The algorithm then proceeds with assigning the activities, one by one, to the resources. Suppose that the first $j-1$ activities have been assigned to resources $1, 2, \ldots, i$. Some of these activities may have been assigned to the same resource. So $i \leq j - 1$. The algorithm then takes the next activity from the list, activity j, and tries to assign it to a resource that already has been utilized before. If this is not possible, i.e., resources $1, \ldots, i$ are all busy at time r_j, then the algorithm assigns activity j to resource $i+1$. The number of resources utilized after activity n has been assigned is the minimum number of resources required.

This last problem turns out to be a special case of a well-known node coloring problem in graph theory. Consider n nodes and let node j correspond to activity j. If there is an (undirected) arc (j, k) connecting nodes j and k, then the processing of activities j and k overlap in time and nodes j and k cannot be given the same color. If the graph can be colored with m (or less) colors, then a feasible schedule exists with m resources. This node coloring problem, which is a feasibility problem that is NP-hard, is more general than the reservation problem in which the number of resources used is minimized. This node coloring problem is actually equivalent to the timetabling problem with operator or tooling constraints described in Section 9.5. The node coloring problem establishes the links between interval scheduling, reservations, and timetabling.

9.3 Reservations with Slack

In the previous section we assumed that there was no slack between the release date and the due date of each activity, i.e.,

9.3 Reservations with Slack

$$p_j = d_j - r_j.$$

A more general version of the reservation model allows for slack in the time window specified, i.e.,

$$p_j \leq d_j - r_j.$$

We again consider first the special case where all release dates and due dates are integer and all processing times are equal to one. The weights of all activities are identical and all M_j sets consist of all m resources. This case is trivial since a schedule can be constructed progressively in time and the maximum number of activities can be assigned.

The more general problem with non-identical durations does not have an easy solution. Maximizing the weighted number of activities assigned is NP-hard, so it is unlikely that there exists an efficient algorithm that would guarantee an optimal solution. We have to rely on heuristics.

The following heuristic is basically a composite dispatching rule as described in Appendix C. It requires, as a first step, the computation of a number of statistics. Let ν_{it} denote the number of activities that may be assigned to resource i during interval $[t-1, t]$. This factor thus corresponds to a potential utilization of resource i in time slot t. The higher this number is, the more flexible resource i is in this time slot. A second factor is the number of resources to which activity j can be assigned, i.e., the number of resources in set M_j, which is denoted by $|M_j|$. The larger this number, the more flexible activity j is. Define for activity j a priority index I_j that is a function of w_j/p_j and $|M_j|$, i.e.,

$$I_j = f(w_j/p_j, |M_j|).$$

The higher w_j/p_j and the smaller $|M_j|$, the lower the index. The activities can now be ordered in increasing order of their indices, i.e.,

$$I_1 \leq I_2 \leq \cdots \leq I_n.$$

The algorithm takes the activity with the lowest index among the remaining activities and attempts to assign it to one of the resources, starting with the resource with the least flexible time intervals. If the activity needs a resource over the period $[t, t+p_j]$, then the selection of resource i depends on a function of the factors $\nu_{i,t+1}, \ldots, \nu_{i,t+p_j}$, i.e., $g(\nu_{i,t+1}, \ldots, \nu_{i,t+p_j})$. Examples of such functions are:

$$g(\nu_{i,t+1}, \ldots, \nu_{i,t+p_j}) = \Big(\sum_{l=1}^{p_j} \nu_{i,t+l}\Big)/p_j;$$

$$g(\nu_{i,t+1}, \ldots, \nu_{i,t+p_j}) = \max(\nu_{i,t+1}, \ldots, \nu_{i,t+p_j}).$$

The heuristic attempts to assign the activity to a resource in a period that has the lowest possible $g(\nu_{i,t+1}, \ldots, \nu_{i,t+p_j})$ value. This one-pass heuristic can be summarized as follows.

Algorithm 9.3.1 (Maximizing Weighted Number of Activities).
Step 1.

   Set $j = 1$

Step 2.

   Take activity j and select, among the resources and time slots available, the resource and time slots with the lowest $g(\nu_{i,t+1}, \ldots, \nu_{i,t+p_j})$ rank.

   Discard activity j if it cannot be assigned to any machine at any time.

Step 3.

   If $j = n$ STOP,

   otherwise set $j = j + 1$ and return to Step 2.

The next example illustrates this heuristic.

Example 9.3.2 (Maximizing Weighted Number of Activities). Consider seven activities and three resources.

activities	1	2	3	4	5	6	7
p_j	3	10	9	4	6	5	3
w_j	2	3	3	2	1	2	3
r_j	5	0	2	3	2	4	5
d_j	12	10	20	15	18	19	14
M_j	{1,3}	{1,2}	{1,2,3}	{2,3}	{1}	{1}	{1,2}

Consider the index function

$$I_j = f(w_j/p_j, M_j) = \frac{|M_j|}{w_j/p_j}.$$

The indices for the activities can be computed and are tabulated below.

activities	1	2	3	4	5	6	7
I_j	3	6.67	9	4	6	2.5	2

The factors ν_{it} are tabulated below.

slot t	0	1	2	3	4	5	6	7	8	9	10	11	12	13	14	15	16	17	18	19
ν_{1t}	1	1	3	3	4	6	6	6	6	6	5	5	4	4	3	3	3	3	2	1
ν_{2t}	1	1	2	3	3	4	4	4	4	4	3	3	3	3	2	1	1	1	1	1
ν_{3t}	0	0	1	2	2	3	3	3	3	3	3	3	2	2	2	1	1	1	1	1

Applying the algorithm using the function

$$g(\nu_{i,t+1}, \ldots, \nu_{i,t+p_j}) = \Big(\sum_{l=1}^{p_j} \nu_{i,t+l}\Big)/p_j$$

yields the schedule depicted in Figure 9.1.

9.4 Timetabling with Workforce Constraints

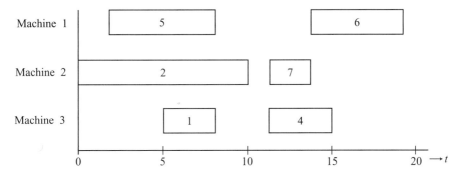

Fig. 9.1. Schedule in Example 9.3.2

Activity	Resource	Period
7	2	11-14
6	1	14-19
1	3	5-8
4	3	11-15
5	1	2-8
2	2	0-10

It turns out that activity 3 (the last activity) does not fit into the schedule. However, in the optimal schedule all activities are assigned (activity 7 starts with resource 1 at time 10 and activity 3 starts with resource 2 at time 11). So the heuristic yields in this case a suboptimal solution.

The function

$$f(w_j/p_j, |M_j|) = \frac{|M_j|^2}{w_j/p_j},$$

yields the same schedule as the one described above (although the sequence in which the activities are put on the resources is slightly different, the final result is still the same).

The function

$$f(w_j/p_j, |M_j|) = \frac{\sqrt{|M_j|}}{w_j/p_j}$$

yields a schedule with activity 3 assigned. However, now activity 5 ends up unassigned. Noting that $w_5 = 1$ and $w_3 = 3$, this last schedule is actually better than the previous one, but still not optimal.

9.4 Timetabling with Workforce Constraints

Consider now an infinite number of identical resources in parallel. There are n activities and all activities have to be done. Activity j can be done by or on any one of the resources, but once the activity has started it has to proceed without interruption until it is completed. There is a workforce that consists of W identical operators. In order to do activity j it is necessary to have W_j operators at hand. If the sum of the requirements of activities j and k is larger than W, i.e., $W_j + W_k > W$, then activities j and k cannot be done at the same time. Actually, the sum of the requirements of any set of activities that are being done at the same time may not exceed W. It is clear that this problem is a special case of the workforce constrained project scheduling problem described in Section 4.6.

Such a model with workforce constraints can be used for workforce scheduling applications. However, workforce scheduling problems in general tend to be more complicated and will be discussed in more detail in Chapter 12.

Example 9.4.1 (Project Management in the Construction Industry). A contractor has to complete n activities. The duration of activity j is p_j and it requires a crew of size W_j. The activities are not subject to precedence constraints. The contractor has W workers at his disposal and his objective is to complete all n activities in minimum time.

Consider now the following special case of the workforce constrained timetabling problem with the number of available resources being unlimited and all activities having the same duration. The activities are not subject to precedence constraints but are subject to workforce constraints and the objective is to minimize the makespan. This problem is equivalent to a famous combinatorial problem known as the *bin packing* problem. In the bin packing problem each bin has capacity W and activity j is equivalent to an item of size W_j. Each bin corresponds to one time slot and the items packed in one bin correspond to the activities done in that time slot. The objective is to pack all the items in a minimum number of bins. This problem has many applications in practice.

Example 9.4.2 (Exam Scheduling). All the exams in a community college have the same duration. The exams have to be held in a gym with W seats. The enrollment in course j is W_j and all W_j students have to take the exam at the same time. The goal is to develop a timetable that schedules all n exams in minimum time.

This workforce constrained scheduling problem with all activities having the same duration is known to be NP-hard (even in the absence of precedence constraints). However, a number of heuristics have been developed that perform reasonably well.

The *First Fit (FF)* heuristic first orders the activities (items) in an arbitrary way. The slots (bins) are numbered $1, 2, 3, \ldots$ The procedure starts at the beginning of the activity list and checks whether the activity fits in slot 1. If it fits, it is inserted there. Otherwise, the procedure checks whether it fits in slot 2, and so on. It has been shown that for any instance of the problem

9.4 Timetabling with Workforce Constraints

$$C_{\max}(FF) \leq \frac{17}{10} C_{\max}(OPT) + 2,$$

where $C_{\max}(FF)$ denotes the makespan under the FF rule and $C_{\max}(OPT)$ denotes the makespan under a (possibly unknown) optimal rule. It is easy to find instances for which

$$\frac{C_{\max}(FF)}{C_{\max}(OPT)} = \frac{5}{3}.$$

Example 9.4.3 (Application of the FF Heuristic). Let $W = 2100$. There are 18 activities.

activities	$1, \ldots, 6$	$7, \ldots, 12$	$13, \ldots, 18$
W_j	301	701	1051

Under the optimal schedule the makespan $C_{\max}$ is 6 and under the FF schedule the makespan is equal to 10. The optimal schedule assigns to each one of the six slots three activities: one of 301, one of 701 and one of 1051.

The FF rule assigns six activities of 301 to slot 1. To each one of the next three slots it assigns two activities of 701. To each one of the last six slots it assigns a single activity of 1051.

This example shows that an FF schedule may be far from optimal when the activities initially are listed in a haphazard way. If the activities are ordered initially in a clever way the First Fit heuristic may perform better. The worst case performance of the next heuristic, that is based on this idea, is significantly better.

The *First Fit Decreasing (FFD)* heuristic first orders the activities in decreasing order of W_j. The slots are again numbered $1, 2, 3$, and so on. The procedure starts at the beginning of the activity list and checks whether the activity fits in slot 1. If it fits, it is inserted. Otherwise, the procedure checks whether the activity fits in slot 2, and so on.

It has been shown that for any instance of the problem

$$C_{\max}(FFD) \leq \frac{11}{9} C_{\max}(OPT) + 4,$$

where $C_{\max}(FFD)$ denotes the makespan under the FFD rule. There are instances for which

$$\frac{C_{\max}(FFD)}{C_{\max}(OPT)} = \frac{11}{9}.$$

The following example shows how this worst case bound can be attained.

Example 9.4.4 (Application of the FFD Heuristic). Let $W = 1000$.

activities	$1, \ldots, 6$	$7, \ldots, 12$	$13, \ldots, 18$	$19, \ldots, 30$
W_j	501	252	251	248

Under the optimal schedule the makespan $C_{\max}$ is 9 and under the FFD schedule the makespan is 11. The optimal schedule assigns to each one of the first six slots three activities: one of 501, one of 251 and one of 248. To each one of the remaining three slots it assigns four activities: two of 252 and two of 248.

The FFD rule assigns to each one of the first six slots one activity of 501 and one activity of 252. To the next two slots it assigns three activities of 251. To each one of the last three slots it assigns four activities of 248.

The two heuristics described above can be applied with some minor modifications to cases where the activities have different release dates. If the activities have due dates and the objective is the minimization of a due date related penalty function, then a different heuristic is required. If the activities have deadlines and the objective is to find a feasible schedule, then also additional modifications are needed.

9.5 Timetabling with Operator or Tooling Constraints

In the previous section we considered models with W identical operators. The operators were basically interchangeable.

In what follows the operators are not identical. Each operator is unique, has his own identity and his own skill. An operator in this model may actually be equivalent to a specific piece of machinery, fixture, or tool. An activity either needs or does not need any given operator or tool. Each activity needs for its execution a specific set of different operators and/or tools.

The difference between timetabling with operator or tooling constraints and timetabling with workforce constraints is significant. In one sense the model in the previous section is more restrictive (there is only one type of operator), and in another sense it is more general (there are a number of that type of operator available).

Each activity now requires one or more different operators or tools. If two activities require the same operator, then they cannot be done at the same time. In the feasibility version of this problem, the goal is to find a schedule or timetable that completes all n activities within the time horizon H. In the optimization version, the objective is to do all the activities and minimize the makespan. It is easy to see that this problem is also a special case of the workforce constrained project scheduling problem discussed in Section 4.6. It is a project scheduling problem with workforce constraints, no precedence constraints, and $W_i = 1$ for $i = 1, \ldots, N$.

Throughout this section we assume that all activity durations are equal to 1. Even the special case with all activity durations being equal does not have an easy solution. In what follows we first focus on the feasibility version when all durations are equal to 1. Finding for this case a conflict-free timetable is structurally equivalent to the node coloring problem described at the end of

9.5 Timetabling with Operator or Tooling Constraints

Section 9.2. In the node coloring problem a graph is constructed by representing each activity as a node. Two nodes are connected by an arc if the two activities require the same operator(s). The two activities, therefore, cannot be scheduled in the same time slot. If the length of the time horizon is H time slots, then the question is: can the nodes in the graph be colored with H different colors in such a way that no two connected nodes receive the same color? This is a feasibility problem. The associated optimization problem is to determine the minimum number of colors needed to color the nodes of the graph in such a way that no two connected nodes have the same color. This minimum number of colors is referred to as the chromatic number of the graph and is equivalent to the makespan in the timetabling problem.

The optimization version of the timetabling problem with all durations being equal to 1 is closely related to the zero slack reservation problem with *arbitrary* durations described at the end of Section 9.2. That this reservation problem (with the number of resources being minimized) is not equivalent to the timetabling problem but rather a special case can be shown as follows: two activities that need the same operator in the timetabling problem are equivalent to two activities that have an overlapping time slot in the reservation problem. If two activities in the reservation problem have an overlapping time slot, then the two nodes are connected. Each color in the coloring process represents a resource and minimizing the number of colors is equivalent to minimizing the number of resources in the reservation problem. That the reservation problem is a special case follows from the fact that the time slots required by an activity in a reservation problem are *adjacent*. However, it may not be possible to order the tools in the timetabling problem in such a way that the tools required for each activity are adjacent to one another. It is this adjacency property that makes the reservation problem easy, while the lack of adjacency makes the timetabling problem with operator constraints hard.

There are a number of heuristics for this timetabling problem with durations equal to 1. In this section we describe only one such procedure. First some graph theory terminology is needed. The degree of a node is the number of arcs connected to a node. In a partially colored graph, the saturation level of a node is the number of differently colored nodes already connected to it. In the coloring process, the first color to be used is labeled Color 1, the second Color 2, and so on.

Algorithm 9.5.1 (Graph Coloring Heuristic).

Step 1.

Arrange the nodes in decreasing order of their degree.

Step 2.

Color a node of maximal degree with Color 1.

Step 3.

Choose an uncolored node with maximal saturation level.

If there is a tie, choose any one of the nodes with maximal degree in the uncolored subgraph.

Step 4.

Color the selected node with the color with the lowest possible number.

Step 5.

If all nodes are colored, STOP. Otherwise go to Step 3.

Example 9.5.2 (Application of the Graph Coloring Heuristic). Gary, Hamilton, Izak and Reha are university professors attending a national conference. During this conference seven one hour meetings have to be scheduled in such a way that each one of the four professors can be present at all the meetings he has to attend. The goal is to schedule all seven meetings in a single afternoon between 2 p.m. and 6 p.m.

meetings	1 2 3 4 5 6 7
Gary	1 0 0 1 1 0 1
Hamilton	1 1 1 0 0 0 0
Izak	0 0 1 0 1 1 0
Reha	1 0 1 1 1 0 0

This problem can be transformed into a timetabling problem with operator constraints by assuming that the seven meetings are activities and the four professors are operators. Consider the following set of data.

activities	1 2 3 4 5 6 7
operator 1	1 0 0 1 1 0 1
operator 2	1 1 1 0 0 0 0
operator 3	0 0 1 0 1 1 0
operator 4	1 0 1 1 1 0 0

If the activities are regarded as nodes, then their degrees can be computed (see Figure 9.2).

activities (nodes)	1 2 3 4 5 6 7
degree	5 2 5 4 5 2 3

Based on the degrees, activity 5 may be colored first, say with the color red (Color 1). The saturation levels of all nodes connected to node 5, i.e., nodes 1, 3, 4, 6, 7, are equal to 1. Of these nodes, nodes 1 and 3 have the highest degrees. Color node 3 blue (Color 2). The saturation levels and the degrees in the uncolored subgraph are presented in the table below.

activities (nodes)	1 2 3 4 5 6 7
saturation level	2 1 - 2 - 2 1
degree	3 1 - 2 - 0 2

9.5 Timetabling with Operator or Tooling Constraints

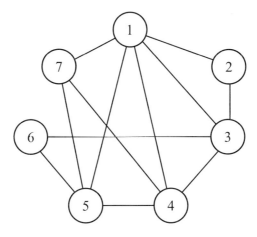

Fig. 9.2. Graph in Example 9.5.2

Based on these numbers node 1 is selected as the node to be colored next, say yellow (Color 3). Node 4 is selected after that and colored green (Color 4). Node 7 follows and is colored with the color that has the lowest number, Color 2 (blue). Node 6 is colored last and colored yellow. Since four colors were needed to color the graph, the makespan of the corresponding schedule is equal to 4. It can easily be seen that this schedule is optimal. Both operators 1 and 4 are needed for 4 activities.

To see why it makes sense to schedule the activity with the highest degree first, consider scheduling the activity with the lowest degree first. Activities 2, 6, and 7 then have to be done in the same time slot. However, activities 4, 1, 3, and 5 are scheduled afterwards in four different time slots and the makespan is 5. If the high degree activities are not scheduled early on, they often end up requiring new colors at the end of the process.

The next example illustrates the relationship between the reservation problem and the timetabling problem.

Example 9.5.3 (Timetabling Compared to Reservations). Consider the following timetabling problem.

activities	1 2 3 4 5 6 7
p_j	1 1 1 1 1 1 1
operator 1	1 0 1 1 1 0 1
operator 2	1 1 1 0 0 0 0
operator 3	0 0 1 0 1 1 0
operator 4	1 0 1 1 1 0 0

Note that the only difference between this example and the previous one is that activity 3 now needs all four operators. The operators can be transformed into time slots as follows. Operators 3 and 4 are equivalent to time slots 1 and 2 and operators 1 and 2 are equivalent to time slots 3 and 4. The time slots required by each activity are now contigious (i.e., adjacent) and the problem is equivalent to a reservation problem.

Consider now a more general timetabling model with all activities again having duration 1. There are a number of feasible slots. However, there is an aversion cost c'_{jt} for assigning activity j to slot t. There is also a proximity cost for scheduling two conflicting activities (that require the same operator) too close to one another. The penalty for scheduling two conflicting activities ℓ slots apart is $\psi(\ell)$, where $\psi(\ell)$ is decreasing in ℓ. The objective function is, for any schedule, the sum of these two costs for every occurrence. The following multi-pass heuristic is applicable to this more general timetabling problem.

Algorithm 9.5.4 (Minimizing Timetabling Costs).

Step 1.

Take activity j from the set of activities not yet scheduled.

Step 2.

Find all feasible slots where activity j can be assigned, i.e., where there is no operator conflict.

If no such slot is found go to Step 4.

Step 3.

For each feasible slot, compute the increase in the cost function (aversion as well as proximity costs);

Assign activity j to the slot with the lowest cost.

Go to Step 1.

Step 4.

Activity j conflicts in every possible slot with other activities.

Find the slots in which activity j can be scheduled by rescheduling all activities that conflict with j.

If there are no such slots go to Step 6. Otherwise go to Step 5.

Step 5.

For each slot calculate the cost of rescheduling all conflicting activities.

Assign activity j to that slot with the lowest rescheduling cost.

Step 6.

If there are no slots for which conflicting activities can be rescheduled without conflict, count for each slot the number of activities that cannot be rescheduled.

Assign activity j to that slot with the least number of such conflicts.

Reschedule as many conflicts as possible and bump the others back on the list of unscheduled activities.

If the number of times activity k is bumped by the same activity j reaches N, then activity k is dropped and considered unschedulable.

The last step of the algorithm can be viewed as a bounded backtracking mechanism that allows it to reconsider earlier decisions. If a pair of activities is difficult to schedule, then they most likely will bump each other relatively early in the backtracking process. The next example illustrates the manner in which the backtracking mechanism works when the algorithm is applied to the more specific model described earlier in this section.

Example 9.5.5 (Minimizing Timetabling Costs). Consider the instance discussed in Example 9.5.2. Assume that the number of feasible slots is 4 (this implies that the algorithm will either find the optimal schedule or it will conclude that there is no feasible schedule). Assume that all aversion costs and proximity costs are zero.

Since Algorithm 9.5.4 does not specify the order in which the unscheduled activities are taken, we assume here that the activities are going to be considered in the order

$$2, 6, 7, 4, 1, 3, 5.$$

Going through Steps 1, 2, and 3 a number of times results in activities 2, 6, and 7 being done in slot [0, 1], activity 4 in slot [1, 2], activity 1 in [2, 3], and activity 3 in [3, 4]. However, when the algorithm attempts to insert the last activity, activity 5, in one of the four slots, then there are conflicts in each one of them. If activity 5 is put in the slot of activity 4, then activity 4 has to be rescheduled, but activity 4 cannot be rescheduled in any of the other three slots. The same thing happens if activity 5 is assigned to the third or fourth slot. So the first slot remains to be checked. If activity 5 is inserted in the first slot, then activities 6 and 7 are in conflict and have to be rescheduled. Activity 6 can be scheduled together with activity 4 in the second slot and activity 7 can be scheduled together with activity 3 in the fourth slot. So the optimal schedule has been obtained.

9.6 Assigning Classes to Rooms at U.C. Berkeley

The University of California at Berkeley enrolls about 30,000 students in over 80 academic departments. Each semester, all departments provide the scheduling office an estimated enrollment, a requested meeting time, and special requirements (e.g., with regard to audiovisual equipment) for each section of each course. The scheduling office must assign 4000 classes to about 250 classrooms. The office consists of three schedulers (one for each academic unit) and one supervisor.

The assignment has to take a number of objectives into consideration. A room with fewer seats than students is undesirable, as is a room that is much too large. In addition, some courses require special equipment. The location of the room is also important. From a professor's point of view, it is nice to have a room that is close to his or her office. From a students' point of view it is convenient to have consecutive classes close together.

It is not easy to state a formal objective for this optimization problem, since there are often no clear priorities. For example, if there is no room to accomodate both Chemistry 201 and Russian 101 at the same time, then it is not easy to make a choice based on some general principle. Fortunately, some policy guidelines had been established by a campus committee. The policy guidelines are based on standard time patterns for offering courses. The nine-hour day, starting at 8 a.m., is partitioned into nine one hour blocks and, at the same time, also into 6 one and one-half-hour blocks. Classes may be scheduled only for whole blocks. Certain time blocks are defined as prime time and departments may not request more than 60% of their courses during prime time. Standard courses have priority over nonstandard courses. These policies provide some means of deciding which courses should not be assigned during overloaded time blocks. However, they still do not provide a watertight method for resolving conflicts between departments.

This classroom assignment problem can be formulated as a large $0-1$ integer programming problem. The objective function of this integer program is rather complicated and contains many terms. First, there is a penalty associated for not assigning a class at all. By making this penalty large relative to the other terms in the objective, the total number of unassigned classes is minimized. The cost terms in the objective associated with the assignment variables account for distances, overutilized facilities, and empty seats.

Since the integer program is huge (approximately 500,000 variables and 30,000 constraints), it is solved heuristically even though the problem does not have to be solved in real time. The heuristic works in a sequential manner and is based on the principle of always solving the hardest remaining subproblem next.

Some notation is needed in order to describe the heuristic. Let J denote the set of all classes to be scheduled. Let t denote a timeslot and let J_t denote the set of all classes for timeslot t. Let M denote the set of all classrooms and let M_j denote the set of all classrooms that can accomodate class j.

The heuristic can be summarized in the following four steps.

Algorithm 9.6.1 (Room Assignment Heuristic).

Step 1. *(Select Time Slot)*

   Select, among time slots not yet considered, slot t with the smallest supply/demand ratio.

Step 2. *(Greedy Algorithm)*

   Rank all classes j in J_t in decreasing order of class size.

9.6 Assigning Classes to Rooms at U.C. Berkeley

Go in a single pass through the list of classes, and assign class j to the (still vacant) room in M_j with lowest cost.

Step 3. *(Improvement Phase)*

Rank all classes j in J_t in decreasing order of current cost.

Go in a single pass through the list of classes and do the following: If class j is not assigned, find all feasible interchanges in which class j moves into an occupied room displacing the assigned class k into a vacant room; if this set is not empty, make the interchange with maximum cost reduction.

If class j is assigned, find the set of feasible assignment interchanges for class j that reduce total cost; if this set is not empty, apply the interchange with the maximum cost reduction.

Step 4. *(Stopping Criterion)*

If Step 3 has resulted in a reduction of the total cost return to Step 3; otherwise delete from the unscheduled list all classes scheduled for the current time slot.

If all time slots have been considered STOP, otherwise go to Step 1.

The heuristic has proven to be a very fast method for generating near-optimal solutions. Combining the rule of "selecting the hardest subproblem next" with a dynamic recalculation of the costs of wasted resources seems very effective. The decision support system is designed so that it is easy to use interactively and it is flexible enough to accomodate future policy modifications without extensive reprogramming.

The system is used in the following manner. Approximately six months before the start of the semester, departments submit room request forms that list all classes scheduled for the semester. Within a couple of weeks, a preliminary schedule is generated, showing those classes that could not be assigned to rooms. Departments then submit revised requests and negotiate with the scheduling office about possible pre-assignments. The system is then run again, with the unchanged standard lectures that had already been assigned flagged as pre-assigned. The resulting set of assignments is then published in time for pre-enrollment. Using the system, the scheduling office is able to complete its part of the cycle several weeks earlier than with the original manual procedure.

The system has been used for a number of years. A number of factors contributed to the success of the system. The most important one is a flexible user interface. While an optimization model is being used, its behavior is easily altered to explore different trade-off strategies. Furthermore, the system is designed so that partial solutions, in the form of easily generated pre-assignments, can be incorporated, while allowing the heuristic to generate a solution for the remaining problem. Special needs that were not anticipated when the model was designed can be accomodated and the schedulers can evaluate their own heuristics.

This classroom assignment problem is very similar to the reservation problems with slack and without slack described in Sections 9.2 and 9.3. The meeting rooms are the resources and activity j can only be assigned to a resource belonging to the subset M_j. However, in the classroom assignment problem *all* activities have to be scheduled. The multi-pass heuristic implemented here is clearly more sophisticated than the one-pass heuristic described in Algorithm 9.3.1.

9.7 Discussion

Even though the four sections of this chapter deal with four different types of scheduling and timetabling problems, it is not hard to imagine real world scheduling problems that have all the features discussed in this chapter, i.e., release dates and due dates (with or without slack), workforce constraints as well as operator constraints. The objective may also be a combination of the minimization of the makespan and the maximization of the (weighted) number of activities done. The separate analyses of these four different aspects give an indication of how hard real world problems can be.

The interval scheduling models and reservation models discussed in this chapter are relatively simple. They only give a flavor of the thinking behind these problems. A company that has to deal with these types of problems typically relies on models that are significantly more complicated. First of all, the models have to be dynamic rather than static. Calls for reservations come in continuously and the decision making process embedded in the company's systems depends heavily on forecasts of future demands. Second, the reservation systems depend on the existing pricing structure and the pricing structure depends on the current occupancy as well as on forecast demand. A significant amount of research has been done recently on reservation models that include a pricing mechanism. These models are beyond the scope of this book.

A natural generalization of timetabling problems with operator constraints and timetabling problems with workforce constraints is the following: Suppose there are a number of different types of operators, say N. A limited number W_i of operators of type i, $i = 1, \ldots, N$, are available. Doing activity j requires W_{ij} operators of type i. The activities have to be scheduled subject to the operator availability constraints. The FF and FFD rules can be adapted to this situation with multiple types of operators. This problem is basically equivalent to the problem discussed in Section 4.6 without precedence constraints.

In this chapter we have not considered any preemptions. When the durations of the activities are not equal to 1, preemptions may improve the performance measures. However, heuristics for preeemptive models may be very different from those for nonpreemptive models.

Exercises

9.1. Consider the following reservation problem with 10 activities and zero slack. There are three identical resources in parallel.

activities	1	2	3	4	5	6	7	8	9	10
p_j	6	1	4	2	3	3	6	2	1	3
r_j	2	7	5	2	1	0	4	8	0	0
d_j	8	8	9	4	4	3	10	10	1	3

(a) Apply Algorithm 9.2.1 to find the schedule with the maximum number of activities done.

(b) Find the schedule that maximizes the total amount of processing (i.e., the sum of the durations of the activities done).

(c) What is the minimum number of resources needed to satisfy the total demand?

9.2. Consider a car rental agency where the following 10 reservations have been made.

reservations	1	2	3	4	5	6	7	8	9	10
p_j	3	1	4	2	3	1	1	3	3	3
r_j	1	3	0	1	0	0	2	3	4	2
d_j	4	4	4	3	3	1	3	6	7	5

Determine the minimum number of cars needed to satisfy the demand.

9.3. Consider the following instance with eight activities and three resources.

activities	1	2	3	4	5	6	7	8
p_j	4	3	10	9	4	6	5	3
w_j	3	2	3	3	2	1	2	3
r_j	8	5	0	2	3	2	4	5
d_j	12	12	10	20	15	18	19	14
M_j	{2}	{1,3}	{1,2}	{1,2,3}	{2,3}	{1}	{1}	{1,2}

(a) Select appropriate index functions I_j and $g(\nu_{i,t+1}, \ldots, \nu_{i,t+p_j})$ and apply Algorithm 9.3.1.

(b) Is the solution obtained optimal? If not, can you modify the algorithm to obtain an optimal solution?

9.4. Consider a hotel with two types of rooms: suites and regular rooms. There are n_1 suites and n_2 regular rooms. If someone wants a suite, then the hotel makes a profit of w_1 dollars per night. If someone wants a regular room, the hotel makes a profit of w_2 dollars per night ($w_2 < w_1$). If a person wants a regular room and all regular rooms are taken, then the hotel can put that person up in a suite. However, the hotel makes then only w_2 dollars per night.

(a) Assume that the hotel cannot ask the guest to change rooms in the middle of his stay; once assigned to a room or suite the guest will stay there until he leaves (i.e., preemptions are not allowed). Explain how this problem fits the framework described in Section 9.2.

(b) Assume that the hotel can ask a guest to change rooms in the middle of his stay. How does this affect the problem? Can the hotel rent out more rooms this way? Illustrate your answer with a numerical example.

9.5. Design a heuristic for the reservation problem described in part (a) of the previous exercise. If guest j requests a regular room then the weight is $p_j w_2$ (even if he is put up in a suite); if he requests a suite his weight is $p_j w_1$. Apply your heuristic to the data set below.

guests	1	2	3	4	5	6	7	8	9
p_j	4	3	10	9	4	6	5	3	4
type	1	1	2	2	2	2	2	2	2
r_j	8	5	0	2	3	2	4	5	7
d_j	12	12	10	15	15	15	13	14	13

There are 3 regular rooms and 1 suite. The value $w_1 = 3$ and the value $w_2 = 2$. Check if your heuristic yields the optimal solution.

9.6. Consider a hotel with m identical rooms. Arriving guests request either a single room or two rooms for a certain period. If the hotel cannot provide two rooms to families that request two rooms, then they go elsewhere. Design a heuristic that has as goal to maximize the number of rooms rented out over time.

9.7. Consider again the instance described in Exercise 9.1. Assume now that it is possible to do an activity either before its release date or after its due date. Doing an activity for τ time units before its actual release date costs $c_1\tau$, where $c_1 = 1$, and doing an activity for τ time units after its due date costs $c_2\tau$, where $c_2 = 3$. Assume that now all the activities *must* be done.

(a) Develop a heuristic to find a low cost schedule.
(b) Apply the heuristic to the instance of Exercise 9.1 with three resources.
(c) Apply the heuristic to the instance of Exercise 9.1 with two resources.

9.8. Consider a timetabling problem with all processing times equal to 1 and tooling constraints. Let T_j denote the set of tools that are required for the processing of job j. Assume that the T_j sets are nested. That is, for any pair of sets T_j and T_k one and only one of the following three statements hold:

(i) Sets T_j and T_k are identical.
(ii) Set T_j is a subset of set T_k.
(iii) Set T_k is a subset of T_j.

(a) Develop a polynomial time algorithm for this problem.

(b) Does the equivalent reservation problem have any special properties?

9.9. Consider the following timetabling problem with tool sets.

jobs	1 2 3 4 5
p_j	1 1 1 1 1
tool 1	1 0 0 1 0
tool 2	0 1 1 0 0
tool 3	1 1 0 1 1
tool 4	0 1 1 1 1

(a) Can the tool sets be numbered in such a way that all the tools needed by each job are adjacent?

(b) Develop an algorithm for verifying whether the tool sets can be numbered in that way.

9.10. Consider the following timetabling problem with two types of personnel. The total number of personnel of type 1 is $W_1 = 3$ and the total number of personnel of type 2 is $W_2 = 4$.

activities	1 2 3 4 5 6 7
p_j	1 1 1 1 1 1 1
W_{1j}	2 0 1 2 1 2 1
W_{2j}	2 4 0 2 3 1 2

(a) Determine first which type of personnel is the most critical (the tightest).

(b) Use the information under (a) to develop a heuristic for this problem with two types of personnel (your heuristic may be a generalization of the FFD heuristic described in Section 9.5).

(c) Give a numerical example of bad behavior of your heuristic.

9.11. Consider the workforce constrained timetabling problem with the activities having arbitrary processing times. In this problem the makespan has to be minimized subject to the workforce constraint W. Consider now the dual of this problem. All activities have to be completed by a fixed time C_{max}. However, the maximum number of people used for these activities has to be minimized. Prove or disprove any mathematical relationship between these two problems.

Comments and References

The first part of Section 9.2 is based primarily on the paper by Bouzina and Emmons (1996). The last part of Section 9.2 that focuses on the processing of all activities with a minimum number of resources is (as described at the end of the section) a special case of a node coloring problem on an arbitrary graph. This problem is actually equivalent to a node coloring problem on a graph with a certain special property; the graphs that satisfy this special property are usually referred to as interval graphs.

A fair amount of research has been done on single and parallel machine models with release dates and due dates or deadlines. A partial list includes McNaughton (1959), Moore (1968), Emmons (1969), Sahni and Cho (1979), Potts (1980), Garey, Johnson, Simons, and Tarjan (1981), Martel (1982a, 1982b), Posner (1985), Potts and van Wassenhove (1988), Ow and Morton (1989), Hall and Posner (1991), Hall, Kubiak and Sethi (1991), and Dondeti and Emmons (1992).

Section 4 of this chapter is based on the research that has been done in bin packing. A very good overview of bin packing results is presented in Garey and Johnson (1979).

A significant amount of research has been done on the timetabling problem with operator or tooling constraints and the problems that are equivalent, i.e., graph coloring and examination timetabling. Brelaz (1979) presents a number of methods to color the nodes of a graph. Algorithm 9.5.4 is due to Laporte and Desroches (1984). Carter (1986) presents a survey of practical applications of examination timetabling algorithms. For excellent overviews of the recent developments in timetabling, see Burke and Ross (1996), Burke and Carter (1998), Burke and Erben (2001) and Burke and De Causmaecker (2003). For an in-depth analysis of local search techniques applied to course timetabling and examination timetabling problems, see Di Gaspero (2003).

The class assignment system developed at UC Berkeley is described in detail by Glassey and Mizrach (1986). Mulvey (1982), Gosselin and Truchon (1986), and Carter and Tovey (1992) also consider the classroom assignment problem.

Chapter 10

Scheduling and Timetabling in Sports and Entertainment

10.1 Introduction 229
10.2 Scheduling and Timetabling in Sport
 Tournaments 230
10.3 Tournament Scheduling and Constraint
 Programming 237
10.4 Tournament Scheduling and Local Search 240
10.5 Scheduling Network Television Programs 243
10.6 Scheduling a College Basketball Conference 245
10.7 Discussion 248

10.1 Introduction

The previous chapter covered the basics of interval scheduling and timetabling. The models considered were relatively simple and their main goal was to provide some insights. In practice, there are many, more complicated applications of interval scheduling and timetabling. For example, there are important applications in sports as well as in entertainment, e.g., the scheduling of games in tournaments and the scheduling of commercials on network television.

This chapter covers basically two topics, namely tournament scheduling and the scheduling of programs on network television. These two topics turn out to be somewhat related. The next section focuses on some theoretical properties of tournament schedules that are prevalent in U.S. college basketball, major league baseball, and European soccer; this section also presents a general framework for tackling the associated optimization problems via integer programming. The third section describes a completely different procedure for dealing with the same problem, namely the constraint programming approach. The fourth section looks at two tournament scheduling problems that are slightly different from the one discussed in the second and third section, and the solution techniques used are based on local search. The fifth section considers a scheduling problem in network television, i.e., how to schedule the programs in order to maximize overall ratings. The subsequent section contains a case study on tournament scheduling; the tournament considered being a college basketball conference. This particular tournament scheduling problem has been tackled with integer programming as well as with constraint programming techniques. The last section discusses the similarities and differences between the different models and approaches considered in this chapter.

10.2 Scheduling and Timetabling in Sport Tournaments

Many tournament schedules are constrained in time; that is, the number of rounds or slots in which games are played is equal to the number of games each team must play plus some extra rounds or slots that are typically required in leagues with an odd number of teams. For example, in a so-called single round robin tournament each team has to play every other team once, either at home or away. Such a tournament among n teams with n being even requires $n-1$ rounds. If the number of teams is odd, then the number of rounds is n (due to the fact that in every round one of the teams has to remain idle). In a double round robin tournament each team has to play every other team twice, once at home and once away. It turns out that such a tournament among n teams requires either $2n-2$ or $2n$ rounds (dependent upon whether n is even or odd).

In order to formulate the most basic version of a tournament scheduling problem certain assumptions have to be made. Assume for the time being that the number of teams, n, is even. (It happens to be the case that tournament scheduling with an even number of teams is slightly easier than with an odd number of teams.) Consider a single round robin tournament in which each team has to play every other team exactly once, i.e., each team plays $n-1$ games. Because of the fact that there are an even number of teams it is possible to create for such a tournament a schedule that consists exactly of $n-1$ rounds with each round having $n/2$ games.

More formally, let t denote a round (i.e., a date or a time slot) in the competition. The $0-1$ variable x_{ijt} is 1 if team i plays at home against team j in round t; the variable x_{ijt} is 0 otherwise. Of all the x_{ijt} variables a total of $(n/2)(n-1)$ are 1; the remaining are 0. The following constraints have to be satisfied:

$$\sum_{i=1}^{n}(x_{ijt}+x_{jit})=1 \quad j=1,\ldots,n;\ t=1,\ldots,n-1,$$

$$\sum_{t=1}^{n-1}(x_{ijt}+x_{jit})=1 \quad \text{for all } i \neq j.$$

10.2 Scheduling and Timetabling in Sport Tournaments

In practice, there are usually many additional constraints concerning the pairing of teams and the sequencing of games; examples of such constraints are described in the case study at the end of this chapter. When there are a large number of constraints, one may just want to end up with a *feasible* schedule. Finding a feasible schedule may already be hard.

However, it may at times also occur that one would like to optimize an objective. In order to formulate one of the more common objective functions in tournament scheduling some terminology is needed. If one considers the sequence of games played by a given team, each game can be characterized as either a *Home (H)* game or as an *Away (A)* game. The pattern of games played by a given team can thus be characterized by a string of H's and A's, e.g., $HAHAA$. There is typically a desire to have for any given team the home games and the away games alternate. That is, if a team plays one game at home, it is preferable to have the next game away and vice versa. If a team plays in rounds t and $t+1$ either two consecutive games at home or two consecutive games away, then the team is said to have a *break* in round $t+1$. A common objective in tournament scheduling is to minimize the total number of breaks. It has been shown in the literature that in any timetable for a single round robin tournament with n teams (n being even), the minimum number of breaks is $n-2$. The algorithm that generates a schedule with this minimum number of breaks is constructive and very efficient.

Example 10.2.1 (Breaks in a Single Round Robin Tournament). Consider 6 teams and 5 rounds.

	round 1	round 2	round 3	round 4	round 5
team 1	(-6)	**(-3)	**(-5)	(2)	(-4)
team 2	(-5)	(6)	**(4)	(-1)	(3)
team 3	(4)	**(1)	(-6)	(5)	(-2)
team 4	(-3)	(5)	(-2)	(6)	**(1)
team 5	(2)	(-4)	(1)	(-3)	**(-6)
team 6	(1)	(-2)	(3)	(-4)	(5)

When team i plays in round t against team j and the game is entered in the table as (j), then team i plays at the site of team j. If it is entered as $(-j)$, then team i plays at home. The timetable shown above has 6 breaks, each of them marked with a **. Since there are 6 teams, it may be possible to find for this tournament a schedule with 4 breaks (see Exercise 10.4).

Assume now that the number of teams is odd. If n is odd, then the minimum number of rounds in a single round robin tournament is larger than $n-1$. If the number of teams is odd, then one team has to remain idle in each round. When a team does not play in one round, it is referred to as a *Bye (B)*. So when the number of teams is odd, the sequence of games that have to be played by a given team is a string of H's, A's, and one or more B's, e.g., $HAHABA$. With these more complicated types of patterns a break can be defined in several ways. An HBH substring may or may not be considered a break; if the B is considered equivalent to an A, then there is no break. If the B is considered equivalent to an H, then there is a break (actually, then there are two breaks). However, one can argue that an HBH pattern is less bad than an HHH pattern; one can even argue that it is less bad than an HH pattern. So, as far as penalties or costs are concerned, the cost of an HBH pattern may actually be less than the cost of a single break.

It turns out that a single round robin tournament problem with arbitrary n (n either even or odd) can be described as a graph coloring problem. This equivalence is somewhat similar to the relationships between timetabling problems and graph coloring problems described in the previous chapter; it provides some additional insight into the tournament scheduling problem as well. Consider a single round robin tournament in which each club has to face every other club once and only once; the game is either a home game (H) or an away game (A). A directed graph $G = (N, B)$ can be constructed in which set N consists of n nodes and each node corresponds to one team. Each node is linked via an arc to each other node. The arcs are initially undirected. In Figure 10.1 the n nodes are positioned in such a way that they form a polygon. If in a graph each node is connected to every other node, it is referred to as a clique or as a complete graph.

A well-known graph coloring problem concerns the coloring of the arcs in a graph; the coloring has to be done in such a way that all the arcs that are linked to any given node have different colors and the total number of colors is minimized. It is a well-known fact that a clique with n nodes can be colored this way with n colors (which is often referred to as its *chromatic number*). Each subgraph that receives a specific color consists of one arc that lies on the boundary of the polygon and a number of internal arcs (see Figure 10.1).

The equivalence between the graph coloring problem and the single round robin tournament scheduling problem is based on the fact that each round in the tournament corresponds to a subgraph with a different color. The coloring of the arcs for the different rounds thus determines a schedule for a single round robin tournament in which each team plays every other team only once. One question is how to partition the arcs into a number of subsets with each subset having a different color. A second question has to be addressed as well: when one team plays another it has to be decided at which one of the two sites the game is played, i.e., which team plays at home and which team will be away. In order to determine this, each arc in the graph has to be directed; if a game between teams i and j takes place at team j's site, then the arc linking nodes i and j emanates from i and goes to j. In order to avoid breaks in two consecutive rounds of a schedule, the arcs have to be directed in such a way that the two subgraphs corresponding to the two consecutive rounds in the timetable constitute a so-called *directed Hamiltonian path*. A directed Hamiltonian path is a path that goes from one node to another (see Figure 10.2) with each node having at most one outgoing arc and at most one incoming arc (see Exercise 10.3).

10.2 Scheduling and Timetabling in Sport Tournaments

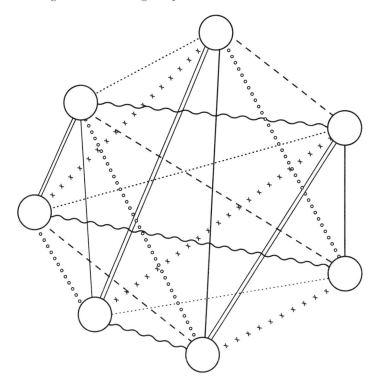

Fig. 10.1. Coloring of a Complete Graph

Many approaches for developing tournament schedules are based on a standard framework for the search for good feasible schedules. In this framework a pattern is equivalent to a string consisting of H's (Home games), A's (Aways) and B's (Byes), for example $HABAHHA$. For a single round robin tournament the length of a string is $n-1$ when the number of teams is even and n when the number of teams is odd. These strings are often referred to as Home Away Patterns (HAPs). The following three step algorithm provides a framework for generating single round robin schedules.

Algorithm 10.2.2 (Scheduling Single Round Robin Tournaments).

Step 1. *(Assemble a Collection of HAPs)*

  Find a collection of n different HAPs.

  This set of HAPs is referred to as the pattern set.

Step 2. *(Create a Timetable)*

  Assign a game to each entry in the pattern set.

  The resulting assignment is referred to as a timetable.

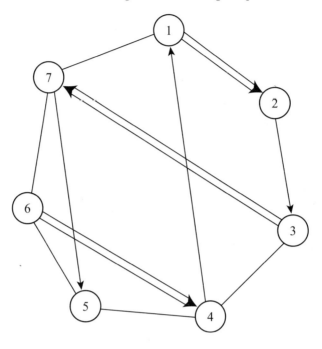

```
      1st   ROUND          2nd   ROUND
          1 ⇒ 2                  2 → 3
          7 ⇐ 3                  1 ← 4
          6 ⇒ 4                  7 → 5
          5   free               6   free
```

Hamiltonian path: 6 ⇒ 4 → 1 ⇒ 2 → 3 ⇒ 7 → 5

Fig. 10.2. Hamiltonian Path in a Graph

Step 3. *(Assign Teams to Patterns)*

Assign a team to each pattern.

Together with the timetable, this creates a single round robin schedule.

A schedule for a double round robin tournament can be created in a similar fashion. First, a single round robin tournament schedule is generated. Then a fourth step is added, which is typically referred to as the *mirroring* step. The single round robin schedule is extended by attaching immediately behind it a schedule that is exactly the same but with the home and away games reversed. The following example illustrates the use of Algorithm 10.2.2.

10.2 Scheduling and Timetabling in Sport Tournaments

Example 10.2.3 (Scheduling Round Robin Tournaments). Consider a four team single round robin tournament with teams a, b, c, and d. In Step 1 a pattern set is selected:

team 1: AHA
team 2: HAH
team 3: HHA
team 4: AAH

This means that team 1 plays the first game away, the second at home, and the third away. Teams 1, 2, 3, and 4 are in the literature referred to as *placeholders*. However, at this point it has not been specified yet which one of teams $\{a, b, c, d\}$ is team 1.

Step 2 assigns games consistent with the pattern set to get a timetable.

team 1:	(3)	(-4)	(2)
team 2:	(-4)	(3)	(-1)
team 3:	(-1)	(-2)	(4)
team 4:	(2)	(1)	(-3)

In the table above a (j) in row i means that team i is visiting team j; a $(-j)$ in row i means that team i is playing at home against team j. Note that there are two breaks in the schedule and, since there are 4 teams, this is the minimum number of breaks.

Step 3 assigns teams to patterns based on, say, their preferences for being at home in the given slots; the resulting schedule is the following.

team d:	(a)	$(-c)$	(b)
team b:	$(-c)$	(a)	$(-d)$
team a:	$(-d)$	$(-b)$	(c)
team c:	(b)	(d)	$(-a)$

This single round robin schedule can be extended to a double round robin schedule by mirroring. The final result is a schedule for the double round robin tournament.

team d:	(a)	$(-c)$	(b)	$(-a)$	(c)	$(-b)$
team b:	$(-c)$	(a)	$(-d)$	(c)	$(-a)$	(d)
team a:	$(-d)$	$(-b)$	(c)	(d)	(b)	$(-c)$
team c:	(b)	(d)	$(-a)$	$(-b)$	$(-d)$	(a)

Note that that this schedule has six breaks. Note also that the distance between two games in which the same two teams face one another is always three (i.e., before a team faces the same team again, it has to play two games against other teams).

In practice, the framework of Algorithm 10.2.2 is often used somewhat differently. In Step 1 usually more than n different HAPs are generated and based on this larger collection of HAPs more than one pattern set is created. Additional pattern sets give more choices and flexibility in the creation of timetables and schedules.

If the tournament under consideration has a large number of teams, then each one of the steps in Algorithm 10.2.2 requires a certain computational effort. There are actually various approaches that can be used for each step in Algorithm 10.2.2. Each step can be implemented following either an optimization approach or a constraint programming approach. The remaining part of this section describes the use of optimization techniques in each step of the framework; the next section focuses on the use of constraint programming techniques in each step.

In Step 1 several pattern sets can be generated by first listing all the preferred patterns (with alternating H's and A's and one B) of appropriate length. There may not be that many of such preferred patterns. A list of some of the less preferred patterns (say, with one or two breaks) is created as well. It is not likely that a set that consists only of preferred patterns will ultimately lead to an acceptable schedule. Because of this, additional pattern sets are created that contain, for example, $n - 2$ preferred patterns and two patterns that are less preferred. If we allow only a small number of less preferred patterns in a pattern set, then the number of pattern sets that can be generated is still relatively small.

Step 2 creates timetables for different teams. Determining the timetables can also be done through integer programming. It is clear that every pattern in each one of its rounds is linked to another pattern. Let S denote a set of n patterns and let T denote the set of rounds. The binary variable $x_{k\ell t}$ is 1 if the team associated with pattern k plays in round t at the site of the team associated with pattern ℓ. Of course, this variable is only defined if the kth pattern has an A in position t and the ℓth pattern has an H in position t. Let F denote the set of all feasible (k, ℓ, t) triplets. In order to find for a single round robin tournament a solution that satisfies all the constraints the following integer program can be formulated.

$$\text{minimize} \sum_{(k,\ell,t) \in F} x_{k\ell t}$$

subject to

$$\sum_{t:(k,\ell,t) \in F} x_{k\ell t} + \sum_{t:(\ell,k,t) \in F} x_{\ell k t} = 1 \qquad \text{for all } k \in S,\ \ell \in S,\ k \neq \ell$$

$$\sum_{\ell:(k,\ell,t) \in F} x_{k\ell t} + \sum_{\ell:(\ell,k,t) \in F} x_{\ell k t} \leq 1 \qquad \text{for all } k \in S,\ t \in T$$

$$x_{k\ell t} \in \{0,1\} \qquad \text{for all } (k,\ell,t) \in F$$

The first set of constraints specifies that during the tournament there will be exactly one game between teams represented by patterns k and ℓ. The second set of constraints specifies that pattern k plays at most one game in round t. (In a single round robin with an even number of teams this inequality constraint becomes an equality constraint.) The objective function for this integer program is somewhat arbitrary, because the only goal is to find a solution that satisfies all constraints.

Step 3 assigns teams to patterns. This step may in certain situations also be formulated as an integer program. Let y_{ik} denote a $0-1$ variable taking value 1 if team i is assigned to HAP k and 0 otherwise. Let c_{ik} denote the relative cost of such an assignment (this relative cost is estimated by taking all stated preferences into account). A timetable for the competition can be constructed as follows.

$$\text{minimize} \sum_{i=1}^{n} \sum_{k=1}^{n} c_{ik} y_{ik}$$

subject to

$$\sum_{i=1}^{n} y_{ik} = 1 \qquad \text{for } k = 1, \ldots, n$$

$$\sum_{k=1}^{n} y_{ik} = 1 \qquad \text{for } i = 1, \ldots, n$$

Of course, each team is assigned to one HAP and each HAP is assigned to one team. In practice, the mathematical program formulated for Step 3 is often more complicated. It is usually of a form that is referred to as a Quadratic Assignment Problem.

This approach, in which each step is based on an integer programming technique, is used in the case that is discussed later on in this chapter.

10.3 Tournament Scheduling and Constraint Programming

A completely different approach for generating tournament schedules is based on constraint programming. The reason why in practice constraint programming appears to be very suitable for tournament scheduling problems is due to the fact that acceptable schedules have to satisfy many constraints. The different types of constraints that a schedule has to adhere to may include:

(i) *Break Constraints.* No team is allowed to play three or more consecutive games at Home or three or more consecutive games Away.

(ii) *First Rounds.* Each team must have Home games or Byes in at least two of the first four rounds.

(iii) *No Two Final Aways.* No team is allowed to play Away in both of the two last rounds.

In addition to these fairly general constraints, there may be many more constraints that are either game-specific or team-specific. For example, certain popular games may have to be played within given time periods and there may even be fixed game assignments. There may also be opponent sequencing constraints; that is, if two teams in the league are considered to be very strong, then none of the other teams should have to face these two teams in consecutive rounds. In a constraint programming implementation all these constraints have to be formulated properly and stored in a constraint store (see Appendix D).

In what follows we consider a double round robin tournament with the number of teams n being odd. The total number of rounds is therefore $2n$. Each one of the steps in Algorithm 10.2.2 can be implemented using a constraint programming approach. Step 1 generates feasible pattern sets. In order to formulate the constraints needed for generating pattern sets, let h_t, a_t, and b_t denote $0-1$ variables. If a team plays at home in round t, then $h_t = 1$, and $a_t = b_t = 0$. The following constraints can be formulated with respect to potential patterns for any given team i:

$$h_t + a_t + b_t = 1 \quad \text{for } t = 1, \ldots, 2n$$
$$a_t + a_{t+1} + a_{t+2} \leq 2 \quad \text{for } t = 1, \ldots, 2n-2$$
$$h_t + h_{t+1} + h_{t+2} \leq 2 \quad \text{for } t = 1, \ldots, 2n-2$$
$$a_{n-1} + a_n \leq 1$$
$$h_1 + h_2 + h_3 + h_4$$
$$+ b_1 + b_2 + b_3 + b_4 \geq 2$$

A very basic constraint program can generate all allowable HAPs with little computational effort. Let ν denote the total number of feasible HAPs found.

Step 1 also has to aggregate these patterns into feasible pattern sets with each set containing n different HAPs. Generating these pattern sets can also be done following a contraint programming approach. In order to generate feasible pattern sets, let **H**, **A**, and **B** denote three $\nu \times 2n$ matrices with all entries being either 0 or 1; the entry h_{kt} in matrix **H** being 1 indicates that pattern k has a home game in round t, and so on. In the generation of a pattern set, each pattern k, $k = 1, \ldots, \nu$, has a $0-1$ variable x_k that indicates whether this pattern is in the pattern set or not. The following constraints have to be satisfied:

$$\sum_{k=1}^{\nu} x_k = n$$

10.3 Tournament Scheduling and Constraint Programming

$$\sum_{k=1}^{\nu} h_{kt} x_k = \lfloor n/2 \rfloor \qquad \text{for } t = 1, \ldots, n$$

$$\sum_{k=1}^{\nu} a_{kt} x_k = \lfloor n/2 \rfloor \qquad \text{for } t = 1, \ldots, n$$

$$\sum_{k=1}^{\nu} b_{kt} x_k = 1 \qquad \text{for } t = 1, \ldots, n$$

The performance of the constraint program for generating feasible pattern sets can be improved by excluding pairs of patterns that have no possible meeting date for the two corresponding teams (see Exercise 10.2).

Applying a constraint programming approach in Step 2 of Algorithm 10.2.2 (the generation of feasible timetables) requires the formulation of a different set of constraints. In order to formulate these constraints, let $\mathbf{H}^*$, $\mathbf{A}^*$, and $\mathbf{B}^*$ denote three $n \times 2n$ matrices of which the entries h_{kt}^*, a_{kt}^*, and b_{kt}^* indicate Home, Away, and Bye games in round t for pattern k. Let $\mathbf{H}^{**}$, $\mathbf{A}^{**}$, and $\mathbf{B}^{**}$ denote three $n \times 2n$ matrices of $0-1$ variables of which the entries indicate when team i plays at Home, Away, or has a Bye in round t. The variable ϕ_i, $i = 1, \ldots, n$, has a range over the integers 1 to n. If team i plays according to the pattern in row ϕ_i of $\mathbf{H}^*$, $\mathbf{A}^*$, and $\mathbf{B}^*$, then

$$H_{it}^{**} = H_{\phi_i, t}^*.$$

Let the $n \times 2n$ matrix $\mathbf{T}$ denote the target timetable. The entries g_{it} in this matrix range over the integers $0, \ldots, n$; the value of g_{it} specifies the opponent of team i in round t.

To specify the constraints on all the matrices, let the $0-1$ variable $I(x = y)$ equal 1 when $x = y$ and the $0-1$ variable $I(x \in D)$ equal 1 if x is an element of set D. The constraint

$$\text{alldifferent } (x_1, \ldots, x_m)$$

ensures that the variables $x_1, \ldots, x_m$ are distinct integers; this constraint is an example of a so-called global constraint. The constraint

$$\text{element } (k, \bar{v}, \ell)$$

applies to integers k and ℓ, and vector $\bar{v}$; it specifies that ℓ obtains the value $\bar{v}_k$, i.e., $\ell = \bar{v}_k$. Furthermore, let $\mathbf{H}_t^*$ denote the t-th column of matrix $\mathbf{H}^*$, i.e., $\mathbf{H}_t^*$ is a vector. The following constraints now have to be included in the constraint store.

$$h_{it}^{**} + a_{it}^{**} + b_{it}^{**} = 1 \qquad \text{for } i = 1, \ldots, n; \ t = 1, \ldots, n$$
$$I(g_{it} = j) = I(g_{jt} = i) \qquad \text{for } i = 1, \ldots, n; \ j = 1, \ldots, n$$

and

$$alldifferent(g_{1t},\ldots,g_{nt}) \quad \text{for } t=1,\ldots,n$$
$$element(\phi_i, \mathbf{H}_t^*, \mathbf{H}_{it}^{**}) \quad \text{for } t=1,\ldots,n;\ i=1,\ldots,n$$
$$element(\phi_i, \mathbf{A}_t^*, \mathbf{A}_{it}^{**}) \quad \text{for } t=1,\ldots,n;\ i=1,\ldots,n$$
$$element(\phi_i, \mathbf{B}_t^*, \mathbf{B}_{it}^{**}) \quad \text{for } t=1,\ldots,n;\ i=1,\ldots,n$$

The matrices $\mathbf{H}^*$, $\mathbf{A}^*$, and $\mathbf{B}^*$ are fixed and the search strategy enumerates over all the variables $\phi_1,\ldots,\phi_n$. The matrices $\mathbf{H}^{**}$, $\mathbf{A}^{**}$, and $\mathbf{B}^{**}$ are then gradually determined. By consistently propagating all the constraints that are generated by the variables that already have been fixed the search tree can be examined in a fairly short time. Finally, the g_{it} entries in $\mathbf{T}$ are enumerated. In this search also, a considerable amount of pruning of the tree can be achieved by consistently propagating all the constraints.

10.4 Tournament Scheduling and Local Search

The previous sections make it clear that the basic tournament scheduling problem is not that easy. Many heuristic techniques have been developed for this problem, including various local search techniques. This section focuses on the application of local search techniques to two variants of the basic tournament scheduling problem.

We first consider a version of the tournament scheduling problem that is only slightly different from the single round robin tournament scheduling problem discussed in the previous sections. Each team must play every other team exactly once. We assume that the number of teams n is even, implying that there are $n-1$ rounds. The difference in the model lies in the following feature: instead of a round having the Home-Away-Bye feature, a round consists now of $n/2$ different periods and each period has to be assigned one game. Suppose there are 8 teams and each round lasts a week with 4 games in each week. However, because of television broadcasting it is preferable not to schedule all four games at the same time. It makes sense to have four different *periods* in a week (e.g., friday evening, saturday afternoon, saturday evening, and sunday afternoon) in which the four different games can be played. This type of scheduling allows television viewers to see more than one game in real time.

However, the fact that a round now consists of several periods leads to new types of preferences, constraints, and objectives. For example, there may be a constraint that no team plays more than twice in the same period over the entire season; that is, the games any given team plays during the tournament should be spread out evenly over the different periods. Instead of the break minimization objective described in the previous sections, a different objective can now be formulated: let N_{iu} denote the number of games team i plays in period u. Recall that there is a preference to keep this number less than or equal to 2. The value of the objective function is now

10.4 Tournament Scheduling and Local Search

$$\sum_{i=1}^{n}\sum_{u=1}^{n/2}\max(N_{iu}-2,0)$$

Example 10.4.1. Consider an example with $n=8$ teams. The tournament lasts 7 weeks and has 4 periods in each week. A valid tournament schedule is presented in the table below.

	week 1	week 2	week 3	week 4	week 5	week 6	week 7
period 1	1-2	1-3	5-8	4-7	4-8	2-6	3-5
period 2	3-4	2-8	1-4	6-8	2-5	1-7	6-7
period 3	5-6	4-6	2-7	1-5	3-7	3-8	1-8
period 4	7-8	5-7	3-6	2-3	1-6	4-5	2-8

It can be verified easily that the schedule is feasible since each team plays at most twice in each period.

There are $(n/2)\times(n-1)$ games that have to be scheduled. Since a schedule can be thought of as being a permutation of these matches, the size of the search space is $((n/2)(n-1))!$. For any local search mechanism it is necessary to specify (i) a schedule representation, (ii) a neighbourhood design, and (iii) a search process within the neighbourhood (see Appendix C).

For this problem it is easy to design a schedule representation. The design of the neighbourhood is considerably more complicated. One type of neighbourhood can be designed by considering all games in a given week t'. All the games and the corresponding periods in that week are identified in which one of the participating teams plays more than twice in the same period over the duration of the tournament. For each period in which such a violation occurs it is also determined which team(s) play only one game in that period over the entire duration of the tournament. Using this type of information one can consider pairwise interchanges within week t' in order to reduce the value of the objective function.

Example 10.4.2. The following two tables give an example of a pairwise interchange between two games in the same week.

	week 1	week 2	week 3	week 4	week 5
period 1	1-2	2-6	3-4	5-6	↓ (2-4) ↓
period 2	4-6	1-3	2-5	1-4	3-6
period 3	3-5	4-5	1-6	2-3	↑ (1-5) ↑

Because team 2 plays three times in period 1 and team 5 plays three times in period 3 the value of the objective function is 2. Consider week 5, i.e., $t'=5$. Note that team 1 plays once in period 1 and team 4 plays once in period 3. Interchanging game (2-4) in period 1 of week 5 with game (1-5) in period 3 of week 5 results in the following timetable.

	week 1	week 2	week 3	week 4	week 5
period 1	1-2	2-6	3-4	5-6	1-5
period 2	4-6	1-3	2-5	1-4	3-6
period 3	3-5	4-5	1-6	2-3	2-4

The value of the objective function is now 0 since each team plays at most twice in every period.

A second variant of the basic tournament scheduling problem is often referred to as the Travelling Tournament Problem. In the basic tournament scheduling problem described in the previous sections the travel times were not considered significant. However, in many leagues, e.g., Major League Baseball (MLB) in the U.S., the times in between consecutive games are short while the travel distances are significant. It is important then to minimize the travel times since excessive travel may cause player fatigue. However, if this is the case, then the alternating home away patterns are not suitable. To reduce travel times a team must visit more than one team on each trip.

Consider again a double round-robin tournament that has $2n-2$ rounds. Let τ_{ij} denote the travel time between the homes of teams i and j. The objective is to minimize the sum of the travel times of each one of the teams. There are two sets of so-called "soft" constraints, namely

(i) *Atmost Constraints*: no team is allowed more than three consecutive home games or three consecutive away games.
(ii) *Nonrepeat Constraints*: a game between teams i and j at team i's home cannot be followed by a game between i and j at team j's home.

It is desirable, but not mandatory, that these constraints are satisfied. If they are not satisfied a penalty is incurred that is added to the total cost of the schedule. So the total cost of the schedule is the total travel time plus a function of the number of violations of the soft constraints. If $\mathcal{T}$ denotes the total travel time of all teams and $\mathcal{V}$ denotes the total number of violations of the soft constraints, then an appropriate objective function is

$$\sqrt{\mathcal{T}^2 + (\alpha\, f(\mathcal{V}))^2},$$

where α is a weight and the function f is increasing concave in the number of violations. The reason why the function f should be increasing concave is that the first violation costs more than a subsequent one; adding one violation to a schedule that already has five violations does not make much of a difference.

Example 10.4.3. Consider the following tournament with 6 teams in Table 10.1. This schedule has team 1 first playing against team 6 at home, then against team 2 away, followed by team 4 at home, team 3 at home, team 5 away, team 4 away, team 3 away, team 5 at home, team 2 at home, and team 6 away. The travel time of team 1 is

10.5 Scheduling Network Television Programs

	slot 1	slot 2	slot 3	slot 4	slot 5	slot 6	slot 7	slot 8	slot 9	slot 10
team 1	-6	2	-4	-3	5	4	3	-5	-2	6
team 2	-5	-1	3	6	-4	-3	6	4	1	5
team 3	4	-5	-2	1	-6	2	-1	-6	5	-4
team 4	-3	6	1	5	2	-1	-5	-2	6	3
team 5	2	3	-6	-4	-1	6	4	1	-3	-2
team 6	1	-4	5	-2	3	-5	-2	3	-4	-1

Table 10.1. Schedule for Example 10.4.3.

$$\tau_{12} + \tau_{21} + \tau_{15} + \tau_{54} + \tau_{43} + \tau_{31} + \tau_{16} + \tau_{61}$$

Note that long stretches of games at home do not contribute to the total travel time but are limited by the atmost constraints. However, the schedule above has two violations of the atmost constraints. Teams 2 and 4 both play a stretch of four games away, i.e., $\mathcal{V} = 2$.

In order to apply a local search routine to the travelling tournament problem a neighbourhood of a schedule has to be defined. A neighboring schedule can be obtained by applying one of several types of moves, namely

(i) SwapSlots
(ii) SwapTeams
(iii) SwapHomes

A SwapSlot move simply swaps slots t and u, i.e., two columns in the schedule above are interchanged. A SwapTeam move simply swaps the schedules of teams i and j, i.e., two rows in the schedule above are interchanged. A SwapHome is a little bit more complicated. This move swaps the Home/Away roles of teams i and j. If team i plays against team j at Home in slot t and again against team j Away in slot u, then a SwapHome move makes team i play against team j Away in slot t and at Home in slot u. The rest of the schedule remains the same.

In the literature on tournament scheduling more complicated neighbourhoods have been considered and various types of local search procedures, including simulated annealing, tabu-search and genetic algorithms, have been implemented.

10.5 Scheduling Network Television Programs

The topic in this section, namely the scheduling of network television programs, is somewhat different from tournament scheduling. However, scheduling network television programs does have a number of similarities with tournament scheduling. The scheduling horizon is typically one week and the week consists of a fixed number of time slots. A number of shows are available for broadcasting and these shows have to be assigned to the different time slots in such a way that a certain objective function is optimized. Moreover, the assignment of shows to slots is subject to a variety of conditions and constraints. For example, assigning a show to one slot may affect the contribution to the objective function of another show in a different slot. The integer programming formulations are, therefore, somewhat similar to the integer programming formulations for tournament scheduling.

Major television networks typically have a number of shows available for broadcasting. Some of these belong to series of half hour shows, whereas others belong to series of one hour shows. There are shows of other lengths as well. There are a fixed number of 30 minute time slots, implying that some shows require one time slot, whereas others need two consecutive time slots. If a particular show is assigned to a given time slot, then a certain rating can be expected. The forecasted ratings may be based on past experience with the show and/or the time slot; it may be based on lead-in effects due to the shows immediately preceding it, and it may also be based on shows that competing networks assign to that same slot. The profits of the network depend very much on the ratings. So one of the main objectives of the network is to maximize its average ratings.

If the length of program j is exactly half an hour, then the binary decision variable x_{jt} is 1 if program j is assigned to slot t; if the length of program j is longer than half an hour, then the decision variable x_{jt} is 1 if the *first* half hour of program j is assigned to slot t (i.e., broadcasting program j may require both slots t and $t+1$, but only the decision variable associated with slot t is 1 while the one associated with slot $t+1$ remains zero). Let π_{jt} denote the total profit (or the total ratings) obtained by assigning program j to time slot t. If program j occupies more than one slot, then π_{jt} denotes the profit generated over all slots the program covers. Let A denote the set of all feasible assignments (j,t). Let the binary variable b_{jtv} be 1 if time slot v is filled by program j or by part of program j because of the assignment (j,t) and 0 otherwise. Clearly, $b_{jtt} = 1$ and b_{jtv} can only be nonzero for $v > t$. The following integer program can be formulated to maximize the total profit.

$$\text{maximize} \sum_{(j,t) \in A} \pi_{jt} x_{jt}$$

subject to

$$\sum_{t:(j,t) \in A} x_{jt} \leq 1 \qquad \text{for } j = 1, \ldots, n$$

$$\sum_{(j,t) \in A} x_{jt} b_{jtv} = 1 \qquad \text{for } v = 1, \ldots, H$$

$$x_{jt} \in \{0,1\} \qquad \text{for } (j,t) \in A$$

This integer program takes into account the fact that there are shows of different durations. However, the formulation above is still too simple to be of any practical use. One important issue in television broadcasting revolves around so-called lead-in effects. These effects may have a considerable impact on the ratings (and the profits) of the shows. If a very popular show is followed by a new show for which it would be hard to forecast the ratings, then the high ratings of the popular show may have a spill-over effect on the new show; the ratings of the new show may be enhanced by the ratings of the popular show. Incorporating lead-in effects in the formulation described above can be done in several ways. One way can be described as follows: let (j, t, k, u) refer to a lead-in condition that involves show j starting in slot t and show k starting in slot u. Let $\mathcal{L}$ denote the set of all possible lead-in conditions. The binary decision variable y_{jtku} is 1 if in a schedule the lead-in condition (j, t, k, u) is indeed in effect and 0 otherwise. Let π'_{jtku} denote the additional contribution to the objective function if the lead-in condition is satisfied. The objective function in the formulation above has to be expanded with the term

$$\sum_{(j,t,k,u) \in \mathcal{L}} \pi'_{jtku} y_{jtku}$$

and the following constraints have to be added:

$$\begin{aligned} y_{jtku} - x_{jt} &\leq 0 & \text{for } (j,t,k,u) \in \mathcal{L} \\ y_{jtku} - x_{ku} &\leq 0 & \text{for } (j,t,k,u) \in \mathcal{L} \\ -y_{jtku} + x_{jt} + x_{ku} &\leq 1 & \text{for } (j,t,k,u) \in \mathcal{L} \\ y_{jtku} &\in \{0,1\} & \text{for } (j,t,k,u) \in \mathcal{L} \end{aligned}$$

The first set of constraints ensures that y_{jtku} can never be 1 when x_{jt} is zero. The second set of constraints is similar. The third set of constraints ensures that y_{jtku} never can be 0 when both x_{jt} and x_{ku} are equal to 1.

10.6 Scheduling a College Basketball Conference

The Atlantic Coast Conference (ACC) is a group of nine universities in the southeastern United States that compete against each other in a number of sports. From a revenue point of view, the most important sport is basketball. Most of the revenues come from television networks that broadcast the games and from gate receipts. The tournament schedule has an impact on the revenue stream. Television networks need a regular stream of quality games and spectators want neither too few nor too many home games in any period.

The ACC consists of nine universities: Clemson (Clem), Duke (Duke), Florida State (FSU), Georgia Tech (GT), Maryland (UMD), North Carolina (NC), North Carolina State (NCSt), Virginia (UVA) and Wake Forest (Wake). Every year, their basketball teams play a double round robin tournament in the first two months of the year. Each team plays every other team twice, once at home and once away. Usually, a team plays twice a week, often on wednesday and on saturday (these two slots are referred to as the weekday and the weekend slot). Because the total number of teams is odd, there will be in each slot one team with a Bye. In each slot there are four conference games. The entire schedule consists therefore of 18 slots, which implies that the length of the schedule is 9 weeks. Every team plays 8 slots at Home, 8 Away, and has two Byes.

There are numerous restrictions in the form of pattern constraints, game count constraints, and team pairing constraints. The patterns of Home games and Away games is important because of wear and tear on the teams, issues of missing class time, and spectator preferences. No team should play more than two Away games consecutively, nor more than two Home games consecutively. A Bye is usually regarded as an Away game. Similar rules apply to weekend slots (no more than two at Home in consecutive weekends).

In addition, the first five weekends are used for recruiting future student-athletes, so each team must have at least two Home or one Home and one Bye weekend among the first five. A Bye is acceptable here because the open slot could be used to schedule a non-conference Home game.

The last week of the season is of great importance to all teams, so no team can be Away in both slots of the final week. The final weekend of the season is the most important slot, and is reserved for "rival" pairings; the games Duke-UNC, Clem-GT, NCSt-Wake, and UMD-UVA are usually played on that day. Duke-UNC is the most critical pairing in the schedule. It must occur in slot 17 and also in slot 10.

Since every team plays two games against every other team, the conference prefers to have the two games somewhat apart. A separation of nine slots can be achieved by mirroring a single round robin schedule. Because of certain fixed game assignments, it turns out that a perfect mirror in this case is not possible. However, a similar idea is used in order to ensure large separations.

An approach similar to Algorithm 10.2.2 was adopted to create a schedule. However, the mirroring of the schedule was already done in Step 1 of this algorithm rather than as an additional step after Step 3. Step 1 created HAPs of length 18. A number of different pattern sets were created in Step 1 by solving a series of integer programs. Generating the timetables in Step 2 can also be done through an integer program of the type described in Section 10.2. Step 3 requires also a significant computational effort. For each timetable there are

$$9! = 362,880$$

assignments of teams to patterns. Each of these is checked for feasibility aspects (e.g., are the final games the right games) and for preference aspects (e.g., the number of prime TV slots and the number of slots that are not prime). The schedule generation process is illustrated by the flow chart in Figure 10.3 that also summarizes which constraints are enforced in each step.

10.6 Scheduling a College Basketball Conference   247

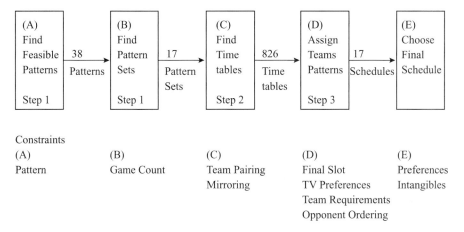

Fig. 10.3. Algorithm Flow Chart

Following this procedure the official schedule for the 1996-1997 ACC Basketball Tournament was generated. The first half of the tournament is presented below.

	slot 1	slot 2	slot 3	slot 4	slot 5	slot 6	slot 7	slot 8	slot 9
Clem	UVA	—	-Duke	FSU	UMD	-NCSt	-Wake	UNC	-GT
Duke	-FSU	GT	Clem	-Wake	—	-UVA	NCSt	UMD	-UNC
FSU	Duke	-NCSt	-UVA	-Clem	GT	—	-UNC	Wake	-UMD
GT	UMD	-Duke	-Wake	—	-FSU	UNC	UVA	-NCSt	Clem
UMD	-GT	-UVA	UNC	NCSt	-Clem	Wake	—	-Duke	FSU
UNC	—	Wake	-UMD	UVA	-NCSt	-GT	FSU	-Clem	Duke
NCSt	-Wake	FSU	—	-UMD	UNC	Clem	-Duke	GT	UVA
UVA	-Clem	UMD	FSU	-UNC	Wake	Duke	-GT	—	-NCSt
Wake	NCSt	-UNC	GT	Duke	-UVA	-UMD	Clem	-FSU	—

The second half of the tournament is as follows.

	slot 10	slot 11	slot 12	slot 13	slot 14	slot 15	slot 16	slot 17	slot 18
Clem	NCSt	—	UMD	Wake	-UVA	Duke	-FSU	-UNC	GT
Duke	-GT	Wake	-NCSt	UVA	FSU	-Clem	—	-UMD	UNC
FSU	UVA	UNC	-GT	UMD	-Duke	—	Clem	NCSt	-Wake
GT	Duke	-UVA	FSU	—	-UNC	-UMD	NCSt	Wake	-Clem
UMD	-Wake	-NCSt	Clem	-FSU	—	GT	-UNC	Duke	UVA
UNC	—	-FSU	-UVA	NCSt	GT	-Wake	UMD	Clem	-Duke
NCSt	-Clem	UMD	Duke	-UNC	Wake	-UVA	-GT	-FSU	—
UVA	-FSU	GT	UNC	-Duke	Clem	NCSt	-Wake	—	-UMD
Wake	UMD	-Duke	—	-Clem	-NCSt	UNC	UVA	-GT	FSU

The slots with an even number are weekend slots and the slots with an odd number are weekday slots. The schedule presented above has the following characteristics:
  (i) the minimum difference between repeating games is 4;
  (ii) the number of strings with three or more consecutive home games (considering a Bye as an Away) is 2;
  (iii) the number of strings with three or more consecutive home games (considering a Bye as a Home) is 10;
  (iv) the number of strings with three or more consecutive away games (considering a Bye as a Home) is 0;
  (v) the number of strings with three or more consecutive away games (considering a Bye as an Away) is 1.

This same scheduling problem has also been solved using a constraint programming approach. The approach followed was based on the one described in Section 10.3. As far as Steps 1 and 2 are concerned, the computational effort needed using a constraint programming approach seems to be comparable to the computational effort needed using an integer programming approach. However, as far as Step 3 is concerned the constraint programming technique seems to have a clear edge. Since the integer programming approach in Step 3 is basically equivalent to complete enumeration, it is not surprising that constraint programming can do better.

10.7 Discussion

In this chapter tournament scheduling problems have been dealt with through integer programming techniques, constraint programming techniques, and local search techniques. It seems that in the college basketball conference example the constraint programming technique has been more effective than the integer programming technique. However, it still would be interesting to find out when it is appropriate to use an optimization technique, when a constraint programming technique, and when a local search technique. Optimization techniques may be more suitable when there is a clearly defined objective function and the set of feasible solutions is large. Another question may be of interest as well: are there cases in which it would make sense to use a hybrid approach that includes optimization, constraint programming, and local search? What is the best way to incorporate the three approaches within a single framework? More research is needed in order to get some insight in how the techniques have to be combined in order to maximize the overall effectiveness.

The framework of Algorithm 10.2.2 is based on an approach that first decides on the HAPs and then assigns teams to the different HAPs. A completely different approach first assigns teams to games, and then determines the home-away assignments while attempting to minimize the number of breaks (see Exercise 10.7).

10.7 Discussion

Since each framework for dealing with a tournament scheduling problem consists of multiple steps, it may be the case that in some steps it is appropriate to use a customized heuristic. There have been a number of applications of customized heuristics (e.g., local search) in tournament scheduling.

In practice, there are many side-constraints in tournament scheduling. Basically, the dates on which some of the games can be played are fixed in advance. They may also be influenced by outside considerations such as national holidays, long weekends, and so on. Also, the schedule of one team may depend very strongly on the schedule of another team. For example, in Europe a number of big cities have more than one top team in their national soccer league (e.g., Milan has AC and Inter, Madrid has Real and Atletico, and so on) and these teams may have fans in common. It is desirable that a schedule does not have the home games of two such teams in the same weekend. Ideally, when one club plays at home, the other should be away.

There are similarities as well as differences between tournament scheduling and the scheduling of television programs. In tournaments, games have to be assigned to slots and in television programs shows have to be assigned to slots as well. However, the reason why tournament scheduling tends to be a harder problem is based on the fact that a game involves two teams and the schedules of both teams have to be taken into consideration. On the other hand, the assignment of a television show to a slot is somewhat (but not completely) independent of the assignment of other shows to other slots. There may be some dependency because of lead-in effects. The dependency between the ratings of two consecutive shows can be compared to having one team face two very strong teams in a row.

Besides the scheduling of the shows that have to be broadcasted, there are also scheduling problems concerning the commercials from advertisers. Large companies that are major advertisers buy hundreds of time slots from a network to air commercials during a broadcast season. The actual commercials to be assigned to these slots are determined at a later stage. During the broadcast season the clients ship the videotapes of the commercials to be aired in the slots they purchased. The advertisers often specify the following guideline: whenever a commercial has to be aired multiple times within a given period, say a month, the advertisements should be spaced out as evenly as possible over that period. The question then arises: how to assign the commercials to the slots in such a way that airings of the same commercial are spaced as evenly as possible.

There are also some similarities between tournament scheduling and the scheduling of commercials. For example, commercials from the same advertiser have to be distributed as evenly as possible over the planning horizon and games between the same pairs of teams have to be spaced out as evenly as possible over the duration of the tournament.

Exercises

10.1. Explain why it is necessary to generate n different HAPs in Step 1 of Algorithm 10.2.2. That is, why is it not possible for two teams to have the same HAP?

10.2. Give an example of a pair of different HAPs (each one containing one Bye) that cannot appear together in the same pattern set. Explain why. Can you give an example of a pair of HAPs (with neither one containing any Byes) that cannot appear in the same pattern set? Explain why or why not.

10.3. Show why in the graph coloring problem described in Section 10.2 the two subgraphs that correspond to two consecutive rounds have a Hamiltonian path when there is no break in the second one of the two rounds. Discuss the structure of the path for the case when n is even as well as for the case when n is odd.

10.4. Consider the tournament schedule generated in Example 10.2.1. The number of breaks in the final schedule is 6. Can you create a schedule with less breaks?

10.5. Consider the tournament schedule in Example 10.2.3. The minimum distance between any two games between the same pair of teams is 3. Show that it is impossible to create a schedule in which the minimum distance is more than 3.

10.6. Describe the single round robin tournament scheduling problem as a workforce constrained project scheduling problem.

10.7. Consider a situation in which some of the games have already been assigned to specific time slots. However, of those games that already have been assigned to given slots, it has not been determined yet which team plays at home and which team plays away.

slots	1	2	3	4	5
team a:	(b)	(f)		(c)	
team b:	(a)		(f)		
team c:	(d)			(e)	(a)
team d:	(c)	(e)			
team e:	(f)	(d)	(c)		
team f:	(e)	(a)	(b)		

(a) Complete the partial schedule presented above without taking into consideration whether games are at home or away. Is this type of *completion problem* an easy problem or a hard problem?

(b) With the schedule of games developed under (a), assign to each game a home team and an away team in such a way that the total number of breaks is minimized.

10.8. Consider the Travelling Tournament problem described in Example 10.4.3. The travel time matrix between the sites of the 6 teams is symmetric and given below.

teams	1 2 3 4 5 6
team 1:	- 3 7 2 0 5
team 2:	3 - 4 6 3 5
team 3:	7 4 - 7 1 4
team 4:	2 6 7 - 5 1
team 5:	0 3 1 5 - 2
team 6:	5 5 4 1 2 -

(a) Compute the value of the objective function of the schedule presented in Example 10.4.3 assuming that $\alpha = 10$.

(b) Apply each one of the three types of moves described (i.e., SwapSlots, SwapTeams, and SwapHomes) to the solution given in Example 10.4.3. Describe how the new schedules are generated and compute the new values of the objective function.

10.9. Consider the network television scheduling problem described in Section 10.5.

(a) Assume that each show takes exactly one slot (half an hour) and that there are no lead-in effects. Show that the problem then reduces to an assignment problem.

(b) Consider now the case where some shows take two slots and the remaining shows take one slot (again, no lead-in effects). Is the problem still an assignment problem?

Comments and References

A significant amount of work has been done on timetabling of sport tournaments. De Werra (1988) studied the graph theoretic properties of tournament scheduling. The break minimization problem has been studied by Miyashiro and Matsui (2003). The approach described in Section 10.2 which first selects the HAPs and then assigns teams to HAPs has been used by Nemhauser and Trick (1998) and Schreuder (1992). An alternative approach, which first assigns teams to games and then determines the home-away assignment, has been studied by Trick (2001).

Section 10.3 is based on the work by Henz (2001). Many other researchers have also applied constraint programming techniques to tournament scheduling; see, for example, McAloon, Tretkoff and Wetzel (1997), Schaerf (1999), Régin (1999), Henz, Müller, and Thiel (2004), Aggoun and Vazacopoulos (2004).

The first part of Section 10.4 is based on the work by Hamiez and Hao (2001); they applied tabu-search to tournament scheduling. The second part of Section 10.4, the Traveling Tournament Problem, is based on the work by Anagnostopoulos, Michel, Van Hentenryck, and Vergados (2003). A number of other researchers have applied local search techniques to tournament scheduling problems; see, for example, Schönberger, Mattfeld, and Kopfer (2004) who applied various types of local search algorithms to the timetabling of non-commercial sport leagues.

Section 10.5, which focused on the timetabling of television shows, is based on the work by Horen (1980) and Reddy, Aronson, and Stam (1998). For other work concerning scheduling in broadcast networks, see Hall, Liu and Sidney (1998), Bollapragada, Cheng, Phillips, Garbiras, Scholes, Gibbs, and Humphreville (2002). For work concerning the scheduling of commercials in broadcast television, see Bollapragada and Garbiras (2004), Bollapragada, Bussieck, and Mallik (2004) and Hägele, Dúnlaing, and Riis (2001).

The scheduling of the college basketball conference described in Section 10.6 is based on the papers by Nemhauser and Trick (1998) and Henz (2001). An interesting case concerning the scheduling of soccer teams in the Netherlands was analyzed by Schreuder (1992). Bartsch, Drexl and Kröger (2004) developed schedules for the professional soccer leagues of Austria and Germany.

Chapter 11

Planning, Scheduling, and Timetabling in Transportation

11.1 Introduction 253
11.2 Tanker Scheduling 254
11.3 Aircraft Routing and Scheduling 258
11.4 Train Timetabling 272
11.5 Carmen Systems: Designs and Implementations . 279
11.6 Discussion 283

11.1 Introduction

In the transportation industry planning and scheduling problems abound. The variety in the problems is due to the many modes of transportation, e.g., shipping, airlines, and railroads. Each mode of transportation has its own set of characteristics. The equipment and resources involved, i.e.,

(i) ships and ports,
(ii) planes and airports, and
(iii) trains, tracks, and railway stations,

have different cost characteristics, different levels of flexibilities, and different planning horizons.

The second section of this chapter focuses on oil tanker scheduling. These models are used in practice in a rolling horizon manner. Among all models discussed in this chapter, this one is the easiest to formulate. The subsequent section considers aircraft routing and scheduling. In aircraft routing and scheduling the goal is to create a periodic (daily) timetable. In a certain sense this model is an extension of the model for oil tanker scheduling. The integer programming formulation of the aircraft routing and scheduling problem is very similar to the formulation described for the oil tanker scheduling problem; however, in the airline case there are additional constraints that enforce periodicity. The fourth section discusses timetabling of trains. Track capacity constraints in railway operations specify that one train can pass another only at a station, and not in between stations. The fifth section describes the airline routing and scheduling systems designed and implemented by Carmen Systems in Sweden. The discussion section focuses on the similarities and differences between tanker scheduling, airline routing and scheduling, and train timetabling.

11.2 Tanker Scheduling

Companies that own and operate tanker fleets typically make a distinction between two types of ships. One type of ship is company-owned and the other type of ship is chartered. The operating cost of a company-owned ship is different from the cost of a charter that is typically determined on the spot market. Each ship has a specific capacity, a given draught, a range of possible speeds and fuel consumptions, and a given location and time at which the ship is ready to start a new trip.

Each port also has its own characteristics. Port restrictions take the form of limits on the deadweight, draught, length, beam and other physical characteristics of the ships. There may be some additional government rules in effect; for example, the Nigerian government imposes a so-called 90% rule which states that all tankers must be loaded to more than 90% of capacity before sailing.

A cargo that has to be transported is characterized by its type (e.g., type of crude), quantity, load port, delivery port, time window constraints on the load and delivery times, and the load and unload times. A schedule for a ship defines a complete itinerary, listing in sequence the ports visited within the time horizon, the time of entry at each port and the cargoes loaded or delivered at each port.

The objective typically is to minimize the total cost of transporting all cargoes. This total cost consists of a number of elements, namely the operating costs for the company-owned ships, the spot charter rates, the fuel costs, and the port charges. Port charges vary greatly between ports and within a given port charges typically vary proportionally with the deadweight of the ship.

In order to present a formal description of the problem the following notation is used. Let n denote the number of cargoes to be transported, T the number of company-owned tankers, and p the number of ports. Let $\mathcal{S}_i$ denote the set of all possible schedules for ship i. Schedule l for ship i, $l \in \mathcal{S}_i$, is represented by the column vector

$$\begin{array}{c} a_{i1}^l \\ a_{i2}^l \\ \vdots \\ a_{in}^l \end{array}$$

11.2 Tanker Scheduling

The constant a_{ij}^l is 1 if under schedule l ship i transports cargo j and 0 otherwise. Let c_i^l denote the incremental cost of operating a company-owned ship i under schedule l versus keeping ship i idle over the entire planning horizon. The operating cost can be computed once schedule l has been specified, since it may depend in various ways on the characteristics of the ship and of the schedule, including the distance travelled, the time the ship is used, and the ports visited. The cost c_j^* denotes the amount that has to be paid on the spot market to transport cargo j on a ship that is not company owned.

Let

$$\pi_i^l = \sum_{j=1}^n a_{ij}^l \, c_j^* - c_i^l$$

denote the "profit" (i.e., the amount of money that does not have to be paid on the spot market) by operating ship i according to schedule l. The decision variable x_i^l is 1 if ship i follows schedule l and zero otherwise.

The Tanker Scheduling Problem can now be formulated as follows:

$$\text{maximize} \quad \sum_{i=1}^T \sum_{l \in \mathcal{S}_i} \pi_i^l \, x_i^l$$

subject to

$$\sum_{i=1}^T \sum_{l \in \mathcal{S}_i} a_{ij}^l \, x_i^l \leq 1 \qquad j = 1, \ldots, n$$

$$\sum_{l \in \mathcal{S}_i} x_i^l \leq 1 \qquad i = 1, \ldots, T$$

$$x_i^l \in \{0, 1\} \qquad l \in \mathcal{S}_i, \ i = 1, \ldots, T$$

The objective function specifies that the total profit has to be maximized. The first set of constraints imply that each cargo can be assigned to at most one tanker. The second set of constraints specifies that each tanker can be assigned at most one schedule. The remaining constraints imply that all decision variables have to be binary $0-1$. This optimization problem is typically referred to as a set-packing problem.

The algorithm used to solve this problem is a branch-and-bound procedure. However, before the branch-and-bound procedure is applied, a collection of candidate schedules have to be generated for each ship in the fleet. As stated before, such a schedule specifies an itinerary for a ship, listing the ports visited and the cargoes loaded or delivered at each port. The generation of an initial collection of candidate schedules has to be done by a separate ad-hoc heuristic that is especially designed for this purpose. The collection of candidate schedules should include enough schedules so that potentially optimal schedules are not ignored, but not so many that the set-packing problem becomes intractable. Physical constraints such as ship capacity and speed, port depth and time windows limit the number of feasible candidate schedules considerably. Schedules that have a negative profit coefficient in the objective function of the set-packing formulation can be omitted as well.

The branch-and bound method for solving the problem is typically based on customized branching and bounding procedures. Since the problem is a maximization problem a good schedule generated by a clever heuristic (or a manual method) provides a lower bound for the value of the optimal solution. When considering a particular node in the branching tree, it is necessary to develop an upper bound for the collection of schedules that correspond to all the descendants of this particular node; if this upper bound is less than the lower bound on the optimum provided by the best schedule currently available, then this node can be fathomed.

There are a variety of suitable branching mechanisms for the branch-and-bound tree. The simplest mechanism is just the most basic $0 - 1$ branching. Select at a node a variable x_i^l which has not been fixed yet at a higher level node and generate branches to two nodes at the next level down: one branch for $x_i^l = 0$ and one for $x_i^l = 1$. The selection of the variable x_i^l may depend on the solution of the LP relaxation at that node; the most suitable x_i^l may be the one with a value closest to 0.5 in the solution of the LP relaxation. If at a node a variable x_i^l is not equal to 1 for ship i, then certain schedules for other ships can be ruled out for all the descendants of this node; that is, the schedules for other ships that have a cargo in common with schedule l for ship i do not have to be considered any more.

Another way of branching can be done as follows: Select at a given node a ship i that has not been selected yet at a higher level node and generate for each schedule l in $\mathcal{S}_i$ a branch to a node at the next level down. In the branch corresponding to schedule l the variable $x_i^l = 1$. Using this branching mechanism, one still has to decide at each node which ship i to select. One could select the i based on several criteria. For example, a ship that transports many cargoes or a ship that may be responsible for a large profit. Another way is to select an i that has a highly fractional solution in the LP relaxation of the problem (e.g., there may be a ship i with a solution $x_i^l = 1/K$ for K different schedules with K being a fairly large number).

An upper bound at a node can be obtained by solving the linear relaxation of the set-packing problem corresponding to that node, i.e., the integrality constraints on x_i^l are replaced by the nonnegativity constraints $x_i^l \geq 0$. This problem may be referred to as the *continuous* set-packing problem. The value of the optimal solution is an upper bound for the values of all possible solutions of the set-packing problem at that node. It is nowadays possible to find with little computational effort optimal solutions (or at least good upper bounds) for very large continuous set-packing problems, making such a bounding mechanism quite effective.

11.2 Tanker Scheduling

Example 11.2.1 (Oil Tanker Scheduling). Consider three ships and 12 cargoes that have to be transported. A feasibility analysis shows that for each one of the ships there are five feasible schedules. The 15 columns in the table below represent the 15 feasible schedules for the three ships.

Schedules	a_{1j}^1	a_{1j}^2	a_{1j}^3	a_{1j}^4	a_{1j}^5	a_{2j}^1	a_{2j}^2	a_{2j}^3	a_{2j}^4	a_{2j}^5	a_{3j}^1	a_{3j}^2	a_{3j}^3	a_{3j}^4	a_{3j}^5
cargo 1	1	0	0	1	1	0	1	0	0	0	0	0	0	1	0
cargo 2	1	0	0	0	0	1	0	0	0	0	0	1	0	1	1
cargo 3	0	0	1	0	1	0	0	0	1	1	0	0	0	0	0
cargo 4	0	1	1	1	0	1	0	1	0	0	0	0	0	0	0
cargo 5	1	1	0	0	0	0	0	0	1	0	0	0	1	0	1
cargo 6	0	0	0	1	1	0	1	0	0	1	1	0	0	0	0
cargo 7	0	0	0	0	0	0	0	1	1	0	0	0	0	0	1
cargo 8	0	1	0	0	0	1	0	1	1	1	0	0	0	0	0
cargo 9	0	0	1	0	0	0	1	0	0	1	1	1	1	0	0
cargo 10	0	1	0	0	0	1	0	0	0	0	1	1	0	0	0
cargo 11	0	0	0	0	0	0	1	1	0	0	0	1	1	1	0
cargo 12	0	0	0	1	0	0	0	0	0	0	1	0	1	1	1

If a cargo is transported by a charter, then a charter cost is incurred.

Cargoes	1	2	3	4	5	6	7	8	9	10	11	12
Charter Costs	1429	1323	1208	512	2173	2217	1775	1885	2468	1928	1634	741

The operating costs of the tankers under each one of the schedules are tabulated below:

Schedule l	1	2	3	4	5
cost of tanker 1 (c_1^l)	5658	5033	2722	3505	3996
cost of tanker 2 (c_2^l)	4019	6914	4693	7910	6868
cost of tanker 3 (c_3^l)	5829	5588	8284	3338	4715

The profits of each schedule can now be computed.

Schedule l	1	2	3	4	5
profit of tanker 1 (π_1^l)	−733	1465	1466	1394	858
profit of tanker 2 (π_2^l)	1629	834	1113	−869	910
profit of tanker 3 (π_3^l)	1525	1765	−1268	1789	1297

The integer program can now be formulated as follows:

maximize

$$-773x_1^1 + 1465x_1^2 + 1466x_1^3 + 1394x_1^4 + 858x_1^5$$
$$+1629x_2^1 + 834x_2^2 + 1113x_2^3 - 869x_2^4 + 910x_2^5$$
$$+1525x_3^1 + 1765x_3^2 - 1268x_3^3 + 1789x_3^4 + 1297x_3^5$$

subject to

$$x_1^1 + x_1^4 + x_1^5 + x_2^2 + x_3^4 \le 1$$
$$x_1^1 + x_2^1 + x_3^2 + x_3^4 + x_3^5 \le 1$$
$$x_1^3 + x_1^5 + x_2^4 + x_2^5 \le 1$$
$$x_1^2 + x_1^3 + x_1^4 + x_2^1 + x_2^3 \le 1$$
$$x_1^1 + x_1^2 + x_2^4 + x_3^3 + x_3^5 \le 1$$
$$x_1^4 + x_1^5 + x_2^2 + x_2^5 + x_3^1 \le 1$$
$$x_2^3 + x_2^4 + x_3^5 \le 1$$
$$x_1^2 + x_2^1 + x_2^3 + x_2^4 + x_2^5 \le 1$$
$$x_1^3 + x_2^2 + x_2^5 + x_3^1 + x_3^2 + x_3^3 \le 1$$
$$x_1^2 + x_2^1 + x_3^1 + x_3^2 \le 1$$
$$x_2^2 + x_2^3 + x_3^2 + x_3^3 + x_3^4 \le 1$$
$$x_1^4 + x_3^1 + x_3^3 + x_3^4 + x_3^5 \le 1$$

$$x_1^1 + x_1^2 + x_1^3 + x_1^4 + x_1^5 \le 1$$
$$x_2^1 + x_2^2 + x_2^3 + x_2^4 + x_2^5 \le 1$$
$$x_3^1 + x_3^2 + x_3^3 + x_3^4 + x_3^5 \le 1$$
$$x_i^l \in \{0, 1\}$$

An initial upper bound can be obtained by solving the linear relaxation of the integer program, i.e., allowing x_i^l to assume any value between 0 and 1. The solution of the linear program is $x_1^2 = x_1^3 = x_1^5 = 1/3$, $x_2^1 = x_2^5 = 1/3$, and $x_3^1 = 1/3$, $x_3^4 = 2/3$. The value of the solution (i.e., the upper bound) is 3810.33.

Solving this integer program via branch-and-bound results in the tree shown in Figure 11.1. The optimal solution of the integer program assigns schedule 3 to ship 1 (i.e., $x_1^3 = 1$), and schedule 4 to ship 3 (i.e., $x_3^4 = 1$). Ship 2 remains idle and cargoes 5,6,7,8, and 10 are transported by charters. The value of this solution is 3255.

In contrast to the transportation problems analyzed in the next two sections, schedules for tankers (oil, natural gas, bulk cargo in general) are usually not cyclic. The scheduling process is based on a rolling horizon procedure.

11.3 Aircraft Routing and Scheduling

A major problem faced by every airline is to construct a daily schedule for a heterogeneous aircraft fleet. A plane schedule consists of a sequence of flight legs that have to be flown by a plane with the exact times at which the legs

11.3 Aircraft Routing and Scheduling

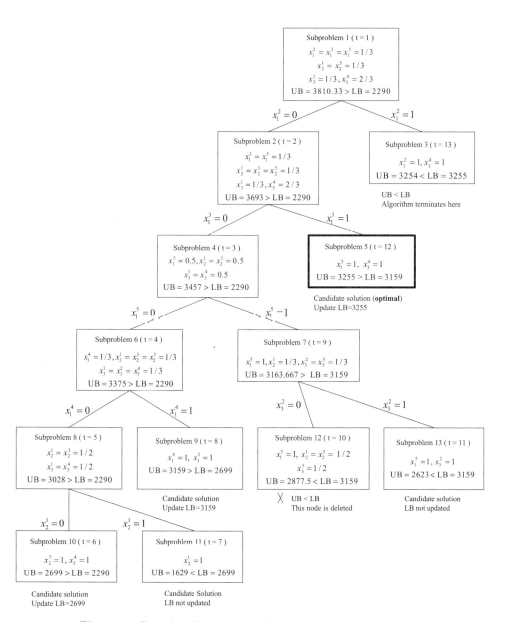

Fig. 11.1. Branch-and-bound tree for tanker scheduling problem

must start and finish at the respective airports. The first part of the problem (determining the sequence of flight legs) is basically a routing problem, whereas the second part of the problem (determining the exact times) is a scheduling problem. The fleet schedule is important, since the total revenue of the airline can be estimated if the demand function of each leg is known. Moreover, the fleet schedule also determines the total cost incurred by the airline, including the cost of fuel and the salaries of the crews.

An airline typically has, from past experience and through marketing research, estimates of customer demands for specific flight legs (a flight leg is characterized by its point of origin, its departure time and its destination). It can be assumed that a minor shift in the departure time of the flight leg does not have an effect on the demand. So an airline has for each leg a (narrow) time window in which it can depart. An airline also has estimates for the revenue derived from a specific flight leg as a function of the type of plane utilized, and of the costs involved.

The Daily Aircraft Routing and Scheduling Problem can now be formulated as follows: Given a heterogeneous aircraft fleet, a collection of flight legs that have to be flown in a one-day period with departure time windows, durations, and cost/revenues corresponding to the aircraft type for each leg, a fleet schedule has to be generated that maximizes the airline's profits (possibly subject to certain additional constraints).

Some of the additional constraints that often have to be taken into account in an Aircraft Routing and Scheduling Problem are the number of available planes of each type, the restrictions on certain aircraft types at certain times and at certain airports, the required connections between flight legs (the so-called "thrus") imposed by the airline and the limits on the daily service at certain airports. Also, the collection of flight legs may have to be balanced, i.e., at each airport there must be, for each airplane type, as many arrivals as departures. One must further impose at each airport the availability of an equal number of aircraft of each type at the beginning and at the end of the day.

In the formulation of the problem the following notation is used: L denotes the set of flight legs, T denotes the number of different aircraft types, and m_i denotes the number of available aircraft of type i, $i = 1, \ldots, T$. So the total number of aircraft available is

$$\sum_{i=1}^{T} m_i.$$

(Note that this is in contrast to the previous section where T was the total number of tankers.) Some flight legs may be flown by more than one type of aircraft. Let L_i denote the set of flight legs that can be flown by an aircraft of type i and let $\mathcal{S}_i$ denote the set of feasible schedules for an aircraft of type i. This set includes the empty schedule (0); an aircraft assigned to this schedule is simply not being used. Let π_{ij} denote the profit generated by covering flight leg j with an aircraft of type i. With each schedule $l \in \mathcal{S}_i$ there is a

11.3 Aircraft Routing and Scheduling

total anticipated profit

$$\pi_i^l = \sum_{j \in L_i} \pi_{ij} a_{ij}^l,$$

where a_{ij}^l is 1 if schedule l covers leg j and 0 otherwise. If an aircraft has been assigned to an empty schedule, then the profit is π_i^0. The profit π_i^0 may be either negative or positive. It may be negative when there is a high fixed cost with keeping a plane for a day; it may be positive when there is a benefit with having a plane idle (some airlines want to have idle planes that can serve as stand by). Let $\mathcal{P}$ denote the set of airports, and $\mathcal{P}_i$ be the subset of airports that have facilities to accomodate aircraft of type i. Let o_{ip}^l be equal to 1 if the origin of schedule l, $l \in \mathcal{S}_i$, is airport p, and 0 otherwise; let d_{ip}^l be equal to 1 if the final destination of schedule l is airport p, and 0 otherwise.

The binary decision variable x_i^l takes the value 1 if schedule l is assigned to an aircraft of type i, and 0 otherwise; the integer decision variable x_i^0 denotes the number of unused aircraft of type i, i.e., the aircraft that have been assigned to an empty schedule.

The Daily Aircraft Routing and Scheduling Problem can now be formulated as follows:

$$\text{maximize} \quad \sum_{i=1}^{T} \sum_{l \in \mathcal{S}_i} \pi_i^l x_i^l$$

subject to

$$\sum_{i=1}^{T} \sum_{l \in \mathcal{S}_i} a_{ij}^l x_i^l = 1 \qquad j \in L$$

$$\sum_{l \in \mathcal{S}_i} x_i^l = m_i \qquad i = 1, \ldots, T$$

$$\sum_{l \in \mathcal{S}_i} (d_{ip}^l - o_{ip}^l) x_i^l = 0 \qquad i = 1, \ldots, T, \; p \in \mathcal{P}_i$$

$$x_i^l \in \{0, 1\} \qquad i = 1, \ldots, T, \; l \in \mathcal{S}_i$$

The objective function specifies that the total anticipated profit has to be maximized. The first set of constraints imply that each flight leg has to be covered exactly once. (This set of constraints is somewhat similar to the first set of constraints in the formulation of the tanker scheduling problem.) The second set of constraints specifies the maximum number of aircraft of each type that can be used. The third set of constraints correspond to the flow conservation constraints at the beginning and at the end of the day at each airport for each aircraft type. The remaining constraints imply that all decision variables have to be binary $0 - 1$.

This model is basically a Set Partitioning Problem with additional constraints; the tanker scheduling model of the previous section is a Set Packing problem because of the fact that the first set of constraints are inequality ($\leq$) constraints (see Appendix A). The algorithm to solve this problem is also based on branch-and-bound; the version of branch-and-bound is typically referred to as branch-and-price (see Appendix B). We first describe the mechanism for generating upper bounds and then describe the branching strategies.

Upper bounds can be obtained by using a so-called column generation procedure that solves the linear relaxation of the integer program formulated above (i.e., in the linear relaxation the integrality constraints on the decision variables x_i^l are replaced by nonnegativity constraints $x_i^l \geq 0$; because the first set of constraints already ensure that the decision variables x_i^l have to be less than or equal to 1).

The column generation procedure is used to avoid the necessity of generating all possible schedules. The procedure divides the linear program into a restricted master problem and a subproblem. The restricted master problem is basically a linear program defined over a relatively small number of candidate schedules for the aircraft. The decision variable x_i^l that corresponds to a column that has not been included is assumed to be 0. The idea is to find an optimal solution for the current restricted master problem and compute the dual variables associated with this solution. These dual variables represent the imputed unit costs for the "resources", such as the legs, the aircraft and the airports. These variables are used in the subproblem to compute the potential profit of other candidate schedules. The subproblem is basically used to test whether the solution of the current restricted master problem is optimal over *all* possible schedules, i.e., for the unrestricted (but still linear) master problem. The subproblem, which uses the dual variables of the optimal solution for the current restricted master problem, turns out to be equivalent to a longest path problem with time windows; it can be solved by dynamic programming. If the current solution is not optimal over all schedules (i.e., there are schedules with a positive potential profit in the subproblem), then the subproblem must provide one or more new aircraft schedules to be added to the set of candidate schedules in the restricted master problem.

The subproblem for planes of type i can be formulated as follows. Consider a directed graph

$$G_i = (N_i, B_i)$$

that is used to generate new feasible candidate schedules for an aircraft of type i, see Figure 11.2. There are five types of nodes: one source, one sink, origination airport nodes, termination airport nodes, and flight leg nodes. There are five types of arcs in B_i: source arcs, sink arcs, schedule origination arcs, schedule termination arcs, and turn arcs. A source arc goes from the source to an origination airport node. A sink arc goes from a termination airport node to the sink. A schedule origination arc emanates from an origination airport node and goes to a flight leg node (the flight has to start at

11.3 Aircraft Routing and Scheduling

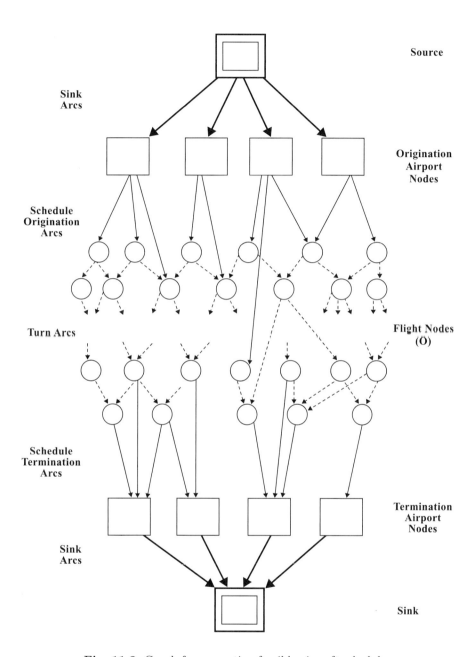

Fig. 11.2. Graph for generating feasible aircraft schedules

the airport associated with the origination airport node). A schedule termination arc begins at a flight leg node and goes to a termination airport node (the flight has to end at the airport associated with the termination airport node). A turn arc connects two flight leg nodes and simply represents a connection between these two flight legs. Such an arc exists between two flight leg nodes only if the flight legs can be flown by the same aircraft consecutively while respecting the time windows of each flight leg and the necessary time it takes to turn the aircraft around in between the flight legs. (A minimum turnaround time is needed in order to deplane one group of passengers, clean the plane and board the departing passengers). So the existence of a turn arc between two flight nodes depends on the given time windows for the departure times of the two flight legs as well as on the minimum turnaround time.

The objective of the subproblem is to find a feasible plane schedule with maximum marginal profit. To be feasible, a schedule must satisfy the time window constraints concerning departure times. If e_{ij} is the earliest possible departure time of flight leg j and ℓ_{ij} is the latest possible departure time, then the value of the actual departure time variable must lie within the time window [e_{ij}, ℓ_{ij}]. If τ_{ij} is the duration of leg j and δ_{ijk} is the minimum turnaround time between legs j and k, then for a turn arc to exist between legs j and k,

$$e_{ij} + \tau_{ij} + \delta_{ijk} \leq \ell_{ik}$$

Each path from the source node to the sink node corresponds to a feasible schedule. The marginal profit of a schedule can now be computed using the revenue and the cost of each activity (i.e., the profit) as well as the dual variables associated with the current optimal solution of the restricted master problem. The arcs in graph G_i are assigned values corresponding to the profits of the activities and the dual variables associated with the optimal solution of the current restricted master problem. Every arc leaving flight node j is assigned a profit π_{ij} (which is a known quantity). In addition to the profits, the dual variables have to be placed on specific arcs in the network. Let α_j, β_i and γ_{ip} denote the dual variables associated with the first, second and third set of constraints in the restricted master problem. A dual variable represents the increase in the total profit of the current solution of the restricted master problem assuming that the right hand side of the corresponding constraint is increased by one unit. So if the number of aircraft of type i is increased by 1, then the profit goes up by β_i. If there are a number of unused aircraft in the solution of the current master problem, then $\beta_i = 0$. Another interpretation of the dual variable β_i is the following: if in the current solution of the restricted master problem all aircraft are being used, then assigning a schedule l to an aircraft implies a cost β_i (since that is the current opportunity cost of a plane). The potential profit $\bar{\pi}_i^l$ of schedule $l \in \mathcal{S}_i$ with respect to the current solution can now be computed as it is the profit of the corresponding path in the network (schedule l may or may not be part of the restricted master

11.3 Aircraft Routing and Scheduling

problem). The potential profit is

$$\bar{\pi}_i^l = \sum_{j \in L_i}(\pi_{ij} - \alpha_j)a_{ij}^l - \beta_i - \sum_{p \in \mathcal{P}_i} \gamma_{ip}(d_{ip}^l - o_{ip}^l),$$

where π_{ij} is the profit generated by flying leg j with an aircraft of type i. The α_j is the current cost incurred by operating leg j, the β_i is the current cost incurred by using an additional aircraft of type i, and the γ_{ip} is the current cost incurred by allowing an imbalance of one airplane of type i at airport p. In the graph G_i, the potential profit is equal to the sum of the profits on the arcs forming the path corresponding to schedule l (negative profits being costs). Summarizing, in order to compute the potential profit of a schedule $l \in \mathcal{S}_i$, the following values must be assigned to the arcs of G_i:

arc type	profit
source arcs	$-\beta_i$
sink arcs	0
origination airport arcs leaving origination airport p	$+\gamma_{ip}$
termination airport arcs leaving flight node j	$\pi_{ij} - \alpha_j - \gamma_{ip}$
turn arcs emanating from flight node j	$\pi_{ij} - \alpha_j$

By regarding the profits as "distances", a longest path algorithm can now be used to find the path with the largest potential profit in network G_i. For this particular version of the longest path problem it is easy to develop an efficient algorithm (since the time windows in the network make sure that the paths cannot cycle). If the largest potential profit is positive, then this flight schedule is included in the restricted master problem, which is then solved again. If for none of the aircraft types there is a path with a positive potential profit, then the current solution is optimal over all schedules and the generation of schedules terminates.

After this description of the mechanism to generate upper bounds we consider the branching procedure. There are various types of branching strategies. The simplest mechanism is again the basic $0-1$ branching. Select at a node a variable x_i^l which has not been fixed yet at a higher level and generate branches to two nodes at the next level down: one branch for $x_i^l = 1$ and the other branch for $x_i^l = 0$. Setting $x_i^l = 1$ implies that in the corresponding node all the flight legs covered by that schedule can be deleted in G_i, for all $i = 1, \ldots, T$. Moreover, the constraints in the master problem that cover the corresponding flights can be deleted also; other constraints may be adjusted accordingly. Setting $x_i^l = 0$ implies that at that node the corresponding column must be ignored. There are ways to ensure that in the subproblem another path is selected.

Another issue that has to be resolved in the branch-and-bound procedure is the selection of columns that should be considered as candidate schedules at any particular node. One may want to keep all columns that have been generated so far at all nodes previously analyzed. Or, one may want to select through some heuristic a collection of promising schedules with a positive potential profit.

Example 11.3.1 (Aircraft Routing and Scheduling). Consider two types of planes, i.e., $T = 2$. Type 1 planes are widebodies and $m_1 = 2$; type 2 planes are narrowbodies and $m_2 = 2$. Twelve legs have to be flown between four airports. The four airports are:

$$p = 1 : \text{San Francisco (SFO)}$$
$$p = 2 : \text{Los Angeles (LAX)}$$
$$p = 3 : \text{New York (NYC)}$$
$$p = 4 : \text{Seattle (SEA)}$$

For each flight the pair of cities is given as well as a time window in which the plane has to fly and the type(s) of plane that can be used. In this example each leg can be flown by either type of plane and the time it takes is independent of the type. However, the profit made on a flight depends on the type assigned.

Leg j	1	2	3	4	5	6	7	8	9	10	11	12
cities	1→2	1→2	2→1	2→1	1→4	1→4	4→1	4→1	3→1	3→1	1→3	1→3
τ_{1j}	1.5	1.5	1.5	1.5	3	3	3	3	6	6	6	6
time	a.m.	p.m.	a.m.	p.m.	a.m.	p.m.	a.m.	p.m.	a.m.	p.m.	a.m.	p.m.

An a.m. flight must take off after 5 a.m. and land before 1 p.m.; a p.m. flight must take off after 1 p.m. and land before 5 a.m. the next day. Based on these flight data it can be checked easily whether a specific trip for a plane is feasible. Initial sets of candidate schedules can be generated for both types of planes. An initial subset $\mathcal{S}_1$ is presented below.

Schedules	a_{1j}^1	a_{1j}^2	a_{1j}^3	a_{1j}^4	a_{1j}^5	a_{1j}^6	a_{1j}^7	a_{1j}^8	a_{1j}^9	a_{1j}^{10}
flight 1	0	0	1	1	0	0	0	0	0	0
flight 2	0	0	0	0	0	0	0	1	0	1
flight 3	0	0	0	1	0	0	0	0	0	0
flight 4	0	0	1	0	0	0	0	1	0	0
flight 5	0	1	1	1	0	0	0	0	0	0
flight 6	0	0	0	0	1	0	0	1	0	1
flight 7	0	1	1	0	1	0	1	0	0	0
flight 8	0	0	0	1	0	0	0	0	0	1
flight 9	0	0	0	0	0	1	0	1	0	1
flight 10	0	1	0	0	1	1	1	0	1	0
flight 11	0	0	0	0	1	1	1	0	1	0
flight 12	0	1	0	0	0	1	1	0	1	0

An initial subset $\mathcal{S}_2$ is given in Table 11.1.
Note that the first candidate schedule in both $\mathcal{S}_1$ and $\mathcal{S}_2$ is the empty schedule in which the plane is kept idle. The o_{ip}^l and d_{ip}^l values of each one of these schedules can be found in Section 5 of the accompanying CD-ROM.

11.3 Aircraft Routing and Scheduling

Schedules	a_{2j}^1	a_{2j}^2	a_{2j}^3	a_{2j}^4	a_{2j}^5	a_{2j}^6	a_{2j}^7	a_{2j}^8	a_{2j}^9
flight 1	0	1	0	1	0	0	0	0	0
flight 2	0	0	0	0	0	0	1	0	1
flight 3	0	0	0	1	0	0	0	1	0
flight 4	0	1	0	0	0	0	1	0	0
flight 5	0	1	0	0	0	0	0	0	0
flight 6	0	0	1	1	0	0	1	0	1
flight 7	0	1	1	0	0	1	0	0	0
flight 8	0	0	0	1	0	0	0	0	1
flight 9	0	0	0	0	1	0	1	0	1
flight 10	0	0	1	0	1	1	0	1	0
flight 11	0	0	1	0	1	1	0	1	0
flight 12	0	0	0	0	1	1	0	1	0

Table 11.1. Table for Example 11.3.1.

The profits generated by assigning an aircraft of type i to flight leg j is π_{ij}. The π_{ij} are tabulated below.

Leg j	1	2	3	4	5	6	7	8	9	10	11	12
π_{1j}	450	300	500	400	900	900	900	900	1500	1500	1500	1500
π_{2j}	450	450	500	500	1000	1000	1000	1000	1350	1350	1350	1350

The profit associated with the assignment of schedule (or round trip) l to a plane of type i is π_i^l. For the large planes of type 1 the profits are tabulated below.

schedule l	1	2	3	4	5	6	7	8	9	10
π_1^l	-375	4800	2650	2750	4800	6000	5400	3100	4500	3600

Note that keeping an aircraft of type 1 idle actually costs money. For the small planes of type 2 the profits are tabulated below.

schedule l	1	2	3	4	5	6	7	8	9
π_2^l	0	2950	4700	2950	5400	5050	3300	4550	3800

Solving the linear relaxation of the restricted master problem results in an optimal solution that sets $x_1^4 = x_1^7 = x_1^8 = 2/3$ and $x_2^1 = 1$ and $x_2^2 = x_2^8 = x_2^9 = 1/3$. The optimal value of the objective function is 11,267.

By applying a standard branch-and-bound procedure to the integer program (see Figure 11.3), it can be verified that the optimal integer solution for this restricted master problem has a total profit of 10,725 and that the optimal solution has $x_1^1 = x_1^4 = 1$ and $x_2^6 = x_2^7 = 1$.

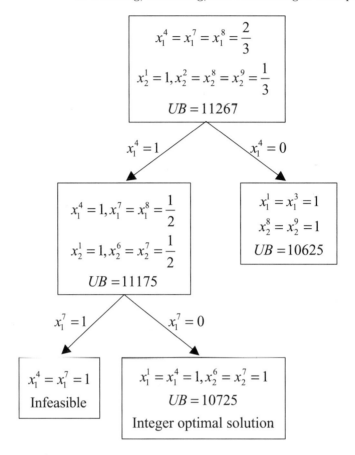

Fig. 11.3. Branch-and-bound tree for aircraft scheduling problem

In order to determine which schedules (round trips) are feasible and have a potential profit if they were included in the restricted master problem, the subproblem has to be formulated. The dual variables (or shadow prices) generated in the linear programming solution of the restricted master problem are the following:

Leg j	1	2	3	4	5	6	7	8	9	10	11	12
α_j	0	-616.7	216.7	2650	2550	2733.3	-2250	0	1500	4333.3	0	183.3

The dual variables corresponding to the second set of constraints are $\beta_1 = -16.67$ and $\beta_2 = 0$. (It was to be expected that the β_2 would be zero since one of the small planes is kept idle in the optimal solution.) The dual variables corresponding to the third set of constraints are tabulated below:

11.3 Aircraft Routing and Scheduling

airport p	1	2	3	4
γ_{1p}	0	0	0	3150
γ_{2p}	0	0	183.33	2966.67

Note that the dual variables associated with the optimal solution of the linear programming solution are not necessarily a unique set. The dual variables generated may depend very much on the type of linear programming implementation, see Exercise 11.4.

In the network G_1 for widebodies a number of candidate schedules can be generated. These candidate schedules involve flight legs 1, 2, 3, 4, 5, 7, 12. The durations of these flight legs and the time windows during which they have to take place are given.

Leg j	1	2	3	4	5	7	12
τ_{1j} (hours)	1.5	1.5	1.5	1.5	3	3	6
e_{1j}	5 a.m.	1 p.m.	5 a.m.	1 p.m.	5 a.m.	5 a.m.	1 p.m.
ℓ_{1j}	1 p.m.	5 a.m.	1 p.m.	5 a.m.	1 p.m.	1 p.m.	5 a.m.

The candidate schedules in the table below for the large planes are feasible and of each one of these candidate schedules the potential profit can be computed.

Schedules	a_{1j}^{11}	a_{1j}^{12}	a_{1j}^{13}	a_{1j}^{14}	a_{1j}^{15}
flight 1	0	1	0	1	1
flight 2	0	0	0	0	0
flight 3	1	1	0	1	1
flight 4	0	0	0	0	0
flight 5	0	0	0	0	0
flight 6	0	1	1	1	1
flight 7	0	0	1	1	1
flight 8	0	1	0	0	0
flight 9	0	0	0	0	0
flight 10	1	0	1	0	1
flight 11	1	0	1	0	0
flight 12	1	0	0	0	1
profit π_1^l	5000	2750	4800	2750	5750
dual variables	4716.67	2933.34	4800	683.34	5200
potential profit $\bar{\pi}_1^l$	283.33	-183.33	0	2066.67	550

Graph G_1 with schedules 11 and 12 is shown in Figure 11.4. A plane of type 1 that follows schedule 11 starts out in Los Angeles and ends up in New York after covering legs 3, 11, 10, 12 (in that order). That is, the plane flies from Los Angeles to San Francisco, then to New York, back to San Francisco and again to New York. The potential profit of schedule 11 is

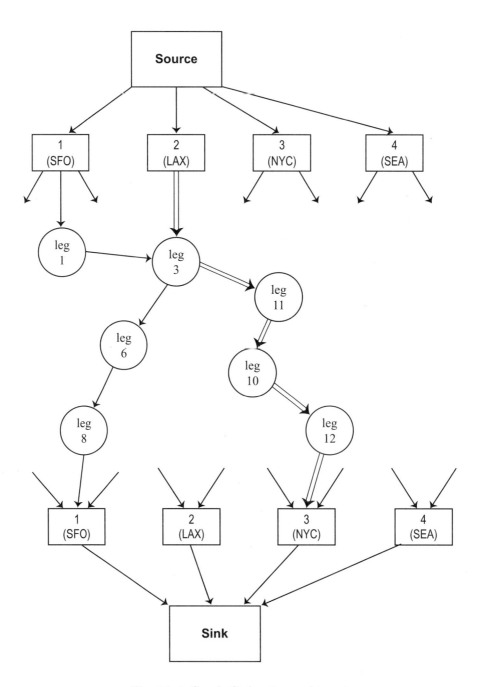

Fig. 11.4. Graph G_1 for planes of type 1

11.3 Aircraft Routing and Scheduling

$$\bar{\pi}_1^{11} = (500 - 216.67) + (1500 - 0) + (1500 - 4333.33)$$
$$+ (1500 - 183.33) - 16.67 - 0$$
$$= 283.33$$

A plane that follows schedule 12 starts out in San Francisco and ends up back in San Francisco after covering legs 1,3,6,8 (in that order). However, the potential profit of this schedule is negative:

$$\bar{\pi}_1^{12} = (450-0)+(500-216.67)+(900-2733.33)+(900-0)-16.67-0 = -183.33$$

From the table it follows that schedules 11, 14, and 15 should be included in $\mathcal{S}_1$ for the restricted master problem. So in the new restricted master problem $\mathcal{S}_1$ has $10 + 3 = 13$ schedules.

For the narrowbodies a network G_2 has to be set up and candidate schedules have to be examined. There are 6 additional schedules for narrowbodies that have to investigated. The results are tabulated below.

Schedules	a_{2j}^{10}	a_{2j}^{11}	a_{2j}^{12}	a_{2j}^{13}	a_{2j}^{14}	a_{2j}^{15}
flight 1	0	1	0	1	1	1
flight 2	1	1	0	0	0	1
flight 3	0	1	0	1	1	1
flight 4	1	1	0	0	0	1
flight 5	1	1	0	0	0	0
flight 6	0	0	1	1	1	0
flight 7	1	0	1	1	1	0
flight 8	0	1	0	0	0	0
flight 9	0	0	0	0	0	0
flight 10	0	0	1	0	1	0
flight 11	0	0	1	0	0	0
flight 12	0	0	0	0	1	0
profit π_2^l	2950	3900	4700	2950	5650	1900
dual variables	2233.33	4700	4816.67	700	5216.67	2250
potential profit $\bar{\pi}_2^l$	716.67	-800	-116.67	2250	433.33	-350

It follows that schedules 10, 13, and 14 have to be included in $\mathcal{S}_2$. In the new restricted master problem $\mathcal{S}_1$ has $10+3 = 13$ columns (schedules) and $\mathcal{S}_2$ has $9 + 3 = 12$ columns.

Solving the new restricted master problem via linear programming yields the solution $x_1^4 = x_1^6 = x_1^7 = x_1^8 = 0.5$ and $x_2^1 = 1$ and $x_2^4 = x_2^{10} = 0.5$. Again, one of the narrowbodies is kept idle. The value of the objective function is 11,575 (which, of course, has to be at least as high as the value 11,267 of the first linear program, since now a larger number of schedules is taken into consideration).

Imposing the integrality constraints on this mathematical program and solving this restricted master problem as an integer program via branch-and-bound (see Figure 11.5.) yields the solution $x_1^{10} = x_1^{11} = 1$ and $x_2^1 = x_2^2 = 1$. The value of the objective function is 11,550.

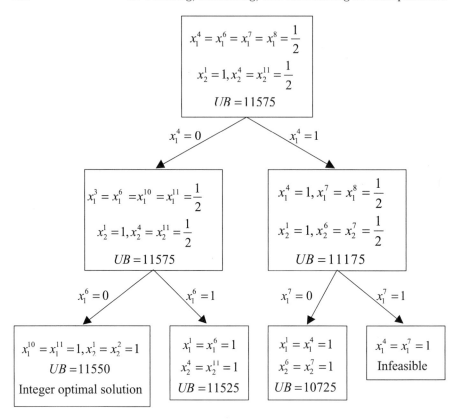

Fig. 11.5. Optimal solution for aircraft scheduling problem

11.4 Train Timetabling

The most common train timetabling problem focuses on a single, one way track that links two major stations with a number of smaller stations in between. A train may or may not stop at a smaller station. This case is of interest because railway networks usually contain a number of important lines, referred to as corridors, that connect major stations. These corridors are made up of two independent one-way tracks that carry traffic in opposite directions. Once the timetables for the trains in the corridors have been determined, it is relatively easy to find a suitable timetable for the trains on the other lines in the network.

The time is typically measured in minutes, from 1 to q, where q represents the length of the given period, e.g., from 1 to 1440 when the period is one day (similar to the airline industry). Link j connects station $j-1$ with station j. There are L consecutive links (numbered 1 to L) and there are $L+1$ stations (numbered 0 to L). Stations 0 and L are the first and last station. Let T denote the set of trains that are candidates to run every period. Let T_j

11.4 Train Timetabling

denote the set of trains that intend to pass through link j. The set T_j may be a subset of T because a train may start out from an intermediate station and/or end up at an intermediate station. Track capacity constraints ensure that one train cannot pass another on the single track between two stations. A train can overtake another only at a station, when the train that is being overtaken makes a stop. Train schedules are usually depicted in a so-called time-space diagram (see Figures 11.6 and 11.7). Such a diagram enables the user to detect conflicts, such as insufficient headways.

For each train there is an ideal timetable, which is the most desirable timetable for that train. This ideal timetable is determined by analyzing passenger behavior and preferences. However, this timetable may be modified in order to satisfy track capacity constraints. It is possible to slow down a train and/or increase its stopping (dwelling) time at a station. Moreover, one can modify the departure time of each train from its first station, or even cancel a train. The final solution of the timetabling problem is referred to as the actual timetable. An actual timetable specifies for train i, $i \in T$, its departure time from its first station and its arrival time at its last station. The timetable of each train is periodic, i.e., it is kept unchanged every period.

The train timetabling problem involves determining the values of various sets of variables, namely the times of arrivals and departures of train i at all stations. In this section the following notation is used: the (continuous) decision variable

y_{ij} = the time train i enters link j (i.e., the time train i departs from station $j-1$);
z_{ij} = the time train i exits link j (i.e., the time train i arrives at station j).

When a timetable is put together there are usually some predetermined arrival and departure times for certain trains at specific stations and some preferred arrival and departure times for other trains. There is a cost (or revenue loss) associated with deviating from these preferred arrival and departure times. If z_{ij} is the arrival time of train i at station j, then the cost function that specifies the revenue loss due to a deviation from this preferred arrival time is denoted by $c_{ij}^a(z_{ij})$. This function may be piece-wise linear or convex with a minimum at the most preferred departure time, see Figure 11.8. A piece-wise linear function has computational advantages, since the optimization problem can be handled in a way that takes advantage of the linearity.

Similarly, there are costs associated with the departure time of train i from station j, i.e., $y_{i,j+1}$, the travel (trip) time of train i on link j, denoted by

$$\tau_{ij} = z_{ij} - y_{ij} \qquad j = 1, \ldots, L,$$

and the stopping (dwelling) time at station j, denoted by

$$\delta_{ij} = y_{i,j+1} - z_{ij} \qquad j = 1, \ldots, L-1.$$

There are deviation costs associated with each one of these quantities, namely c_{ij}^d, c_{ij}^τ, c_{ij}^δ, which are also piece-wise linear and convex.

274 11 Planning, Scheduling, and Timetabling in Transportation

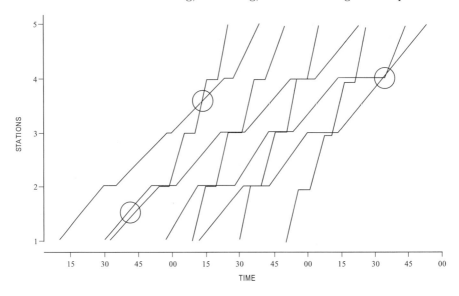

Fig. 11.6. Train Time-Distance (Pathing) Diagrams

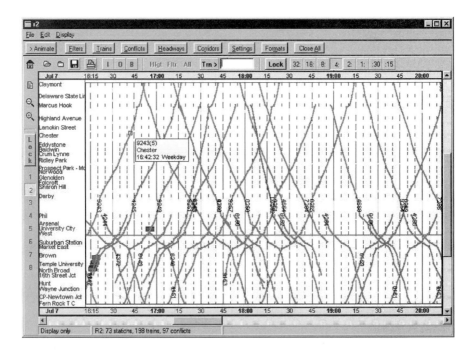

Fig. 11.7. Train diagram graphical user interface

11.4 Train Timetabling

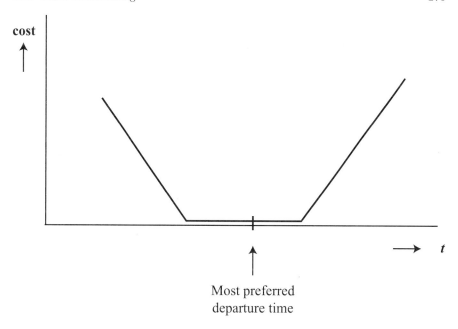

Fig. 11.8. Cost as a Function of Departure Time

The general objective to be minimized in the train timetabling problem is

$$\sum_{i \in T} \sum_{j=1}^{L} \left(c_{ij}^a(z_{ij}) + c_{i,j-1}^d(y_{ij}) + c_{ij}^\tau(z_{ij} - y_{ij}) \right) + \sum_{i \in T} \sum_{j=1}^{L-1} \left(c^\delta(y_{i,j+1} - z_{ij}) \right)$$

The variables are subject to various sets of operational constraints. For example, train i needs at least a minimum time $\tau_{ij}^{\min}$ to traverse link j; train i must stop at station j for a minimum amount of time $\delta_{ij}^{\min}$ to allow passengers to board. For reasons of safety and reliability minimum headways have to be maintained on each link. Let H_{hij}^d be the minimum headway required between the departures $y_{h,j+1}$ and $y_{i,j+1}$ of trains h and i from station j and let H_{hij}^a denote the minimum headways between the arrivals z_{hj} and z_{ij} of trains h and i at station j (ensuring adequate headway between trains h and i when exiting link j). Moreover, there may be upper and lower bounds on all arrival and departure times.

Formulating the timetabling problem as a Mixed Integer Program (MIP) requires a set of $0-1$ variables: the decision variable x_{hij} assumes the value 1 if train h immediately precedes train i on link j, and 0 otherwise. Also, to make the formulation easier, two dummy (artificial) trains i' and i'' are included in set T: train i' has fixed arrival and departure times ensuring that it precedes every other train on each link and train i'' also has fixed arrival and departure times ensuring that it will be the last train on each link. The timetabling problem can now be formulated as the following mixed integer program.

minimize

$$\sum_{i\in T}\sum_{j=1}^{L}\left(c_{ij}^a(z_{ij}) + c_{i,j-1}^d(y_{ij}) + c_{ij}^\tau(z_{ij} - y_{ij})\right) + \sum_{i\in T}\sum_{j=1}^{L-1}\left(c^\delta(y_{i,j+1} - z_{ij})\right)$$

subject to

$$y_{ij} \geq y_{ij}^{\min} \qquad i \in T, \ j = 1,\ldots,L$$
$$y_{ij} \leq y_{ij}^{\max} \qquad i \in T, \ j = 1,\ldots,L$$
$$z_{ij} \geq z_{ij}^{\min} \qquad i \in T, \ j = 1,\ldots,L$$
$$z_{ij} \leq z_{ij}^{\max} \qquad i \in T, \ j = 1,\ldots,L$$
$$z_{ij} - y_{ij} \geq \tau_{ij}^{\min} \qquad i \in T, \ j = 1,\ldots,L$$
$$y_{i,j+1} - z_{ij} \geq \delta_{ij}^{\min} \qquad i \in T, \ j = 1,\ldots,L-1$$
$$y_{i,j+1} - y_{h,j+1} + (1 - x_{hij})M \geq H_{hij}^d \qquad i \in T, \ j = 0,\ldots,L-1$$
$$z_{ij} - z_{hj} + (1 - x_{hij})M \geq H_{hij}^a \qquad i \in T, \ j = 1,\ldots,L$$
$$\sum_{h\in\{T-i\}} x_{hij} = 1 \qquad i \in T, \ j = 1,\ldots,L$$
$$x_{hij} \in \{0,1\}$$

Most of the constraint sets are self-explanatory. However, some of the constraint sets may warrant some explanation. The value M in the seventh and eighth constraint set is made very large for the following reason: if train h is not immediately preceding train i on link j, then $x_{hij} = 0$ and the inequalities with an M on the left hand side are automatically satisfied; if train h precedes train i on link j, then $x_{hij} = 1$ and the headway constraint must be enforced.

Since the integer program above represents only a problem concerning a single link and the railway system typically has to solve a network consisting of many links, it is important to solve the single link problem fast. One heuristic is based on decomposition. According to this heuristic the trains are scheduled one at the time and a solution for the overall problem is generated by solving a series of single train subproblems. In a given iteration it is assumed that in previous iterations the sequences have been determined in which trains belonging to set T_0 go through the different links. This set T_0 includes the two artificial (dummy) trains i' and i'' which are the first and last train on each link. The next iteration has to select a train from set $T - T_0$ for inclusion in set T_0 and determine for each link where this train is inserted in the sequence of trains.

The manner in which the trains are selected for inclusion in set T_0 can affect the quality of the solution generated as well as the speed at which the solution is obtained. One very simple priority rule selects the trains in the order of their desired starting times. Another possible priority rule selects the

11.4 Train Timetabling

trains in decreasing order of their importance; the importance of a particular train may be measured by the train's type, speed, and expected revenue. For example, a descending order of importance may be: express trains without stops, express trains with stops, local trains and freight trains. A third rule would select a train that has little flexibility in its arrival and departure times. A fourth priority rule can be designed by combining the first three rules.

Assume that in a given iteration train k is selected for inclusion in set T_0. So, the trains in set T_0 have already been scheduled in the previous iteration. Even though the computations in the previous iteration determined exact arrival and departure times of the trains at all stations, the only information that is carried forward to the current iteration is the sequence or order in which the trains in set T_0 traverse each link (the exact arrival and departure times are disregarded). Let I_j denote the vector of train indices that specifies the order in which the trains in T_0 are scheduled to traverse link j. It is assumed that in this order I_j train i is immediately followed by train i^*.

The current iteration must schedule train k on each link while maintaining the order I_j in which the trains in T_0 traverse link j, i.e., train k is inserted somewhere in vector I_j. However, even though the orderings $I_j, j = 1, \ldots, L$, are maintained, the exact arrival and departure times of the trains in T_0 are allowed to vary after the inclusion of k. The subproblem that inserts train k in the vectors $I_1, \ldots, I_L$ is a mixed integer program, say MIP(k), which is similar but simpler than the MIP formulated for the original problem.

In order to simplify the formulation of MIP(k), include train k in set T_0. MIP(k) inserts train k in each vector I_j and determines the exact arrival and departure times for all trains in set T_0. The objective function is the same as the objective function of the MIP for the original problem. However, the constraint sets have to be modified. A different $0-1$ variable is used in MIP(k), namely x_{ij} which is 1 if train k is inserted on link j immediately after train i and 0 otherwise. That is, x_{ij} is 1 if train k is scheduled in between train i and the train immediately following train i, which in vector I_j is referred to as train i^*. Let set T_0 now include all trains already scheduled in the previous iterations as well as train k. The integer program MIP(k) can be formulated as follows:

minimize
$$\sum_{i \in T_0} \sum_{j=1}^{L} \left(c_{ij}^a(z_{ij}) + c_{i,j-1}^d(y_{ij}) + c_{ij}^\tau(z_{ij} - y_{ij}) \right) + \sum_{i \in T_0} \sum_{j=1}^{L-1} \left(c^\delta(y_{i,j+1} - z_{ij}) \right)$$

subject to
$$y_{ij} \geq y_{ij}^{\min} \quad i \in T_0, \ j = 1, \ldots, L$$
$$y_{ij} \leq y_{ij}^{\max} \quad i \in T_0, \ j = 1, \ldots, L$$
$$z_{ij} \geq z_{ij}^{\min} \quad i \in T_0, \ j = 1, \ldots, L$$
$$z_{ij} \leq z_{ij}^{\max} \quad i \in T_0, \ j = 1, \ldots, L$$

$$z_{ij} - y_{ij} \geq \tau_{ij}^{\min} \quad i \in T_0, \ j = 1, \ldots, L$$
$$y_{i,j+1} - z_{ij} \geq \delta_{ij}^{\min} \quad i \in T_0, \ j = 1, \ldots, L-1$$
$$y_{i^*,j+1} - y_{i,j+1} \geq H_{ii^*j}^{d} \quad i \in I_j, \ i \neq i'', \ j = 0, \ldots, L-1$$
$$y_{k,j+1} - y_{i,j+1} + (1 - x_{ij})M \geq H_{ikj}^{d} \quad i \in I_j, \ i \neq i'', \ j = 0, \ldots, L-1$$
$$y_{i^*,j+1} - y_{k,j+1} + (1 - x_{ij})M \geq H_{ki^*j}^{d} \quad i \in I_j, \ i \neq i'', \ j = 0, \ldots, L-1$$
$$z_{i^*j} - z_{ij} \geq H_{ii^*j}^{a} \quad i \in I_j, \ i \neq i'', \ j = 1, \ldots, L$$
$$z_{kj} - z_{ij} + (1 - x_{ij})M \geq H_{ikj}^{a} \quad i \in I_j, \ i \neq i'', \ j = 1, \ldots, L$$
$$z_{i^*j} - z_{kj} + (1 - x_{ij})M \geq H_{ki^*j}^{a} \quad i \in I_j, \ i \neq i'', \ j = 1, \ldots, L$$
$$\sum_{i \in \{T_0 - k - i''\}} x_{ij} = 1 \quad j = 1, \ldots, L$$
$$x_{ij} \in \{0, 1\}$$

Thus MIP(k) has the same continuous variables y_{ij} and z_{ij} as the original MIP, but contains far fewer $0-1$ variables than the original MIP. Now the binary $0-1$ variables are x_{ij} rather than x_{hij}.

Before solving MIP(k) it is advantageous to do some preprocessing in order to reduce the number of $0-1$ variables and the number of constraints. Many of the $0-1$ variables x_{ij} are likely to be redundant, due to upper and lower bounds in other constraints on the departure times, arrival times, stopping times, and overall trip times. For example, the variable x_{hj} is not needed if train k cannot be scheduled immediately after train h on link j (i.e., in between trains h and h^*), because of a lack of sufficient time in between trains h and h^* (not allowing for the required headways).

Solving MIP(k) can be done through branch-and-bound. There are several strategies one can follow in the branching process. At a given node in the branching tree, all branches correspond to a certain link j. More specifically, there is a branch for each member of the set of trains already scheduled that train k would be able to follow. This multiple branching (branching dependent on the trains already scheduled) can be combined with fixed order branching that is based on the links. That is, the links are in the order in which they are traversed by the trains.

Summarizing, the following heuristic framework can be used to find a good feasible (or optimal) solution for the original MIP.

Algorithm 11.4.1 (Train Timetabling).

Step 1. *(Initialization)*

  Introduce two "dummy trains" as the first and last trains.

Step 2. *(Select an Unscheduled Train and Solve its Pathing Problem)*

  Select the next train k through the train selection priority rule.

Step 3. *(Set up and Preprocess Mixed Integer Program)*

  Include train k in set T_0.

Set up MIP(k) for the selected train k.

Preprocess MIP(k) to reduce number of $0-1$ variables and constraints.

Step 4. *(Solve Mixed Integer Program)*

Solve MIP(k). If algorithm does not yield feasible solution, STOP.

Otherwise, add train k to the list of already scheduled trains and fix for each link the sequences of all trains in T_0.

Step 5. *(Reschedule All Trains Scheduled Earlier)*

Consider the current partial schedule that includes train k.

For each train $i \in \{T_0 - k\}$ delete it and reschedule it.

Step 6. *(Stopping Criterion)*

If T_0 consists of all trains, then STOP;

otherwise go to Step 2.

Of course, the framework presented above is merely a heuristic. It does not guarantee to yield an optimal solution. The following example illustrates how the heuristic may result in a suboptimal solution.

Example 11.4.2. Consider four trains that have to be scheduled on a single link. Two trains (A and D) are fast trains with travel times τ_1 and two trains (B and C) are slow trains with travel times τ_2. Train A must depart at a fixed time, say 9 a.m., and the other three trains have to either arrive or depart at a time that is as close to 9 a.m. as possible. If the trains are introduced one at a time in the order A, B, C, D, then the resulting schedule is as depicted in Figure 11.9 (a). This solution is a local optimum. The global optimum is as depicted in Figure 11.9 (d).

The decomposition framework described above is in some respects very similar to the shifting bottleneck heuristic described in Chapter 5 for job shop scheduling. First, the train timetabling framework schedules one train at the time. In each iteration, the arrival times and departure times of all the trains in set T_0 are computed; however, the arrival times and departure times are subject to modification in subsequent iterations. Third, after scheduling in an iteration one additional train, it reschedules all the trains that had been scheduled in the previous iterations.

11.5 Carmen Systems: Designs and Implementations

Several software companies specialize in planning and scheduling systems for the transportation industries. Carmen Systems AB, based in Goteborg (Sweden), is one of the more successful companies in this domain. Carmen Systems offers a suite of products designed for the airline industry (see Figure 11.10).

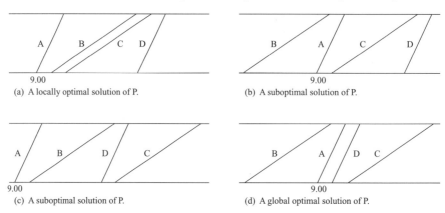

Fig. 11.9. Train Schedules for Example 11.4.2

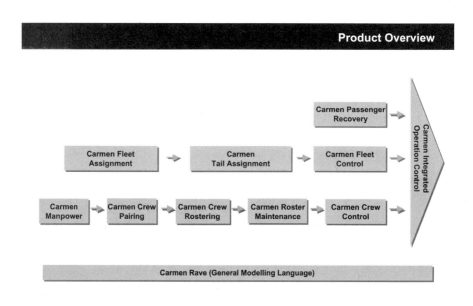

Fig. 11.10. Carmen systems product overview

11.5 Carmen Systems: Designs and Implementations

Carmen's product line is based on Carmen's proprietary modelling language *Rave*. This modelling language makes it easy for the user to describe and implement cost functions, feasibility conditions, and quality constraints. This language supports all Carmen's products and is essential for the fine-tuning of everyday schedules and for conducting simulations. The language is designed in such a way that a program can be linked easily to advanced optimization routines. A planner who uses a system based on Rave can easily implement modifications in the operation principles of the airline by simply switching a rule in the code on or off, or by adjusting the value of a parameter in a rule.

Example 11.5.1. A rule ensuring that it is impossible to have a crew assignment longer than 10 hours can be written as follows:

    rule max-duty =
        %duty% <= 10:00;
        remark "Maximal duty";
    end

The following is an example of a cost function that consists of components for salary, hotel, per diem and flight positioning:

    %pairing__cost% =
        %salary% +
        %hotel% +
        %per__diem% +
        %positioning% +
        ....

Carmen's clients typically have their own rule programmers who ensure that Carmen's products keep on solving the right problems in a changing environment. Carmen Rave also enables users to perform simulation studies and consider what-if scenarios. The rules that are embedded in one module are automatically followed in another module. For example, crew connection rules established in Carmen Crew Pairing are automatically adhered to in Carmen Tail Assignment.

Two of the more important modules in the Carmen product suite are the *Fleet Assignment* module and the *Tail Assignment* module. Fleet Assignment is the process of assigning fleets or aircraft types to the flights in the schedule; this module is mainly concerned with generating a schedule that is balanced (e.g., at any airport, the number of arrivals of aircraft of any given type has to be equal to the number of departures of aircraft of that type), but is not concerned with individual aircraft and their operational constraints, such as maintenance.

The Tail Assignment, on the other hand, is the problem of deciding which individual aircraft (specified by its tail number) should cover which flight. The main focus of the tail assignment is to verify whether the schedule is feasible; it has to deal therefore with individual aircraft and take operational constraints into account. Such operational constraints include:

(i) minimum connection times and maintenance constraints;
(ii) route constraints, e.g., only aircraft equipped to fly overseas can be assigned to such flights;
(iii) that only aircraft with "hush kits" can be sent to noise-restricted airports;
(iv) destination constraints, e.g., only aircraft that do not require an external power supply can be sent to airports where such a service is unavailable.

The Tail Assignment also constrains the crew planning. If a crew remains with the same aircraft on two consecutive legs, then the connection time between the two consecutive flights in a crew roster can be shorter than if the crew had to change aircraft. By minimizing the connection times in a crew roster, a schedule can be created with less crew. Furthermore, a crew remaining with an aircraft makes the schedule more robust and less vulnerable to disturbances that occur during the day of operation. By allowing the crew planning to influence the Tail Assignment, it is possible to reduce costs and increase stability at the same time. This implies an additional operational constraint:

(v) fixed links between flights to protect crew rosters. These constraints can also be modeled as costs incurred when links are broken.

The Tail Assignment usually takes the Fleet Assignment as given. However, many airlines are now looking into the possibility of being more adaptive to actual booking levels and change the fleet assignment close to the day-of-operation in order to accomodate more passengers. This approach is referred to as *Demand Driven Dispatch*. This changes the formulation of the aircraft routing and scheduling problem described in Section 11.3 from a cost minimization problem to a profit maximization problem. A revenue component has to be added to the objective function in order to make sure that the revenue does depend on the type of aircraft that is assigned to the flight. Instead of minimizing costs c_i^l, one has to maximize profits π_i^l; the profit π_i^l is equal to the revenue generated with the assignment of aircraft i to schedule l minus c_i^l. In practice, the problem formulation has more constraints than the formulation in Section 11.3; there are usually additional constraints that ensure that the aircraft assignments are feasible given the already established crew schedules (the crew schedules are often generated before the aircraft assignments are made).

In the pure Tail Assignment application, the optimizer is used to generate aircraft routes that automatically satisfy all feasibility constraints (with regard to minimum turn-around times, maintenance times, and so on). The objective includes maximizing aircraft utilization, prioritizing robust connections with as few critical aircraft changes as possible (i.e., maximizing schedule stability and robustness). The output of the optimization module is the most efficient combination of aircraft rotations. By using the aircraft efficiently it is possible to free up more aircraft for stand-by duty. An aircraft being idle for 2-3 hours may be a waste of a resource, whereas an aircraft being on the ground for 6 hours or more may be equivalent to a plane being available for stand-by duty. By promoting, at the same time, very short and very long connections, the number of planes available for stand-by duty can be increased considerably.

When the system is used according to *Demand Driven Dispatch*, revenue information is used in order to determine if it is advantageous to change aircraft types. Most major airlines have crews that can operate different aircraft types with varying seat capacity. This enables the airline to make late adjustments to the fleet assignments without having to change the crew rosters. The optimizer allocates the aircraft in the most efficient way, balancing the revenues and costs. The user (i.e., the scheduler) determines the flights for which it is permitted to change the fleet assignment. The optimizer then solves the problem for the individual tails with profit maximization as objective. Also, because of Rave's simulation capabilities it is possible to make much more informed decisions when new schedules are being created.

The graphical user interface (GUI) enables the planner to add or modify relevant information. The GUI also provides an overview of solutions and input data in the form of Gantt charts, see Figure 11.11. Through the GUI the user can select flights or only a subset of the plan to modify and improve. He or she can insert maintenance activities for specific planes in the schedule, and so on.

The report generator makes it possible to issue reports at any time (e.g., maintenance reports, airport stand-bys, and crew stability analysis). Reports can be displayed on screen or on paper. The report generator can also export data files to other decision support systems.

Carmen has installed systems for numerous airlines, including British Airways, Northwest Airlines, Singapore Airlines, and AeroMexico. When a system is introduced, savings in the order of 5% are typically achieved immediately. However, airlines expect the long term savings to be significantly higher.

11.6 Discussion

The three transportation settings considered in this chapter have similarities as well as differences. All three problems lead to integer programming formulations. In one case the model results in a set packing formulation and in another case the model results in a set partitioning formulation.

There are various similarities between aircraft scheduling and railway timetabling. Both aircraft scheduling and railway timetabling require cyclic (periodic) schedules; the periods are typically one day (with international flights the periods may be one week). In both aircraft scheduling and railway timetables the input requirements and the constraints on the schedules are determined in advance by marketing departments. The analysis of customer behavior and preferences determine, within time ranges, the ideal departure

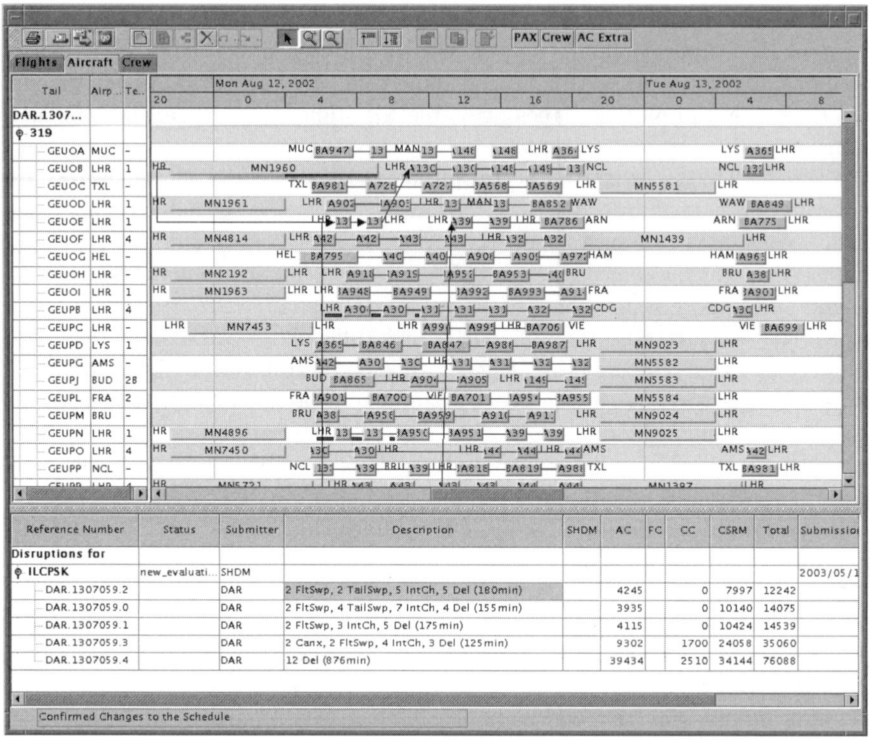

Fig. 11.11. Carmen Systems aircraft scheduling graphical user interface

times of the planes and trains. In railway timetabling as well as in aircraft scheduling weekend days are different from weekdays.

There are of course also some differences between aircraft scheduling and railway timetabling. A flight leg between two airports has to be covered by a plane in the same way as a link between two stations has to be covered by a train. However, flight legs have to be combined with one another into round trips, whereas trains have to move in major corridors.

There are many differences between, on the one hand, tanker scheduling problems and, on the other hand, aircraft scheduling and train timetabling problems. Tanker and shipping schedules are typically not cyclic. Schedules are generated on a rolling horizon basis, planning ahead for three months. It tends to be less complex than the aircraft scheduling problem and the railway timetabling problem. (The sources of complexities in aircraft scheduling are based on the fact that flight legs have to be combined into round trips, and in railway timetabling that tracks have capacity constraints and bypassing is only allowed at stations).

Another important difference between tanker scheduling and aircraft scheduling is based on the fact that in tanker scheduling each ship has its

11.6 Discussion

own identity. In aircraft scheduling, a distinction is made between aircraft types; aircraft that are of the same type are interchangeable. It is important for an airline not to have to make a distinction between aircraft of the same type. This way, the size of the integer program can be reduced considerably.

Another difference between aircraft scheduling and tanker scheduling lies in the objective function. The objective function in the tanker scheduling problem takes into account that certain cargoes may be carried by ships that are not company-owned. The necessary payments are determined on the spot market. Any cargo that is transported on a company ship represents a saving or profit.

There are many similarities between transportation scheduling models and machine scheduling models in manufacturing: The tanker scheduling problem and the aircraft routing and scheduling problem are somewhat similar to parallel machine scheduling problems. Tankers and planes correspond to machines, while cargoes and flight legs correspond to jobs. A fleet of tankers or aircraft may therefore be considered equivalent to a parallel machine environment. Transporting a cargo or covering a leg is similar to processing a job. The jobs have certain processing times and also time windows during which they have to be processed. When a tanker transports one cargo followed by another, a certain type of sequence dependent setup time and setup cost is incurred. The same is true when a given plane flies one leg followed by another. The objectives are, of course, very problem specific. In the aircraft routing problem the objectives are a function of the sum of earliness and the sum of tardiness costs.

If one compares train timetabling with machine scheduling, then the links must represent the machines, and the trains the jobs. The train timetabling problem is then somewhat similar to a flow shop problem, i.e., a number of machines (links) have to process a set of jobs (trains) one after another. Each job has to be processed first on machine 1 (link 1), then on machine 2, and so on. Note that this is different from a permutation flow shop in which each machine processes all the jobs in the same sequence. One machine may process the jobs in a different sequence than another machine, because in the train scheduling problem, it may occur that at a station one train passes another. However, there is another important difference between the train scheduling problem and the flow shop problem. In a flow shop problem, a machine is not allowed to process more than one job at the same time. In the train scheduling problem, a link may be "processing" more than one train at a time (see Exercise 11.8).

In this chapter we did not cover the routing and scheduling of trucks. However, the routing and scheduling of trucks and tanker trucks (which are often referred to as road tankers) is somewhat similar to the routing and scheduling of the oil tankers described in the second section.

In many transportation settings there are other important scheduling problems that have to be dealt with as well. For example, aircraft scheduling affects crew scheduling. The crew of an aircraft is a very important cost component

Exercises

11.1. Consider the tanker scheduling problem in Example 11.2.1. Suppose that c_2^l, for all l is multiplied by a factor K, $K < 1$. At which value of K would tanker 2 be used?

11.2. Consider again the tanker scheduling problem in Example 11.2.1. Suppose that the charter cost of cargo 6 is $2217K$, where K is a factor greater than 1. At what value of of K does it make sense to transport cargo 6 with a company-owned ship?

11.3. In the subproblem of the aircraft routing and scheduling problem, the dual variables represent a certain profit. Section 11.3 provides a clear description of what the variable β_i represents. Give a description of what the variables α_j and γ_{ip} represent.

11.4. Consider Example 11.3.1. The set of dual variables in the first linear programming solution of the restricted master problem are not necessarily unique. Another set of dual variables are the following:

Leg j	1	2	3	4	5	6	7	8	9	10	11	12
α_j	0	-616.7	0	0	0	183.3	2950	2766.7	0	0	1683.3	4333.3

The dual variables corresponding to the second set of constraints are $\beta_1 = -16.67$ and $\beta_2 = 0$. The dual variables corresponding to the third set of constraints are:

airport p	1	2	3	4
γ_{1p}	0	-1500	0	-3550
γ_{2p}	0	-1466.7	0	-3733.3

Compute the potential profit of schedules $a_{1j}^{11}, \ldots, a_{1j}^{15}$ and $a_{2j}^{10}, \ldots, a_{2j}^{16}$. Which schedules should be included now in S_1 and S_2?

11.5. In Example 11.3.1 all flights have to be covered by a plane. Suppose the profit on leg 2 is reduced, i.e., both π_{12} and π_{22} are multiplied with a factor $K < 1$. At what level of K does it make sense not to cover leg 2?

11.6. Redo all the steps in Example 11.3.1 assuming that planes of type 1 are not allowed to fly according to schedule 8 in S_1.

11.7. Consider the Mosel (Dash Optimization) code of Example 11.3.1 that is stored in Section 4 of the CD-ROM. Modify either the code or the data files in such a way that an empty schedule is not available for planes of type 2 (i.e., set x_2^1 equal to 0). Compare the optimal solution with the optimal solution presented in the text.

11.8. Describe the similarities and the differences between a flexible flow shop (see Section 2.2) and the train timetabling problem.

11.9. Consider the following variation of the train timetabling problem in Section 11.4. Instead of two tracks between any two stations there is now a single track between any two stations. The single track has to accomodate the traffic in both directions. At the stations trains can bypass one another in both directions. Develop a heuristic that generates the timetables with the same objectives as described in Section 11.4.

11.10. Consider the train timetabling problem described in Section 11.4. Compare the number of variables and the number of constraints in the MIP with the numbers in MIP(k).

Comments and References

No framework has yet been established for planning and scheduling models in transportation. A series of conferences on computer-aided scheduling of public transport has resulted in a number of very interesting proceedings, see Voss and Daduna (2001), Wilson (1999), Daduna, Branco, and Pinto Paixao (1995), Desrochers and Rousseau (1992), and Wren and Daduna (1988). A volume edited by Yu (1998) considers operations research applications in the airline industry; this volume contains several papers on planning and scheduling in the airline industry. Barnhart, Belobaba and Odoni (2003) present a survey of operations research applications in the air transport industry.

The oil tanker scheduling problem described in the second section is based on the paper by Fisher and Rosenwein (1989) and Perakis and Bremer (1992). Additional details concerning Example 11.2.1 can be found on the CD-ROM accompanying this text. For a slightly different model for ocean transportation of crude oil, see Brown, Graves and Ronen (1987) and for another ship routing model, see Christiansen (1999).

The daily aircraft routing and scheduling formulation is from Desaulniers, Desrosiers, Dumas, Solomon, and Soumis (1997). More details with regard to Example 11.3.1 can be found on the CD that accompanies this book. Desrosiers, Lasry, McInnes, Solomon, and Soumis (2000) developed a routing and scheduling system called ALTITUDE and implemented it at the Canadian airline Transat. Stojkovic, Soumis, Desrosiers, and Solomon (2002) present a model for a real time flight scheduling problem. Keskinocak and Tayur (1998) and Martin, Jones and Keskinocak (2003) consider the scheduling of time-shared jet aircraft. For models that integrate fleet assignment with crew pairing, see Barnhart, Lu and Shenoi (1998) and Cordeau, Stojkovic, Soumis, and Desrosiers (2001).

The train timetabling approach described in this chapter is due to Carey and Lockwood (1995). For another, more elaborate approach, see Caprara, Fischetti and Toth (2002).

Chapter 12

Workforce Scheduling

12.1 Introduction 289
12.2 Days-Off Scheduling 290
12.3 Shift Scheduling 296
12.4 The Cyclic Staffing Problem 299
12.5 Applications and Extensions of Cyclic Staffing .. 301
12.6 Crew Scheduling 303
12.7 Operator Scheduling in a Call Center 307
12.8 Discussion 311

12.1 Introduction

Workforce allocation and personnel scheduling deal with the arrangement of work schedules and the assignment of personnel to shifts in order to cover the demand for resources that vary over time. These problems are very important in service industries, e.g., telephone operators, hospital nurses, policemen, transportation personnel (plane crews, bus drivers), and so on. In these environments the operations are often prolonged and irregular and the staff requirements fluctuate over time. The schedules are typically subject to various constraints dictated by equipment requirements, union rules, and so on. The resulting problems tend to be combinatorially hard.

In this chapter we first consider a somewhat elementary personnel scheduling problem for which there is a relatively simple solution. We then describe an integer programming framework that encompasses a large class of personnel scheduling problems. We subsequently consider a special class of these integer programming problems, namely the cyclic staffing problems. This class of problems has many applications in practice and is easy from a combinatorial point of view. We then consider several special cases and extensions of cyclic staffing. In the sixth section we discuss the crew scheduling problems that occur in the airline industry. In the subsequent section we describe a case that involves the scheduling of operators in a call center.

12.2 Days-Off Scheduling

We first consider a fairly elementary personnel assignment problem. Each day of the week a number of employees have to be present. The number may differ from day to day, but the set of requirements remains the same from week to week. There is a total number of employees available and each has to be assigned a sequence of days. However, the assignment of days to any given employee may be different from one week to the next. By a week, we mean seven days that start with a Sunday and end with a Saturday. The problem is to find the minimum number of employees to cover a seven day a week operation such that the following constraints are satisfied.

(i) The demand per day, $n_j, j = 1, \ldots, 7$, (n_1 is Sunday and n_7 is Saturday) is met.
(ii) Each employee is given k_1 out of every k_2 weekends off.
(iii) Each employee works exactly 5 out of 7 days (from Sunday to Saturday).
(iv) Each employee works no more than 6 consecutive days.

These constraints can have certain effects on the schedule of an employee. For example, if an employee has one weekend off, then he cannot work six days straight and have his next day off on the following Sunday, because he violates then the third constraint (working exactly 5 days out of seven days that start with a Sunday and end with a Saturday). However, an employee can have a consecutive Saturday, Sunday and Monday off, as long as he works after that for at least 5 days in a row. Actually, he could have a weekend off and take again a single day off after 2 or 3 days of work.

We now describe a method that generates an optimal schedule one week at a time, i.e., after the schedule for week i has been set, the schedule for week $i + 1$ is determined, and so on. It turns out that there exists a cyclic optimal schedule that, after a number of weeks, repeats itself.

There are three simple lower bounds on the minimum size of the workforce, W. First, there is the weekend constraint. The average number of employees available each weekend must be sufficient to meet the maximum weekend demand. In k_2 weeks each employee is available for $k_2 - k_1$ weekends. So, assuming that (as near as possible) the same number of workers get each of the k_2 weekends off:

$$(k_2 - k_1)W \geq k_2 \max(n_1, n_7)$$

and therefore

$$W \geq \lceil \frac{k_2 \max(n_1, n_7)}{k_2 - k_1} \rceil.$$

12.2 Days-Off Scheduling

Second, there is the total demand constraint. The total number of employee days per week must be sufficient to meet the total weekly demand. Since each employee works five days per week,

$$5W \geq \sum_{j=1}^{7} n_j$$

or

$$W \geq \lceil \frac{1}{5} \sum_{j=1}^{7} n_j \rceil.$$

Third, we have the maximum daily demand constraint

$$W \geq \max(n_1, \ldots, n_7).$$

The minimum workforce must be at least as large as the largest of these three lower bounds. In what follows, we present an algorithm that yields a schedule that requires a workforce of a size equal to the largest of these three lower bounds.

The algorithm that solves this problem, i.e., that finds a schedule that satisfies all constraints using the smallest possible workforce, is relatively simple. Let W denote the maximum of the three lower bounds and let n denote the maximum weekend demand, i.e.,

$$n = \max(n_1, n_7).$$

Let $u_j = W - n_j$, for $j = 2, \ldots, 6$, and $u_j = n - n_j$, for $j = 1$ and 7; the u_j is the surplus number of employees with regard to day j. The second lower bound guarantees that

$$\sum_{j=1}^{7} u_j \geq 2n.$$

It is clear that employees should be given days off on those days that have a large surplus of employees. The algorithm that constructs the schedule uses a list of so-called off-day pairs. The pairs in this list are numbered from 1 to n and the list is created as follows: First, choose day k such that

$$u_k = \max(u_1, \ldots, u_7).$$

Second, choose day l, $(l \neq k)$, such that $u_l > 0$; if $u_l = 0$ for all $l \neq k$, then choose $l = k$. Third, add pair (k, l) to the list and decrease both u_k and u_l by 1. Repeat this procedure n times. At the end of the list pairs of the form (k, k) may appear; these pairs are called nondistinct pairs.

Now number the employees from 1 to W. Note that since the maximum demand during a weekend is n, the remaining $W - n$ employees can have that weekend off. Assume that the first day to be scheduled falls on a Saturday, and the first and second days are weekend 1.

Algorithm 12.2.1 (Days-Off Scheduling).

Step 1. *(Schedule the weekends off)*

Assign the first weekend off to the first $W - n$ employees.

Assign the second weekend off to the second $W - n$ employees.

This process is continued cyclically with employee 1 being treated as the next employee after employee W.

Step 2. *(Categorization of employees in week 1)*

In week 1 each employee falls into one of four categories.

Type $T1$: weekend 1 off; 0 off days needed during week 1; weekend 2 off

Type $T2$: weekend 1 off; 1 off days needed during week 1; weekend 2 on

Type $T3$: weekend 1 on; 1 off days needed during week 1; weekend 2 off

Type $T4$: weekend 1 on; 2 off days needed during week 1; weekend 2 on

Since there are exactly n people working each weekend,

$$|T3| + |T4| = n \text{ (because of weekend 1)};$$
$$|T2| + |T4| = n \text{ (because of weekend 2)}.$$

It follows that $|T2| = |T3|$.

Pair each employee of $T2$ with one employee of $T3$.

Step 3. *(Assigning off-day pairs to employees in week 1)*

Assign the n pairs from the top of the list.

First to employees of $T4$: each employee of $T4$ gets both days off.

Second to employees of $T3$: each employee of $T3$ gets from his pair the earlier day off and his companion of $T2$ gets from that same pair the later day off. (So each employee of $T3$ and $T2$ gets one day off in week 1, as required.)

Step 4. *(Assigning off-day pairs to employees in week i)*

Assume a schedule has been created for weeks $1, \ldots, i - 1$.

A categorization of employees can be done for week i in the same way as in Step 2. In order to assign employees to off-day pairs two cases have to be considered.

Case (a): *(All off-day pairs in the list are distinct)*

Employees of $T4$ and $T3$ are associated with the same pairs as those they were associated with in week $i - 1$.

A $T4$ employee gets from his pair both days off.

A $T3$ employee gets from the pair he is associated with the earlier day off and his companion of $T2$ gets from that pair the later day off.

12.2 Days-Off Scheduling

Case (b): (Not all off-day pairs in the list are distinct)

Week i is scheduled in exactly the same way as week 1, independent of week $i-1$.

Set $i = i + 1$ and return to Step 4.

This algorithm needs some motivation. First, it may not be immediately clear that we never will be confronted with the need to schedule a nondistinct pair of days to a type $T4$ worker. In that case a worker needs two weekdays off in a week, and we try to give him the same day twice. It can be shown that the number of $T4$ employees is always smaller than or equal to the number of distinct pairs.

If there are non-distinct pairs, i.e., pairs (k, k), in the off-days list, then week i is independent of week $i - 1$. It can be shown that every pair contains day k and the maximum workstretch from week $i - 1$ to week i is 6 days. The only time that the workstretch is greater than 5 days is when there are nondistinct pairs.

In the next example, there is one distinct pair, one nondistinct pair, and $|T4| = 1$.

Example 12.2.2 (Application of Days-Off Scheduling Algorithm). Consider the problem with the following daily requirements.

day j	1	2	3	4	5	6	7
	Sun.	Mon.	Tues.	Wed.	Thurs.	Fri.	Sat.
Requirement	1	0	3	3	3	3	2

The maximum weekend demand is $n = 2$ and each person requires 1 out of 3 weekends off, i.e., $k_1 = 1$ and $k_2 = 3$. So

$$W \geq \lceil (3 \times 2)/(3 - 1) \rceil = 3,$$
$$W \geq \lceil 15/5 \rceil = 3,$$
$$W \geq 3.$$

So the minimum number of employees W is 3 and $W - n = 1$. We assign a weekend off to one employee each week. This results in the following assignment of weekend days off for the three employees.

	S	S	M	T	W	T	F	S	S	M	T	W	T	F	S	S	M	T	W	T	F	S
1	X	X																				X
2								X	X													
3															X	X						

At this point, there is one surplus employee on Sunday and three on Monday.

day j	1	2	3	4	5	6	7
u_j	1	3	0	0	0	0	0

There are 2 pairs of off days, one distinct and one non-distinct.

Pair 1: Sunday - Monday;
Pair 2: Monday - Monday;

We will use these pairs for each week.

There is one nondistinct pair in the list. The categorization of the employees in the first week results in the following categories: The first pair is assigned to the $T4$ employee (one each week) and the second pair is split between the remaining two employees (types $T2$ and $T3$).

Applying the next step of the algorithm yields the following schedule.

	S	S	M	T	W	T	F	S	S	M	T	W	T	F	S	S	M	T	W	T	F	S
1	X	X	X						X	X							X					X
2			X					X	X	X						X	X					
3		X	X							X					X	X	X					

The schedule produces a six-day workstretch for one employee each week. This cannot be avoided since the solution is unique.

It can be shown that if all off-day pairs are distinct then the maximum workstretch is 5 days. In the next example all off-day pairs are distinct.

Example 12.2.3 (Application of Days-Off Scheduling Algorithm).
Consider the problem with the following daily requirements.

day j	1	2	3	4	5	6	7
	Sun.	Mon.	Tues.	Wed.	Thurs.	Fri.	Sat.
Requirement	3	5	5	5	7	7	3

The maximum weekend demand $n = 3$ and each person requires 3 out of 5 weekends off, i.e., $k_1 = 3$ and $k_2 = 5$. So

$$W \geq \lceil (5 \times 3)/2 \rceil = 8,$$
$$W \geq \lceil 35/5 \rceil = 7,$$
$$W \geq 7.$$

So the minimum number of employees W is 8 and $W - n = 5$. We assign weekends off to 5 employees each week. This results in the following assignment of weekend days off for the eight employees.

12.2 Days-Off Scheduling

	S	S	M	T	W	T	F	S	S	M	T	W	T	F	S	S	M	T	W	T	F	S
1	X	X						X	X													X
2	X	X						X	X													X
3	X	X											X	X								X
4	X	X											X	X								X
5	X	X											X	X								
6				X	X								X	X								
7				X	X								X	X								
8				X	X																	X

At this point, there are 8 people available each weekday, so the surplus u_j is

day j	1	2	3	4	5	6	7
u_j	0	3	3	3	1	1	0

There are a number of ways in which 3 pairs of off days can be chosen. For example,

Pair 1: Monday - Tuesday;
Pair 2: Tuesday - Wednesday;
Pair 3: Tuesday - Wednesday.

We will use these pairs for each week.

There are no nondistinct pairs on the list. The categorization of the employees in the first week results in the following 4 categories:

T1: 1, 2
T2: 3, 4, 5
T3: 6, 7, 8
T4: –

Employee 3 is paired with 6, 4 with 7 and 5 with 8. Thus we need three pairs of weekdays to give off.

Categorization of the employees in the second week yields the following 4 categories.

T1: 6, 7
T2: 1, 2, 8
T3: 3, 4, 5
T4: –

Employee 1 is paired with 3, 2 with 4, and 8 with 5.

Categorization of the employees in the third week results in the following 4 categories.

T1: 3, 4
T2: 5, 6, 7
T3: 1, 2, 8
T4: –

Employee 1 is paired with 6, 2 with 5, and 8 with 7.

Applying the next step of the algorithm results in the schedule shown by the following table:

	S	S	M	T	W	T	F	S	S	M	T	W	T	F	S	S	M	T	W	T	F	S
1	X	X						X	X	→X						X						X
2	X	X						X	X			→X					X←					X
3	X	X		→X						X					X	X						X
4	X	X			→X					X					X	X						X
5	X	X			X					X←					X	X					X	
6			X					X	X						X	X		→X				
7			X					X	X						X	X					X	
8			X←					X	X					X					X←			X

The arrows illustrate how a pair of off-days is shared by two employees. All pairs are distinct, so the maximum workstretch is 5 days. It can be easily verified that the schedule is cyclic and the cycle is 8 weeks.

It can be shown that schedules generated by the algorithm always satisfy the constraints. Because of the way the off-weekends are distributed over the employees (evenly) and because of the first lower bound, it is assured that each employee is given at least k_1 out of k_2 weekends off. That each employee works exactly 5 days out of the week (from Sunday to Saturday) follows immediately from the algorithm. That no employee works more that 6 days in one stretch is a little harder to see. An employee may have a six day workstretch (but not longer) when there are non-distinct pairs (if all pairs are distinct, then the longest workstretch is 5 days). If there are non-distinct pairs (k,k), then day k has to appear in all pairs. In the worst case, an employee can be associated with pair (j,k) in week $i-1$ and pair (k,l) in week i, where $j \leq k \leq l$. In this case, he will receive at least day k off in week $i-1$ as well as in week i which results in a six day workstretch. The stretch is smaller if either $k < j$ or $l < k$, for then he would receive day j or day l off.

It can be shown that there exists an optimal schedule that is cyclic and that the algorithm may yield this schedule. The number of weeks in such a cyclic schedule can be computed fairly easily (see Exercise 12.1).

12.3 Shift Scheduling

In the scheduling problem discussed in the previous section there are various assignment patterns over the cycle. The cost of assigning an employee to a certain work pattern is the same for each pattern and the objective is to minimize the total number of employees. The fact that each assignment pattern has the same cost is one reason why the problem is relatively easy.

In this section we consider a more general personnel scheduling problem and follow a completely different approach. We consider a cycle that is fixed

12.3 Shift Scheduling

in advance. In certain settings the cycle may be a single day, while in others it may be a week or a number of weeks. In contrast to the previous section, each work assignment pattern over a cycle has its own cost and the objective is to minimize the total cost.

The problem can be formulated as follows. The predetermined cycle consists of m time intervals or periods. The lengths of the periods do not necessarily have to be identical. During period i, $i = 1, \ldots, m$, the presence of b_i personnel is required. The number b_i is, of course, an integer. There are n different shift patterns and each employee is assigned to one and only one pattern. Shift pattern j is defined by a vector $(a_{1j}, a_{2j}, \ldots, a_{mj})$. The value a_{ij} is either 0 or 1; it is a 1 if period i is a work period and 0 otherwise. Let c_j denote the cost of assigning a person to shift j and x_j the (integer) decision variable representing the number of people assigned to shift j. The problem of minimizing the total cost of assigning personnel to meet demand can be formulated as the following integer programming problem:

$$\text{minimize} \quad c_1 x_1 + c_2 x_2 + \cdots + c_n x_n$$

subject to

$$a_{11} x_1 + a_{12} x_2 + \cdots + a_{1n} x_n \geq b_1$$
$$a_{21} x_1 + a_{22} x_2 + \cdots + a_{2n} x_n \geq b_2$$
$$\vdots$$
$$a_{m1} x_1 + a_{m2} x_2 + \cdots + a_{mn} x_n \geq b_m$$
$$x_j \geq 0 \qquad \text{for } j = 1, \ldots, n,$$

with $x_1, \ldots, x_n$ integer. In matrix form this integer program is written as follows.

$$\text{minimize} \quad \bar{c}\bar{x}$$

subject to

$$\mathbf{A}\bar{x} \geq \bar{b}$$
$$\bar{x} \geq 0.$$

Such an integer programming problem is known to be strongly NP-hard in general. However, the $\mathbf{A}$ matrix may often exhibit a special structure. For example, shift j, $(a_{1j}, \ldots, a_{mj})$, may contain a *contiguous* set of 1's (a contiguous set of 1's implies that there are no 0's in between 1's). However, the *number* of 1's may often vary from shift to shift, since it is possible that some shifts have to work longer hours or more days than other shifts.

Example 12.3.1 (Shift Scheduling at a Retail Store). Consider a retail store that is open for business from 10 a.m. to 9 p.m. There are five shift patterns.

pattern	Hours of Work	Total Hours	Cost
1	10 a.m. to 6 p.m.	8	$ 50.00
2	1 p.m. to 9 p.m.	8	$ 60.00
3	12 p.m. to 6 p.m.	6	$ 30.00
4	10 a.m. to 1 p.m.	3	$ 15.00
5	6 p.m. to 9 p.m.	3	$ 16.00

Staffing requirements at the store vary from hour to hour.

Hour	Staffing Requirement
10 a.m. to 11 a.m.	3
11 a.m. to 12 a.m.	4
12 a.m. to 1 p.m.	6
1 p.m. to 2 p.m.	4
2 p.m. to 3 p.m.	7
3 p.m. to 4 p.m.	8
4 p.m. to 5 p.m.	7
5 p.m. to 6 p.m.	6
6 p.m. to 7 p.m.	4
7 p.m. to 8 p.m.	7
8 p.m. to 9 p.m.	8

The problem can be formulated as an integer program with the following $\bar{c}$ vector, $\mathbf{A}$ matrix and $\bar{b}$ vector:

$$\bar{c} = (50, 60, 30, 15, 16)$$

$$\mathbf{A} = \begin{bmatrix} 1 & 0 & 0 & 1 & 0 \\ 1 & 0 & 0 & 1 & 0 \\ 1 & 0 & 1 & 1 & 0 \\ 1 & 1 & 1 & 0 & 0 \\ 1 & 1 & 1 & 0 & 0 \\ 1 & 1 & 1 & 0 & 0 \\ 1 & 1 & 1 & 0 & 0 \\ 1 & 1 & 1 & 0 & 0 \\ 0 & 1 & 0 & 0 & 1 \\ 0 & 1 & 0 & 0 & 1 \\ 0 & 1 & 0 & 0 & 1 \end{bmatrix} \quad \bar{b} = \begin{bmatrix} 3 \\ 4 \\ 6 \\ 4 \\ 7 \\ 8 \\ 7 \\ 6 \\ 4 \\ 7 \\ 8 \end{bmatrix}$$

Clearly, an integer solution x_1, x_2, x_3, x_4, x_5 is required. However, solving the linear programming relaxation of this problem yields the solution (0,0,8,4,8). Since this solution is integer, it is clear that it is also optimal for the integer programming formulation.

Even though the integer programming formulation of the general personnel scheduling problem (with an arbitrary $0-1$ **A** matrix) is NP-hard, the special case with each column containing a contiguous set of 1's is easy. It can be shown that the solution of the linear programming relaxation is always integer. There are several other important special cases that are solvable in polynomial time. In the next section we discuss one of them.

12.4 The Cyclic Staffing Problem

A classical personnel scheduling problem is the *cyclic staffing* problem. The objective is to minimize the cost of assigning people to an m period cyclic schedule so that sufficient workers are present during time period i, in order to meet requirement b_i, and each person works a shift of k consecutive periods and is free the other $m-k$ periods. Notice that period m is followed by period 1.

An example is the (5,7)-cyclic staffing problem, where the cycle is seven days and any person works 5 consecutive days followed by two days off. As described in the previous section, this problem can be formulated as an integer program. In the integer programming formulation a column vector of the **A** matrix denotes a possible shift assignment that specifies which two consecutive days are off and which 5 days are workdays. There are 7 possible column vectors. Even though the **A** matrix has a very special structure in this case, the columns do not always have a contiguous set of 1's. So this is not a special case of the example described in the previous section.

$$\mathbf{A} = \begin{bmatrix} 1 & 0 & 0 & 1 & 1 & 1 & 1 \\ 1 & 1 & 0 & 0 & 1 & 1 & 1 \\ 1 & 1 & 1 & 0 & 0 & 1 & 1 \\ 1 & 1 & 1 & 1 & 0 & 0 & 1 \\ 1 & 1 & 1 & 1 & 1 & 0 & 0 \\ 0 & 1 & 1 & 1 & 1 & 1 & 0 \\ 0 & 0 & 1 & 1 & 1 & 1 & 1 \end{bmatrix}$$

The cost c_j of column vector j, i.e., $a_{1j}, \ldots, a_{mj}$, represents the cost of having one person work according to the corresponding schedule. The $\bar{b}$ vector is again the requirements vector, i.e., b_i denotes the number of people that have to be present on day i. The integer decision variable x_j represents the number of people that work according to the schedule defined by column vector j. This results in an integer programming problem with a special structure.

The special structure of this integer programming problem makes it possible to solve it in an efficient manner. Actually, it can be shown that the solution of the linear program relaxation of this problem is very close to the solution of the integer programming problem. Because of this the following algorithm leads to an optimal solution.

Algorithm 12.4.1 (Minimizing Cost in Cyclic Staffing).
Step 1.
Solve the linear relaxation of the original problem to obtain $x'_1, \ldots, x'_n$.
If $x'_1, \ldots, x'_n$ are integer, then it is optimal for the original problem. STOP.
Otherwise go to Step 2.

Step 2.
Form two linear programs LP' and LP'' from the relaxation of the original problem by adding respectively the constraints

$$x_1 + \cdots + x_n = \lfloor x'_1 + \cdots + x'_n \rfloor$$

and

$$x_1 + \cdots + x_n = \lceil x'_1 + \cdots + x'_n \rceil.$$

LP'' always will have an optimal solution that is integer.

If LP' does not have a feasible solution, then the solution of LP'' is an optimal solution to the original problem.

If LP' has a feasible solution, then it has an optimal solution that is integer and the solution to the original problem is the better one of the solutions to LP' and LP''. STOP.

The next example illustrates the use of the algorithm.

Example 12.4.2 (Minimizing Cost in Cyclic Staffing). Consider the (3,5)-cyclic staffing problem

$$\mathbf{A} = \begin{bmatrix} 1 & 0 & 0 & 1 & 1 \\ 1 & 1 & 0 & 0 & 1 \\ 1 & 1 & 1 & 0 & 0 \\ 0 & 1 & 1 & 1 & 0 \\ 0 & 0 & 1 & 1 & 1 \end{bmatrix} \quad \bar{b} = \begin{bmatrix} 3 \\ 4 \\ 6 \\ 4 \\ 7 \end{bmatrix}$$

The cost vector is

$$\bar{c} = (3.6,\ 4.8,\ 5.5,\ 3.7,\ 5.2)$$

Applying Algorithm 12.4.1 leads to the following results.
Step 1. Solving the linear programming relaxation yields

$$\bar{x}' = (1.5,\ 0,\ 4.5,\ 0,\ 2.5).$$

The value of the objective function is 43.15.
Step 2. Adding to the original problem the constraint

$$x_1 + x_2 + x_3 + x_4 + x_5 = 8$$

results in a problem without a feasible solution. Adding to the original problem the constraint
$$x_1 + x_2 + x_3 + x_4 + x_5 = 9$$
yields $\bar{x} = (2, 0, 4, 1, 2)$ with objective value 43.3. So the optimal solution is $\bar{x} = (2, 0, 4, 1, 2)$ with objective value 43.3.

12.5 Applications and Extensions of Cyclic Staffing

In this section we discuss three applications of cyclic staffing.

(i) Days-Off scheduling. Consider the following special case of the problem discussed in Section 12.2. Each employee is guaranteed two days off a week, including every other weekend (a week starts with a Sunday and ends with a Saturday) and is not allowed to work more than 6 days consecutively. This problem can be formulated as an integer program with the following **A** matrix.

$$\begin{bmatrix}
0 & 0 & 0 & 0 & 0 & 0 & \cdots & 1 & 1 & 1 & 1 & 1 & 1 & \cdots \\
- & - & - & - & - & - & \cdots & - & - & - & - & - & - & \cdots \\
0 & 0 & 0 & 0 & 0 & 0 & \cdots & 1 & 0 & 1 & 1 & 0 & 1 & \cdots \\
0 & 0 & 1 & 1 & 1 & 1 & \cdots & 0 & 1 & 1 & 0 & 1 & 1 & \cdots \\
1 & 1 & 0 & 0 & 0 & 1 & \cdots & 1 & 1 & 0 & 1 & 1 & 1 & \cdots \\
1 & 1 & 1 & 1 & 1 & 0 & \cdots & 1 & 1 & 1 & 1 & 1 & 0 & \cdots \\
1 & 1 & 1 & 1 & 1 & 1 & \cdots & 1 & 1 & 1 & 1 & 1 & 1 & \cdots \\
1 & 1 & 1 & 1 & 1 & 1 & \cdots & 1 & 1 & 1 & 1 & 1 & 1 & \cdots \\
1 & 1 & 1 & 1 & 1 & 1 & \cdots & 0 & 0 & 0 & 0 & 0 & 0 & \cdots \\
- & - & - & - & - & - & \cdots & - & - & - & - & - & - & \cdots \\
1 & 0 & 1 & 1 & 0 & 1 & \cdots & 0 & 0 & 0 & 0 & 0 & 0 & \cdots \\
0 & 1 & 1 & 0 & 1 & 1 & \cdots & 0 & 0 & 1 & 1 & 1 & 1 & \cdots \\
1 & 1 & 0 & 1 & 1 & 1 & \cdots & 1 & 1 & 0 & 0 & 0 & 1 & \cdots \\
1 & 1 & 1 & 1 & 1 & 0 & \cdots & 1 & 1 & 1 & 1 & 1 & 0 & \cdots \\
1 & 1 & 1 & 1 & 1 & 1 & \cdots & 1 & 1 & 1 & 1 & 1 & 1 & \cdots \\
1 & 1 & 1 & 1 & 1 & 1 & \cdots & 1 & 1 & 1 & 1 & 1 & 1 & \cdots \\
0 & 0 & 0 & 0 & 0 & 0 & \cdots & 1 & 1 & 1 & 1 & 1 & 1 & \cdots \\
- & - & - & - & - & - & \cdots & - & - & - & - & - & - & \cdots \\
0 & 0 & 0 & 0 & 0 & 0 & \cdots & 1 & 0 & 1 & 1 & 0 & 1 & \cdots
\end{bmatrix}$$

The number of possible patterns tends to be somewhat large, since there are many patterns that satisfy the conditions stated. Each row in the matrix represents a day in the week. The first two rows, the middle two rows and the last two rows represent weekends. The first group of columns correspond to the assignments with the first and the third weekend off and the second group of columns to the assignments with the second weekend off. This problem can be solved by the technique discussed in the previous section.

The general model described in Section 12.2 can also be viewed as a cyclic staffing problem. However, it does not fit the framework described in Section

12.4 that well, because of a number of differences. For starters, the cycle length is not fixed a priori. In addition, even if the cycle length were fixed, it would be hard to describe this problem as an integer programming problem. The number of possible columns is very large and not easy to enumerate.

(ii) Cyclic staffing with overtime. A basic staffing problem occurs in facilities such as hospitals that operate around the clock. Suppose that there are fixed hourly staff requirements b_i, and three basic work shifts, each of 8 hours duration: 8 a.m. to 4 p.m., 4 p.m. to midnight and 12 p.m. to 8 a.m. Overtime of up to an additional 8 hours is possible for each shift. A personnel assignment that meets all staffing requirements at minimum cost has to be found. The constraint matrix $\mathbf{A}$ consists of 9 submatrices, each with 8 rows and 9 columns.

$$\begin{bmatrix} \mathbf{1} & \mathbf{0} & \mathbf{0\backslash 1} \\ \mathbf{0\backslash 1} & \mathbf{1} & \mathbf{0} \\ \mathbf{0} & \mathbf{0\backslash 1} & \mathbf{1} \end{bmatrix}$$

The submatrix $\mathbf{0}$ is a matrix with all entries 0. The submatrix $\mathbf{1}$ is a matrix with all entries 1 and the submatrix $\mathbf{0\backslash 1}$ is the matrix

$$\begin{bmatrix} 0 & 1 & 1 & 1 & 1 & 1 & 1 & 1 \\ 0 & 0 & 1 & 1 & 1 & 1 & 1 & 1 \\ 0 & 0 & 0 & 1 & 1 & 1 & 1 & 1 \\ 0 & 0 & 0 & 0 & 1 & 1 & 1 & 1 \\ 0 & 0 & 0 & 0 & 0 & 1 & 1 & 1 \\ 0 & 0 & 0 & 0 & 0 & 0 & 1 & 1 \\ 0 & 0 & 0 & 0 & 0 & 0 & 0 & 1 \\ 0 & 0 & 0 & 0 & 0 & 0 & 0 & 1 \end{bmatrix}$$

This problem can be solved by linear programming.

(iii) Cyclic Staffing with Linear Penalties for Understaffing and Overstaffing. Suppose the demands for each period are not fixed. There is a linear penalty c'_i for understaffing and a linear penalty c''_i for overstaffing. The penalty c''_i may actually be negative, since there may be some benefits in overstaffing. Let x'_i denote the level of understaffing during period i. The level of overstaffing during period i is then

$$b_i - (a_{i1}x_1 + a_{i2}x_2 + \cdots + a_{in}x_n) - x'_i.$$

The problem can now be formulated as the following linear program.

$$\text{minimize} \quad \bar{c}\bar{x} + \bar{c}'\bar{x}' + \bar{c}''(\bar{b} - \mathbf{A}\bar{x} - \bar{x}')$$

subject to

$$\mathbf{A}\bar{x} + I\bar{x}' \geq \bar{b}$$
$$\bar{x}, \bar{x}' \geq 0 \qquad \bar{x}, \bar{x}' \text{ integer.}$$

If **A** is the matrix for the (k, m) staffing problem (or any other row circular matrix), the problem can be solved by the algorithm described in the previous section, provided that the problem is bounded from below. It turns out that the problem is bounded from below if and only if $c - c''\mathbf{A} \geq 0$ and $c' - c'' \geq 0$.

12.6 Crew Scheduling

Crew scheduling problems are very important in the transportation industry, especially in the airline industry. The underlying model is different from the models considered in the previous sections and so are the solution techniques.

Consider a set of m jobs, e.g., flight legs. A flight leg is characterized by a point of departure and a point of arrival, as well as an approximate time interval during which the flight has to take place. There is a set of n feasible and permissible combinations of flight legs that one crew can handle, e.g., round trips or tours (the number n usually is very large). A round trip may consist of several flight legs, i.e., a plane may leave city A for city B, then go to city C, before returning to city A. Any given flight leg may be part of many round trips. Round trip j, $j = 1, \ldots, n$, has a cost c_j. Setting up a crew schedule is equivalent to determining which round trips should be selected and which ones not. The objective is to choose a set of round trips with a minimum total cost in such a way that each flight leg is covered exactly once by one and only one round trip.

In order to formulate this crew scheduling problem as an integer program some notation is required. If flight leg i is part of round trip j, then a_{ij} is 1, otherwise a_{ij} is 0. Let x_j denote a $0 - 1$ decision variable that assumes the value 1 if round trip j is selected and 0 otherwise. The crew scheduling problem can be formulated as the following integer program.

minimize $\quad c_1 x_1 + c_2 x_2 + \cdots + c_n x_n$

subject to

$$a_{11} x_1 + a_{12} x_2 + \cdots + a_{1n} x_n = 1$$
$$a_{21} x_1 + a_{22} x_2 + \cdots + a_{2n} x_n = 1$$
$$\vdots$$
$$a_{m1} x_1 + a_{m2} x_2 + \cdots + a_{mn} x_n = 1$$
$$x_j \in \{0, 1\} \qquad \text{for } j = 1, \ldots, n.$$

Each column in the **A** matrix is a round trip and each row is a flight leg that must be covered exactly once by one round trip. The optimization problem is then to select, at minimum cost, a set of round trips that satisfies the constraints. The constraints in this problem are often called the partitioning equations and this integer programming problem is referred to as the *Set Partitioning* problem (see Appendix A). For a feasible solution $(x_1,\ldots,x_n)$, the variables that are equal to 1 are referred to as the *partition*. In what follows we denote a partition l by $J^l = \{j \mid x_j^l = 1\}$.

This problem is known to be NP-hard. Many heuristics as well as enumeration schemes (branch-and-bound) have been proposed for this problem. In many of these approaches the concept of *row prices* is used. The vector $\bar{\rho}^l = (\rho_1^l, \rho_2^l, \ldots, \rho_m^l)$ is a set of feasible row prices corresponding to partition J^l satisfying

$$\sum_{i=1}^{m} \rho_i^l a_{ij} = c_j \quad j \in J^l.$$

The price ρ_i^l may be interpreted as an estimate of the cost of covering job (flight leg) i using solution J^l. There are usually many feasible price vectors for any given partition.

The row prices are of crucial importance in computing the change in the value of the objective if a partition J^1 is changed into partition J^2. If Z^1 (Z^2) denotes the value of the objective corresponding to partition 1 (2), then

$$Z^2 = Z^1 - \sum_{j \in J^2} \left(\sum_{i=1}^{m} \rho_i^1 a_{ij} - c_j \right).$$

The quantity

$$\sigma_j = \sum_{i=1}^{m} \rho_i^1 a_{ij} - c_j$$

can be interpreted as the *potential savings* with respect to the first partition to be obtained by including column j. It can be shown that if

$$\sum_{i=1}^{m} \rho_i^1 a_{ij} \leq c_j \quad j = 1, \ldots, n,$$

for any set of feasible row prices $\bar{\rho}^1$ corresponding to partition J^1, then solution J^1 is optimal.

Based on the concept of row prices the following simple heuristic can be used for finding better solutions, given a partition J^1 and a corresponding set of feasible row prices $\bar{\rho}^1$. The goal is to find a better partition J^2. In the heuristic the set N denotes the indices of the columns that are candidates for inclusion in J^2.

Algorithm 12.6.1 (Column Selection in Set Partitioning).
Step 1.
  Set $J^2 = \emptyset$ and $N = \{1, 2, \ldots, n\}$.

12.6 Crew Scheduling

Step 2.

Compute the potential savings

$$\sigma_j = \sum_{i=1}^m \rho_i^1 a_{ij} - c_j \qquad j = 1, \ldots, n.$$

Find the column k in N with the largest potential savings,

$$\sum_{i=1}^m \rho_i^1 a_{ik} - c_k.$$

Step 3.

For $i = 1, \ldots, m$, if $a_{ik} = 1$ set $a_{ij} = 0$ for all $j \neq k$.

Step 4.

Let $J^2 = J^2 \cup \{k\}$ and $N = N - \{k\}$.

Delete from N all j for which $a_{ij} = 0$ for all $i = 1, \ldots, m$

Step 5.

If $N = \emptyset$ STOP, otherwise go to Step 2.

The next example illustrates the heuristic.

Example 12.6.2 (Crew Scheduling and Truck Routing). Consider a central depot and 5 clients, see Figure 12.1. From the depot, a single delivery has to be made to each one of the clients. Assume that each truck can serve at most two clients on a single trip. The objective is to determine which truck should go to which client and the routing of the trucks that minimizes the total distance traveled. Each column in the table below represents one possible truck route and the c_j is equal to the total distance traveled for each trip. For example, column 6 represents a vehicle proceeding from the depot to client 1, then on to client 2, and from there back to the depot. The value of c_6 is 14, which is the total distance traveled during the trip.

Route	1	2	3	4	5	6	7	8	9	10	11	12	13	14	15
c_j	8	10	4	4	2	14	10	8	8	10	11	12	6	6	5
	1	0	0	0	0	1	1	1	1	0	0	0	0	0	0
	0	1	0	0	0	1	0	0	0	1	1	1	0	0	0
	0	0	1	0	0	0	1	0	0	1	0	0	1	1	0
	0	0	0	1	0	0	0	1	0	0	1	0	1	0	1
	0	0	0	0	1	0	0	0	1	0	0	1	0	1	1

Suppose we select $J^1 = 1, 2, 3, 4, 5$ as our first partition. A set of feasible row prices is $\bar{\rho}^1 = (8, 10, 4, 4, 2)$. The corresponding potential savings are

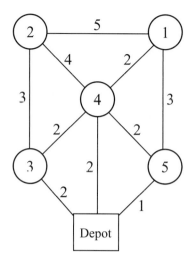

Fig. 12.1. Truck Routing Network

presented in the table below. Applying the heuristic and breaking ties by selecting the column with the lowest index yields the new partition $J^2 = \{6, 13, 5\}$. This new partition has a cost of $Z^2 = 22$ as compared to a cost of $Z_1 = 28$ for the first partition.

Route	1	2	3	4	5	6	7	8	9	10	11	12	13	14	15	
c_j	8	10	4	4	2	14	10	8	8	10	11	12	6	6	5	
	0	0	0	0	0	4	2	4	2	4	3	0	2	0	1	$x_6 = 1$
σ_j	-	-	0	0	0	-	-6	-4	-6	-6	-7	-10	2	0	1	$x_{13} = 1$
	-	-	-	-	0	-	-	-	-	-6	-	-	-10	-	-4	-3	$x_5 = 1$

Many sets of row prices are feasible. The new row prices can be determined using the following procedure. It is clear that

$$\rho_1^2 + \rho_2^2 = 14.$$

Choose

$$\rho_1^2 = \frac{\rho_1^1}{\rho_1^1 + \rho_2^1} \times c_6 = \frac{8}{8+10} \times 14 = 6.222$$

and

$$\rho_2^2 = \frac{\rho_2^1}{\rho_1^1 + \rho_2^1} \times c_6 = \frac{10}{8+10} \times 14 = 7.777.$$

Using the row prices $\bar{\rho}^2 = (6.2, 7.8, 3, 3, 2)$ the potential savings can be computed and the heuristic yields the new partition $J^3 = \{8, 10, 5\}$ with a cost $Z_3 = 20$.

12.7 Operator Scheduling in a Call Center

Route	1	2	3	4	5	6	7	8	9	10	11	12	13	14	15	
c_j	8	10	4	4	2	14	10	8	8	10	11	12	6	6	5	
σ_j	-1.8	-2.2	-1	-1	0	0	-0.8	1.2	0.2	0.8	-0.2	-2.2	0	-1	0	$x_8 = 1$
	-	-2.2	-1	-	0	-6.2	-7	-	-6	0.8	-3.2	-2.2	-3	-1	-3	$x_{10} = 1$
	-	-	-	-	0	-	-	-	-6	-	-	-10	-	-4	-3	$x_5 = 1$

Using the row prices $\bar{\rho}^3 = (5.3, 7.1, 2.9, 2.7, 2)$ we find that all potential savings are negative, so the partition J^3 is optimal.

Route	1	2	3	4	5	6	7	8	9	10	11	12	13	14	15
c_j	8	10	4	4	2	14	10	8	8	10	11	12	6	6	5
σ_j	-2.7	-2.9	-1.1	-1.3	0	-1.6	-1.8	0	-0.7	0	-1.2	-2.9	-0.4	-1.1	-0.3

When the problems become very large, it is necessary to adopt more sophisticated approaches, namely branch-and-bound, branch-and-price, and branch-cut-and-price. The bounding techniques in branch-and-bound are often based on a technique called Lagrangean Relaxation. Branch-cut-and-price combines branching with so-called cutting planes techniques and has been used to solve real world problems arising in the airlines industry with considerable success.

12.7 Operator Scheduling in a Call Center

In every large organization that has a help line or a toll free number, there is a concern with regard to the quality of service. This is measured by the time it takes a caller to get an operator on the line, or, in the case of a conversant system that responds to digital input, by the time it takes the caller to get an appropriate person on the line.

Companies often have detailed statistics with regard to the frequencies of past requests for service. These statistics typically are broken down according to the time of day, the day of the week, and the type of service requested. Figure 12.2 depicts statistics with regard to calls for operators in a call center.

In this section we describe the design and development of an operator scheduling system for a call center. The objective is to determine a set of shifts and assignment of tasks within the shifts, so that the requirement curves are matched as closely as possible by the availability of operators. The shift configurations that are allowed have a certain structure. These structures are often referred to as *tour templates* and are characterized by their start times, finish times, and the number, length and timing of non-work periods. The time unit is 15 minutes (a coffee break is in this case one time unit and a lunch break is anywhere from two to four units). Different templates j and k can be of the same *type* i. A type is characterized by the time between start and finish, but not by the timing of the non-work periods. A sample of the tour template types is depicted in Figure 12.3.

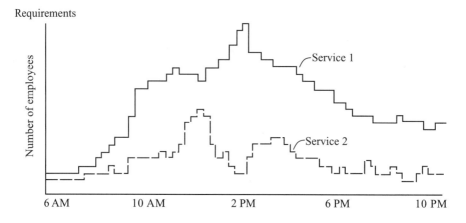

Fig. 12.2. Statistics with Regard to Calls for Operators at a Telephone Company

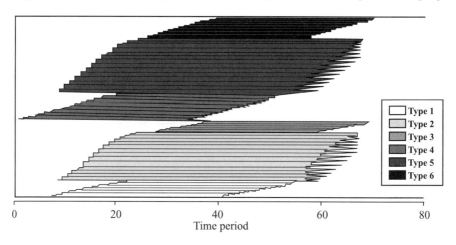

Fig. 12.3. Space of Tour Template Types

The problem is simplified by assuming that none of the variables (e.g., the number of operators available, the call frequencies, and so on) change within any given time unit. So for each 15 minute period the demand as well as the number of operators are fixed. Because of this simplification, each term in any one of the performance measures is a sum rather than an integral.

To describe the objectives formally, we need the following notation. Let x_j denote the number of tours of template j and c_j the cost of a tour of template j. Let a_j and b_j denote the starting time and the finishing time of template j. Let m_i denote the number of tours of type i. Let $y(k,t)$ be 1 if tour k is at work at time t and 0 otherwise and $s(t)$ the number of operators available at time t, i.e., the exact supply at time t. Let $e(t)$ denote the difference between

12.7 Operator Scheduling in a Call Center

the supply and the demand, $e^-(t)$ the negative part (the shortage), i.e.,

$$e^-(t) = \max(0, -e(t)),$$

and $e^+(t)$ the positive part (the surplus), i.e.,

$$e^+(t) = \max(0, e(t)).$$

Based on this, a number of precise performance measures can be defined that take everything into account including coffee breaks. Let $\mathcal{F}$ denote the fitness measure defined as

$$\mathcal{F} = \psi^- \sum_{t=1}^{H} e^-(t) + \psi^+ \sum_{t=1}^{H} e^+(t),$$

where ψ^- and ψ^+ denote the respective penalty costs and H denotes the number of time units. Let $\mathcal{C}$ denote the cost measure

$$\mathcal{C} = \mathcal{F} + \sum_j c_j x_j$$

and $\mathcal{L}$ the smoothness measure

$$\mathcal{L} = \sum_{t=1}^{H} e(t)^2.$$

The overall framework of the approach adopted in the system is depicted in Figure 12.4. The optimization process is divided into a number of modules. The solid tour selection module is based on a mathematical program. In the solid tour selection the breaks in the tours are not taken into consideration. If there are no side constraints then this mathematical program is equivalent to a network flow problem.

The break placement module attempts to minimize $\mathcal{L}$ subject to all break placement rules. This procedure operates in a sequential manner. It finds and places a break in a tour with a maximum decrease in $\mathcal{L}$. This is repeated for all tours and breaks and for all break hierarchies. The target demand modification module attempts to minimize the fitness measure.

A number of side constraints have to be satisfied. There are bounds on tour types and on the total number of tours. Moreover, there are constraints on the tightness of the fit during certain critical periods. With these side constraints, the optimization problem in the solid tour selection module is no longer a simple network flow problem. One way of dealing with this more complicated problem is the following. The original network flow formulation can be replaced by a conventional linear programming formulation and the side constraints can be incorporated as additional constraints. In most cases the linear programming formulation of the solid tour selection yields an integer solution. When it does not, a simple rounding heuristic can restore integrality.

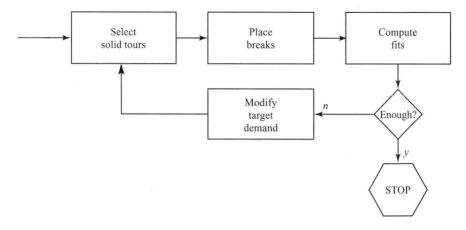

Fig. 12.4. Framework of the Approach in Operator Scheduling

To describe the mathematical programming formulation of the solid tour selection module in more detail additional notation is needed (since in this module the breaks are not taken into account). Let $D(t)$ denote the target demand for solid tours during time t; this target demand for solid tours has to be somewhat higher than the actual demand for operators since at any point in time not all of them may be working. Let $S(t)$ denote the supply of tours during period t (not taking into account that an operator may be on a break). So

$$S(t) = \sum_{j:\ t\in[a_j,b_j]} x_j$$

and

$$S(t) > s(t).$$

Let

$$E(t) = S(t) - D(t),$$
$$E^+(t) = \max(E(t), 0),$$
$$E^-(t) = \max(-E(t), 0).$$

So $E(t)$ denotes the amount of surplus at time t and $E^+(t)$ and $E^-(t)$ denote the positive and the negative part of this surplus. Let $\Psi^+(t)$ and $\Psi^-(t)$ denote the respective penalty costs at time t. The optimization problem can now be formulated as follows.

$$\min_{x, E^+, E^-} \sum_{t=1}^{H} \left(\Psi^-(t)E^-(t) + \Psi^+(t)E^+(t)\right) + \sum_{j=1}^{J} c_j x_j$$

subject to

$$E^+(t) - E^-(t) = \sum_{j:\, t\in[a_j,b_j]} x_j - D(t), \qquad t = 1,\ldots,H$$

$$\sum_{j=1}^{N} x_j \leq U$$

In addition, most of the variables have upper and lower bounds, i.e.,

$$\begin{array}{rcll}
x_j^{\min} \leq & x_j & \leq x_j^{\max} & j = 1,\ldots,N \\
0 \leq & E^-(t) & \leq E^-(t)^{\max} & t = 1,\ldots,H \\
0 \leq & E^+(t) & \leq E^+(t)^{\max} & t = 1,\ldots,H \\
m_i^{\min} \leq & m_i & \leq m_i^{\max} & i = 1,\ldots
\end{array}$$

The approach used here is much faster than implicit tour/break representation approaches, and can handle complex break placement rules.

The operator scheduling problem described here is a more complicated version of the problem described in Section 12.3. The real life version is even harder because of other issues that have to be taken into account. For example, a company may have operators who speak only English and others who speak both English and Spanish. So, some calls can be handled by either type of operator and others by only one type of operator. In the terminology of Chapter 2, the calls are the jobs and the operators are the machines; the M_j sets of the jobs are nested. The problems are further complicated by consideration of labor agreements and personnel policies. An example of such a policy is the FIFO rule, that is, if person A starts his shift earlier than person B, then A's first break cannot start later than B's first break.

12.8 Discussion

It is interesting to compare the models in this chapter with the workforce constrained scheduling and time-tabling models described in Section 9.4. The jobs have to be scheduled in such a way that a certain objective, e.g., the makespan, is minimized and at any point in time the demand for people remains within the limit. So, there is a flexibility in the scheduling of the jobs. In this chapter, the models are somewhat different. There is no flexibility in the requirements, since these are given. However, the size of the workforce and the number of people in each shift are the variables.

In practice, personnel scheduling problems tend to be intertwined with other factory scheduling problems. For example, when it is evident that committed shipping dates cannot be met, extra shifts have to be put in, or overtime has to be scheduled.

In the literature, these more aggregate problems (integrating machine scheduling and personnel scheduling) have not yet been considered. However, a number of scheduling systems, that are currently available on the market, offer machine scheduling features together with shift scheduling features.

Exercises

12.1. Consider the model described in Section 12.2.

a) Explain how algorithm 12.2.1 may result in an optimal schedule that is cyclic.

b) Develop a method to compute the number of weeks in the cycle of an optimal cyclic schedule.

c) Are all optimal schedules cyclic?

12.2. Consider Example 12.2.3. Note that the schedule has the disadvantage that there is a 1-day workstretch (e.g., employee 3 works in week 1 on Monday, while he is off on Sunday and Tuesday). Consider the following list of paired days:

>Pair 1: Monday - Wednesday;
>Pair 2: Tuesday - Thursday;
>Pair 3: Tuesday - Wednesday.

Work out the new schedule and observe that the schedule has minimum 2-day and maximum 5-day workstretches. However, the surplus is less evenly distributed.

12.3. Consider the days-off scheduling model of Section 12.2 and the instance with the following daily requirements.

day j	1	2	3	4	5	6	7
	Sun.	Mon.	Tues.	Wed.	Thurs.	Fri.	Sat.
Requirement	3	5	7	5	5	7	3

Each person requires 3 out of the 5 weekends off. Apply Algorithm 12.2.1 to this instance.

12.4. Consider the days-off scheduling model of Section 12.2 and the instance with the following daily requirements.

day j	1	2	3	4	5	6	7
	Sun.	Mon.	Tues.	Wed.	Thurs.	Fri.	Sat.
Requirement	1	7	7	7	10	11	3

Each person requires at least 1 out of every 2 weekends off. Apply Algorithm 12.2.1 to this instance.

12.5. Consider a retail store that is open for business from 10 a.m. to 8 p.m. There are five shift patterns.

pattern	Hours of Work	Total Hours	Cost
1	10 a.m. to 6 p.m.	8	$ 50.00
2	1 p.m. to 9 p.m.	8	$ 60.00
3	12 p.m. to 6 p.m.	6	$ 30.00
4	10 a.m. to 1 p.m.	3	$ 15.00
5	6 p.m. to 8 p.m.	3	$ 16.00

Staffing requirements at the store varies from hour to hour.

Hour	Staffing Requirement
10 a.m. to 12 a.m.	3
12 a.m. to 2 p.m.	6
2 p.m. to 4 p.m.	7
4 p.m. to 6 p.m.	7
6 p.m. to 8 p.m.	4

a) Formulate this problem as an integer program.

b) Solve the linear programming relaxation of this problem.

c) Is the solution obtained under b) optimal for the original problem?

12.6. Consider the instance described in Exercise 12.4. All the assumptions are still in force with the exception of one. Assume now that each person must work every other weekend and must have every other weekend off. (In Exercise 12.4 it was possible for a person to have every weekend off.)

a) Formulate this instance as an integer program.

b) Solve the linear program relaxation of the integer program formulated. Round off the answer to the nearest integers.

c) Compare the solution obtained under b) with the solution obtained in Exercise 12.4.

12.7. Consider the (5,7)-cyclic staffing problem with the **A** matrix as depicted in Section 12.4. The $\bar{b}$ vector is (4,9,8,8,8,9,4). The first entry corresponds to a Sunday and the last entry corresponds to a Saturday. The cost vector $(c_1, \ldots, c_7)$ is (6,5,6,7,7,7,7), i.e., the least expensive shift is the one that has both Saturday and Sunday off.

Apply Algorithm 12.4.1. to find the optimal solution.

12.8. Consider Application (i) in Section 12.5.

(a) Compute the number of columns in the matrix.

(b) Compute the number of columns if one-day workstretches are not allowed (a one day workstretch is a working day that is preceded and followed by days-off).

(c) Compute the number of columns if one and two day workstretches are not allowed.

12.9. Consider application (ii) in Section 12.5. Assume that the $\bar{c}$ vector is

$$(1, 1.25, 1.5, 1.75, 2, 2.25, 2.5, 2.75, 3,$$
$$1.5, 1.75, 2, 2.25, 2.5, 2.75, 3, 3.25, 3.5,$$
$$2, 2.25, 2.5, 2.75, 3, 3.25, 3.5, 3.75, 4)$$

The requirements vector $\bar{b}$ is

$$10, 10, 10, 10, 10, 10, 10, 10, 8, 8, 8, 8, 8, 8, 8, 8, 5, 5, 5, 5, 5, 5, 5, 5$$

Apply Algorithm 12.4.1 to this instance. (You will need a linear programming code to do this, e.g., LINDO.)

12.10. Consider Example 12.6.2. In the second iteration the row prices $(6.2, 7.8, 3, 3, 2)$ are used. However, this is not the only set of feasible row prices. Consider the set $(7, 7, 3, 3, 2)$, which is also feasible. Perform the next iteration using this set of prices.

12.11. Consider a central depot and 5 clients. From the depot a single delivery has to be made to each one of the clients. The routes that are allowed are shown in the table below. The objective is to determine which truck should go to each client and the routing that minimizes the total distance traveled. Each column in the table represents a possible truck route and the c_j represents the total distance of the route.

Route	1	2	3	4	5	6	7	8	9	10	11	12	13
c_j	8	10	4	4	2	14	10	8	11	12	6	6	5
	1	0	0	0	0	1	1	1	0	0	0	0	0
	0	1	0	0	0	1	0	0	1	1	0	0	0
	0	0	1	0	0	0	1	0	0	0	1	1	0
	0	0	0	1	0	0	0	1	0	1	0	1	1
	0	0	0	0	1	0	0	1	0	1	0	1	1

Apply Algorithm 12.6.1 to this instance.

Comments and References

The elementary textbook by Nanda and Browne (1992) covers some (but not all) of the models discussed in this chapter.

Section 12.2 is taken from Burns and Carter (1985) and is based on the seven days per week, one shift per day model. Burns and Koop (1987) extend this work and look at the seven days per week, multiple shifts per day model. Emmons (1985), Emmons and Burns (1991) and Hung and Emmons (1993) consider related models.

The general integer programming formulation considered in Section 12.3 appears in many handbooks and survey papers; see, for example, the survey papers by Tien and Kamiyama (1982) and Burgess and Busby (1992).

Comments and References

The material presented in Sections 12.4 and 12.5 is primarily based on the paper by Bartholdi, Orlin and Ratliff (1980).

The crew scheduling heuristic presented in Section 12.6 comes from the paper by Cullen, Jarvis and Ratliff (1981). Many papers have focused on crew scheduling problems; see, for example, Marsten and Shepardson (1981), Bodin, Golden, Assad and Ball (1983), and Stojkovic, Soumis, and Desrosiers (1998). A branch-and-cut method applied to crew scheduling is described in Hoffman and Padberg (1993). The airline crew recovery problem is described in Lettovsky, Johnson, and Nemhauser (2000).

A description of the operator scheduling system designed for a long-distance telephone company was presented at a national meeting of the INFORMS society in Washington, D.C., see Gawande (1996).

Part IV

Systems Development and Implementation

13 Systems Design and Implementation 319

14 Advanced Concepts in Systems Design 345

15 What Lies Ahead? 371

Chapter 13

Systems Design and Implementation

13.1 Introduction 319
13.2 Systems Architecture 320
13.3 Databases, Object Bases, and Knowledge-Bases . 322
13.4 Modules for Generating Plans and Schedules ... 327
13.5 User Interfaces and Interactive Optimization ... 330
13.6 Generic Systems vs. Application-Specific
     Systems 336
13.7 Implementation and Maintenance Issues 339

13.1 Introduction

Analyzing a planning or scheduling problem and developing a procedure for dealing with it on a regular basis is, in the real world, only part of the story. The procedure has to be embedded in a system that enables the decision-maker to actually use it. The system has to be integrated into the information system of the organization, which can be a formidable task. This chapter deals with system design and implementation issues.

The next section presents an overview of the infrastructure of the information systems and the architecture of the decision support systems in an enterprise. We focus on planning and scheduling systems in particular. The third section covers database, object base, and knowledge-base issues. The fourth section describes the modules that generate the plans and schedules, while the fifth discusses user interface issues and interactive optimization. The sixth section describes the advantages and disadvantages of generic systems and application-specific systems, while the last section discusses implementation and maintenance issues.

It is, of course, impossible to cover everything concerning the topics mentioned above. Many books have been written on each of these topics. This chapter focuses only on some of the more important issues concerning the design, development and implementation of planning and scheduling systems.

13.2 Systems Architecture

To visualize the information flow through an organization one often uses a *reference* model, which depicts all the relevant information flows in an enterprise. An example of a simplified reference model is presented in Figure 13.1

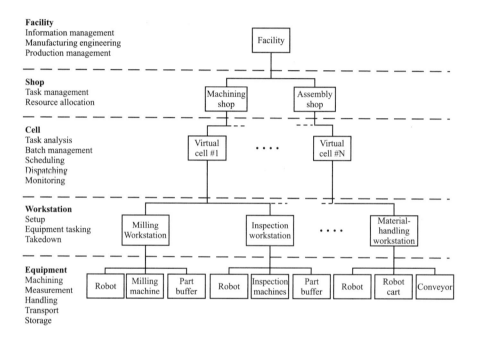

Fig. 13.1. Simplified Version of Reference Model

Nowadays, there are companies that specialize in the design and development of software that can serve as a backbone for an enterprise-wide information system. Each decision support system, on every level in the enterprise, can then be linked to such a backbone. Such a framework facilitates the connectivity of all the modules in an enterprise-wide information system. A company that is very active in this type of software development is SAP, which is headquartered in Walldorf (Germany).

As described in Section 1.2, a planning or scheduling system usually has to interact with a number of different systems in an organization. It may receive information from a higher level system that provides guidelines for the actions

13.2 Systems Architecture

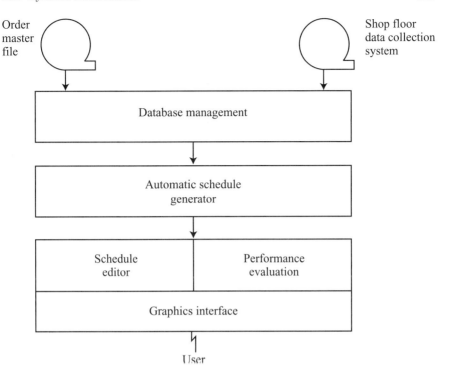

Fig. 13.2. Configuration of a Scheduling System

to be taken with regard to long term plans, medium term plans, short term schedules, workforce allocations, preventive maintenance, and so on. It may interact with a Material Requirements Planning (MRP) system in order to determine proper release dates for the jobs. A system may also interact with a shop floor control system that provides up-to-date information concerning availability of machines, progress of jobs, and so on (see Figures 1.1 and 13.2).

A planning or scheduling system typically consists of a number of different modules. The various types of modules can be categorized as follows:

(i) database, object base, and knowledge-base modules,
(ii) modules that generate the plans, schedules or timetables and
(iii) user interface modules.

These modules play a crucial role in the functionality of the system. Significant effort is required to make a factory's database suitable for input to a planning or scheduling system. Making a database accurate, consistent, and complete often involves the design of a series of tests the data must pass before it can be used. A database management module may also be able to manipulate the data, perform various forms of statistical analysis and allow the decision-maker, through some user interface, to see the data in graphical form. Some systems have a knowledge-base that is specifically designed for planning or scheduling purposes. A knowledge-base may contain, in one format or another, a list of rules that have to be followed in specific situations and maybe also a list of objects representing orders, jobs, and resources. A knowledge-base may at times also take the form of a constraint store which contains the constraints that have to be satisfied by the plans or schedules. However, few systems have a separate knowledge-base; a knowledge-base is usually embedded in the module that generates the plans or schedules.

The module that generates the plans or schedules typically contains a suitable model with objective functions, constraints and rules, as well as heuristics and algorithms.

User interface modules are important, especially with regard to the implementation process. Without an excellent user interface there is a good chance that, regardless of its capabilities, the system will be too unwieldy to use. User interfaces often take the form of an electronic Gantt chart with tables and graphs that enable a user to edit a plan or schedule generated by the system and take last minute information into account (see Figure 13.2). After a user has adjusted the plan or schedule manually, he is usually able to follow the impact of his changes on the performance measures, compare different solutions, and perform "what if" analyses.

13.3 Databases, Object Bases, and Knowledge-Bases

The database management subsystem may be either a custom-made or a commercial system. A number of the commercial database systems available on the market have proven to be useful for planning and scheduling systems. These are usually relational databases incorporating *Stuctured Query Language (SQL)*. Examples of such database management systems are Oracle and Sybase.

Whether a database management subsystem is custom made or commercial, it needs a number of basic functions, which include multiple editing, sorting and searching routines. Before generating a plan or a schedule, a decision-maker may want to see certain segments of an order masterfile and collect some statistics with regard to the orders and the related jobs. Actually, at times, he may not want to feed all the jobs or activities into the planning and scheduling routines, but rather a subset.

Within the database a distinction can be made between *static* and *dynamic* data. Static data include job or activity data and machine or resource data that do *not* depend on the plan or schedule. Some job data may be specified in the customer's order form, such as the ordered product quantity (which is proportional to the processing times of all the operations associated with the job), the committed shipping date (the due date), the time at which all necessary material is available (the release date) and possibly some processing (precedence) constraints. The priorities (weights) of the jobs are also static

13.3 Databases, Object Bases, and Knowledge-Bases

data as they do not depend on the plan or schedule. Having different weights for different jobs is usually a necessity, but determining their values is not that easy. In practice, it is seldom necessary to have more than three priority classes; the weights are then, for example, 1, 2 and 4. The three priority classes are sometimes described as "hot", "very hot" and "hottest" dependent upon the level of manager pushing the job. These weights actually have to be entered manually by the decision-maker into the information system database. To determine the priority level, the person who enters the weight may use his own judgement, or may use a formula that takes into account certain data from the information system (for instance, total annual sales to the customer or some other measure of customer criticality). The weight of a job may also change from one day to another; a job that is not urgent today, may be urgent tomorrow. The decision-maker may have to go into into the file and change the weight of the job before generating a new plan or schedule. Static machine data include machine speeds, scheduled maintenance times, and so on. There may also be static data that are both job and machine dependent, e.g., the setup time between jobs j and k assuming the setup takes place on machine i.

The dynamic data consists of all the data that are dependent upon the plan or schedule: the starting times and completion times of the jobs, the idle times of the machines, the times that a machine is undergoing setups, the sequences in which the jobs are processed on the machines, the number of jobs that are late, the tardinesses of the late jobs, and so on.

The following example illustrates some of these notions.

Example 13.3.1 (Order Master File in a Paper Mill). Consider the paper mill described in Example 1.1.4. The order master file of the rolls on order may contain the following data:

ORDER	CUSTOMER	CMT	WDT	BW	GR	FNSH	QTY	DDT	PRDT
PUR01410	UZSOY CO		16.0	5.0		29.0	55.0	05/25	05/24
PUR01411	UZSOY CO		16.0	4.0		29.0	20.0	05/25	05/25
PUR01412	UZSOY CO		16.0	4.0		29.0	35.0	06/01	
TAM01712	CYLEE LTD	PR	14.0	3.0		21.0	7.5	05/28	05/23
TAM01713	CYLEE LTD		14.0	3.0		21.0	45.0	05/28	05/23
TAM01714	CYLEE LTD		16.0	3.0		21.0	50.0	06/07	
EOR01310	LENSTRA NV	HLD	16.0	3.0		23.0	27.5	06/15	

Each order is characterized by an 8 digit alphanumeric order number. A customer may place a number of different orders, each one representing a specific type of roll that is characterized by a set of parameters. There are three parameters with regard to the type of paper, namely the basis weight (BW), the grade (GR) and the finish (FNS), and one parameter with regard to the size, namely the width (WDT) of the roll. The quantity (QTY) ordered is specified in pounds or tons. Given the quantity and the width the user of the system can compute the number and the diameter of the rolls. He actually has some freedom here: he can either go for a smaller number of rolls with a larger diameter or a larger number of rolls with a smaller diameter. The upper and lower bounds on the diameters of the rolls are determined by equipment limitations. The month and day of the committed shipping date are specified in the DDT column. The month and day of the completion date of the order are specified in the PRDT column; the days specified in this column can be either actual completion dates or scheduled completion dates. The comments (CMT) column is often empty. If a customer calls and puts an order on hold, then HLD is entered in this column and the decision-maker knows that this order should not yet be scheduled. If an order has a high priority, then PR is entered in this column. The weights will be a function of these entries, i.e., a job on hold has a low weight, a priority job has a high weight and the default value may correspond to an average weight.

Setup times may be regarded either as static or as dynamic data, depending on how they are generated. Setup times may be stored in a table so that whenever a particular setup time needs to be known, the necessary table entry is retrieved. However, this method is not very efficient if the set is very large and if relatively few table look-ups are required. The size of the matrix is n^2 and all entries of the matrix have to be computed beforehand, which may require considerable CPU time as well as memory. An alternative way to compute and retrieve setup times, that is more efficient in terms of storage space and that can be more efficient in terms of computation time, is the following. A number of parameters, say

$$a_{ij}^{(1)}, \ldots, a_{ij}^{(l)},$$

may be associated with job j and machine i. These parameters are static data and may be regarded as given machine settings necessary to process job j on machine i. The setup time between jobs j and k on machine i, s_{ijk}, is a known function of the $2l$ parameters

$$a_{ij}^{(1)}, \ldots, a_{ij}^{(l)}, \ a_{ik}^{(1)}, \ldots, a_{ik}^{(l)}.$$

The setup time usually is a function of the differences in machine settings for jobs j and k and is determined by production standards.

Example 13.3.2 (Sequence Dependent Setup Times). Assume that in order to start a job on machine i, three machine settings have to be fixed (for example, the color, the grade and the basis weight of the paper in the paper mill of Example 1.1.4). So the total setup time s_{ijk} depends on the time it takes to perform these three changeovers and is a function of six parameters, i.e.,

$$a_{ij}^{(1)}, a_{ij}^{(2)}, a_{ij}^{(3)}, \ \ a_{ik}^{(1)}, a_{ik}^{(2)}, a_{ik}^{(3)}.$$

If the three changeovers have to be done sequentially, then the total setup time is

13.3 Databases, Object Bases, and Knowledge-Bases

$$s_{ijk} = h_i^{(1)}(a_{ij}^{(1)}, a_{ik}^{(1)}) + h_i^{(2)}(a_{ij}^{(2)}, a_{ik}^{(2)}) + h_i^{(3)}(a_{ij}^{(3)}, a_{ik}^{(3)}).$$

If the three changeovers can be done in parallel, then the total setup time is

$$s_{ijk} = \max\left(h_i^{(1)}(a_{ij}^{(1)}, a_{ik}^{(1)}),\ h_i^{(2)}(a_{ij}^{(2)}, a_{ik}^{(2)}),\ h_i^{(3)}(a_{ij}^{(3)}, a_{ik}^{(3)})\right).$$

Of course, there may be situations where some of the changeovers can be done in parallel while others have to be done in series.

If the setup times are computed this way, they may be considered dynamic data. The total time needed for computing setup times in this manner depends on the type of algorithm. If a dispatching rule is used to determine a good schedule, this method, based on (static) job parameters, is usually more efficient than the table look-up method mentioned earlier. However, if some kind of local search routine is used, the table look-up method will become more time efficient. The decision on which method to use depends on the relative importance of memory versus *CPU* time.

The calendar function is often also part of the database system. It contains information with regard to holidays, number of shifts available, scheduled machine maintenance, and so on. Calendar data are sometimes static, e.g., fixed holidays, and sometimes dynamic, e.g., preventive maintenance shutdowns.

Some of the more modern planning and scheduling systems may rely on an object base in addition to (or instead of) a database. One of the main functions of the object base is to store the definitions of all object types, i.e., it functions as an object library and instantiates the objects when needed. In a conventional relational database, a data type can be defined as a schema of data; for example, a data type "job" can be defined as in Figure 13.3.a and an instance can be as in Figure 13.3.b. Object types and corresponding instances can be defined in the same way. For example, an object type "job" can be defined and corresponding job instances can be created. All the job instances have then the same type of attributes.

There are two crucial relationships between object types, namely, the "is-a" relationship and the "has-a" relationship. An is-a relationship indicates a generalization and the two object types have similar characteristics. The two object types are sometimes referred to as a subtype and a supertype. For example, a "machine" object type may be a special case of a "resource" object type and a "tool" object type may be another special case of a resource object type. A "has-a" relationship is an aggregation relationship; one object type contains a number of other object types. A "workcenter" object may be composed of several machine objects and a "plant" object may comprise a number of workcenter objects. A "routing table" object may consist of job objects as well as of machine objects.

Object types related by is-a or has-a relationships have similar characteristics with regard to their attributes. In other words, all the attributes of a supertype object are used by the corresponding subtypes. For example, a machine object has all the attributes of a resource object and it may also have

| ID | Name | Type | Quantity | Priority | Ready | Due |

(a) Job data type

| 2 | IBM | 4 | 160 | 2 | 10 | 200 |

(b) Job instance

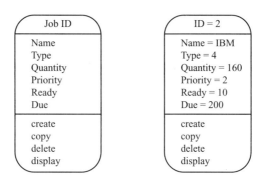

(c) Job object type and a job object

Fig. 13.3. Job Data Type, Job Instance, and Job Object Type

some additional attributes. This is often referred to as inheritance. A hierarchical structure that comprises all object types can be constructed. Objects can be retrieved through commands that are similar to SQL commands in relational databases.

While virtually every planning and scheduling system relies on a database or an object base, not many systems have a module that serves specifically as a knowledge-base. However, knowledge-bases or constraint stores may become more and more important in the future.

The overall architecture of a system, in particular the module that generates the plans or schedules, influences the design of a knowledge-base. The most important aspect of a knowledge-base is the knowledge *representation*. One form of knowledge representation is through *rules*. There are several formats for stating rules. A common format is through an *IF-THEN* statement. That is, *IF* a given condition holds, *THEN* a specific action has to be taken.

Example 13.3.3 (IF-THEN Rule for Parallel Machines). Consider a setting with machines in parallel. The machines have different speeds; some are fast and others are slow. The jobs are subject to sequence dependent setup times that are independent of the machines, i.e., setups take the same amount of time on a fast machine as they take on a slow machine.

Because of the setup times it is advisable to assign the longer jobs to the faster machines while keeping the shorter jobs on the slower machines. One could establish a threshold value and assign the longer jobs to a fast machine.
  IF a job's processing time is longer than a given value,
  THEN the job may be assigned to a fast machine.
It is easy to code such a rule in a programming language such as C++.

Another format for stating rules is through *predicate logic* that is based on propositional calculus. An appropriate programming language for dealing with rules in this format is Prolog.

Example 13.3.4 (Logic Rule for Parallel Machines). Consider the rule in the previous example. A Prolog version of this rule may be:

MACHINEOK(M,L) : − long __ job(L) , fast __ machines(F) , member(M,F).

The M refers to a specific machine, the L to a long job, and the F to a list of all fast machines. The ": −" may be read as "if"and the "," may be read as "and". A translation of the rule would be: machine M is suitable for job L if L is a long job, if set F is the set of fast machines and if machine M is a member of F.

As stated before, the design of the module that generates the plans or the schedules affects the design of a knowledge-base. This is discussed in more detail in the next section.

13.4 Modules for Generating Plans and Schedules

Current planning and scheduling techniques are an amalgamation of several schools of thought that have been converging in recent years. One school of thought, predominantly followed by industrial engineers and operations researchers, is at times referred to as the *algorithmic* or the *optimization* approach. A second school of thought, that is often followed by computer scientists and artificial intelligence experts, include the *knowledge-based* and the *constraint programming* approaches. Recently, the two schools of thought have started to converge and the differences have become blurred. Some hybrid systems combine a knowledge base with fairly sophisticated heuristics; other systems have one segment of the procedure designed according to the optimization approach and another segment according to the constraint programming approach.

Example 13.4.1 (Architecture of a Planning and Scheduling System for a Wafer Fab). A hybrid planning and scheduling system has been designed for a particular semiconductor wafer fabrication unit as follows. The system consists of two levels. The higher level operates according to a knowledge-based approach. The lower level is based on an optimization approach; it consists of a library of algorithms.

The higher level performs the first phase of the planning and scheduling process. At this level, the current status of the environment is analyzed. This analysis takes into consideration due date tightness, bottlenecks, and so on. The rules embedded in this higher level determine for each situation the type of algorithm that has to be used at the lower level.

The algorithmic approach usually requires a mathematical formulation of the problem that includes objectives and constraints. The algorithm could be based on any one or a combination of techniques. The "quality" of the solution is based on the values of the objectives and performance criteria of the given schedule. This form of solution method may consist of three phases. In the first phase, a certain amount of *preprocessing* is done, where the problem instance is analyzed and a number of statistics are compiled, e.g., the average processing time, the maximum processing time, the due date tightness. The second phase consists of the actual algorithms and heuristics, whose structure may depend on the statistics compiled in the first phase (for example, in the way the look-ahead parameter K in the ATC rule may depend on the due date tightness and due date range factors). The third phase may contain a *postprocessor*. The solution that comes out of the second phase is fed into a procedure such as simulated annealing or tabu-search, in order to see if improvements can be obtained. This type of solution method is usually coded in a procedural language such as Fortran, Pascal or C.

The knowledge-based and constraint programming approaches are in various respects different from the algorithmic approach. These approaches are often more concerned with underlying problem structures that cannot easily be described in an analytical format. In order to incorporate the decisionmaker's knowledge into the system, objects, rules or constraints are used. These approaches are often used when it is only necessary to find a *feasible* solution given the many rules or constraints; however, as some plans or schedules are ranked "more preferable" than others, heuristics may be used to obtain a "more preferred" plan or schedule. Through a so-called *inference engine,* such an approach tries to find solutions that do not violate the prescribed rules and satisfy the stated constraints as much as possible. The logic behind the schedule generation process is often a combination of inferencing techniques and search techniques as described in Appendixes C and D. The inferencing techniques are usually so-called *forward chaining* and *backward chaining* algorithms. A forward chaining algorithm is knowledge driven. It first analyzes the data and the rules and, through inferencing techniques, attempts to construct a feasible solution. A backward chaining algorithm is result oriented. It starts out with a promising solution and attempts to verify whether it is feasible. Whenever a satisfactory solution does not appear to exist or when the system's user thinks that it is too difficult to find, the user may want to reformulate the problem by relaxing some of the constraints. The relaxation

13.4 Modules for Generating Plans and Schedules

of constraints may be done either automatically (by the system itself) or by the user. Because of this aspect, the knowledge-based approach has also been referred to as the *reformulative* approach.

The programming style used for the development of knowledge-based systems is different from the ones used for systems based on algorithmic approaches. The programming style may depend on the form of the knowledge representation. If the knowledge is represented in the form of *IF-THEN* rules, then the system can be coded using an expert system shell. The expert system shell contains an inference engine that is capable of doing forward chaining or backward chaining of the rules in order to obtain a feasible solution. This approach may have difficulties with conflict resolution and uncertainty. If the knowledge is represented in the form of logic rules (see Example 13.3.4), then Prolog may be suitable. If the knowledge is represented in the form of frames, then a language with object oriented extensions is required, e.g., C++. These languages emphasize user-defined objects that facilitate a modular programming style. Examples of systems that are designed according to a constraint programming approach are described in Chapter 10 and in Appendix D.

Algorithmic approaches as well as knowledge-based approaches have their advantages and disadvantages. An algorithmic approach has an edge if

(i) the problem allows for a crisp and precise mathematical formulation,
(ii) the number of jobs involved is large,
(iii) the amount of randomness in the environment is minimal,
(iv) some form of optimization has to be done frequently and in real time,
(v) the general rules are consistently being followed without too many exceptions.

A disadvantage of the algorithmic approach is that if the operating environment changes (for example, certain preferences on assignments of jobs to machines), the reprogramming effort may be substantial.

The knowledge-based and constraint programming approaches may have an edge if only feasible plans or schedules are needed. Some system developers believe that changes in the environment or in the rules or constraints can be more easily incorporated in a system that is based on such an approach than in a system that is based on the algorithmic approach. Others, however, believe that the effort required to modify any system is mainly a function of how well the code is organized and written; the effort required to modify does not depend that much on the approach used.

A disadvantage of the knowledge-based and constraint programming approaches is that obtaining reasonable plans or schedules may require in certain settings substantially more computer time than an algorithmic approach. In practice certain planning or scheduling systems have to operate in near-real time (it is very common that plans or schedules must be generated within minutes).

The amount of available computer time is an important factor in the selection of a schedule generation technique. The time allowed to generate a plan or a schedule varies from application to application. Many applications require real time performance: a plan or schedule has to be generated in seconds or minutes on the available computer. This may be the case if rescheduling is required many times a day because of schedule deviations. It would also be true if the planning or scheduling engine runs iteratively, requiring human interaction between iterations (perhaps for adjustments of workcenter capacities). However, some applications do allow overnight number crunching. For example, a user may start a program at the end of the day and expect an output by the time he or she arrives at work the next day. A few applications require extensive number crunching. When, in the airline industry, quarterly flight schedules have to be determined, the investments at stake are such that a week of number crunching on a mainframe is fully justified.

As stated before, the two schools of thought have been converging and many planning and scheduling systems that are currently being designed have elements of both. One language of choice is C++ as it is an easy language for coding algorithmic procedures and it also has object-oriented extensions.

13.5 User Interfaces and Interactive Optimization

The user interfaces are very important parts of the system. The interfaces usually determine whether the system is going to be used or not. Most user interfaces, whether the system is based on a workstation or PC, make extensive use of window mechanisms. The user often wants to see several different sets of information at the same time. This is the case not only for the static data that is stored in the database, but also for the dynamic data that depend on the plan or schedule.

Some user interfaces allow for extensive user interaction. A decision-maker may modify the current status or the current information. Other user interfaces may not allow any modifications. For example, an interface that displays the values of all the relevant performance measures would not allow the user to change any of the numbers. A decision-maker may be allowed to modify the plan or schedule in another interface which then automatically changes the values of the performance measures, but a user may not change performance measures directly.

User interfaces for database modules often take a fairly conventional form and may be determined by the particular database package used. These interfaces must allow for some user interaction, because data such as due dates often have to be changed during a planning or scheduling session.

There are often a number of interfaces that exhibit general data concerning the plant or enterprise. Examples of such interfaces are:

(i) the plant layout interface,
(ii) the resource calendar interface, and
(iii) the routing table interface.

13.5 User Interfaces and Interactive Optimization

The plant layout interface may depict graphically the workcenters and machines in a plant as well as the possible routes between the workcenters. The resource calendar displays shift schedules, holidays and preventive maintenance schedules of the machines. In this interface the user can assign shifts and schedule the servicing of the resources. The routing table typically may show static data associated with the jobs. It specifies the machines and/or the operators who can process a particular job or job type.

The module that generates the plans or schedules may provide the user with a number of computational procedures and algorithms. Such a library of procedures within this module will require its own user interface, enabling the user to select the appropriate algorithm or even design an entirely new procedure.

User interfaces that display information regarding the plans or schedules can take many different forms. Interfaces for adjusting or manipulating the plans or schedules basically determine the character of the system, as these are the ones used most extensively. The various forms of interfaces for manipulating solutions depend on the level of detail as well as on the planning horizon being considered. In what follows four such interfaces are described in more detail, namely:

(i) the Gantt Chart interface,
(ii) the Dispatch List interface,
(iii) the Capacity Buckets interface, and
(iv) the Throughput Diagram interface.

The first, and probably most popular, form of schedule manipulation interface is the Gantt chart (see Figure 13.4). The Gantt chart is the usual horizontal bar chart, with the x-axis representing the time and the y-axis, the various machines. A color and/or pattern code may be used to indicate a characteristic or an attribute of the corresponding job. For example, jobs that are completed after their due date under the current schedule may be colored red. The Gantt chart usually has a number of scroll capabilities that allow the user to go back and forth in time or focus on particular machines, and is usually mouse driven. If the user is not entirely satisfied with the generated schedule, he may wish to perform some manipulations on his own. With the mouse, the user can "click and drag" an operation from one position to another. Providing the interface with a click, drag, and drop capability is not a trivial task for the following reason: after changing the position of a particular operation on a machine, other operations on that machine may have to be pushed either forward or backward in time to maintain feasibility. The fact that other operations have to be processed at different times may have an effect on the schedules of other machines. This is often referred to as *cascading* or *propagation* effects. After the user has repositioned an operation of a job, the system may call a reoptimization procedure that is embedded in the planning or scheduling engine to deal with the cascading effects in a proper manner.

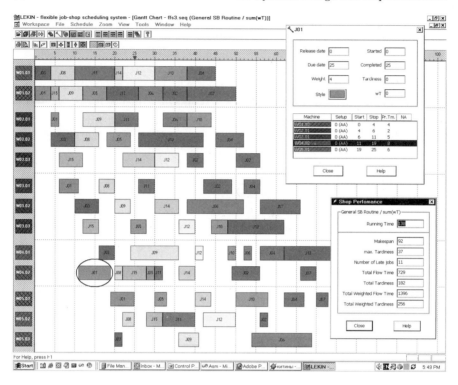

Fig. 13.4. Gantt Chart Interface

Example 13.5.1 (Cascading Effects and Reoptimization). Consider a three machine flow shop with unlimited storage space between the successive machines and therefore no blocking. The objective is to minimize the total weighted tardiness. Consider a schedule with 4 jobs as depicted by the Gantt chart in Figure 13.5.a. If the user swaps jobs 2 and 3 on machine 1, while keeping the order on the two subsequent machines the same, the resulting schedule, because of cascading effects, takes the form depicted in Figure 13.5.b. If the system has reoptimization algorithms at its disposal, the user may decide to reoptimize the operations on machines 2 and 3, while keeping the sequence on machine 1 frozen. A reoptimization algorithm then may generate the schedule depicted in Figure 13.5.c. To obtain appropriate job sequences for machines 2 and 3, the reoptimization algorithm has to solve an instance of the two machine flow shop with the jobs subject to given release dates at the first machine.

Gantt charts do have disadvantages, especially when there are many jobs and machines. It may be hard to recognize which bar or rectangle corresponds to which job. As space on the screen (or on the printout) is rather limited, it is hard to attach text to each bar. Gantt chart interfaces usually provide the

13.5 User Interfaces and Interactive Optimization

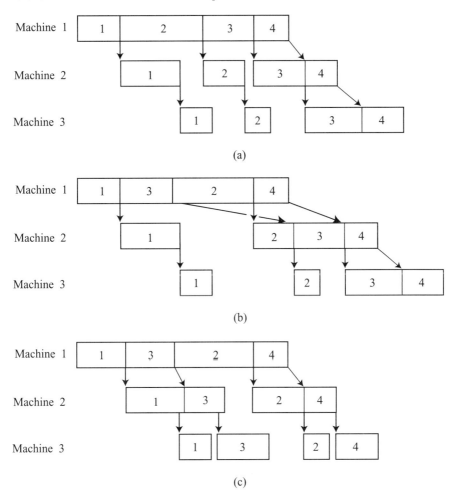

Fig. 13.5. Cascading and reoptimization after swap: (a) original schedule, (b) cascading effects after swap of jobs on machine 1, (c) schedule after reoptimization of machines 2 and 3

capability to click on a given bar and open a window that displays detailed data regarding the corresponding job. Some Gantt charts also have a filter capability, where the user may specify the job(s) that should be exposed on the Gantt chart while disregarding all others. The Gantt chart interface depicted in Figure 13.4 is from the LEKIN system described in Chapter 5.

The second form of user interface displaying schedule information is the *dispatch-list* interface (see Figure 13.6). Schedulers often want to see a list of the jobs to be processed on each machine in the order in which they are to be processed. With this type of display schedulers also want to have editing

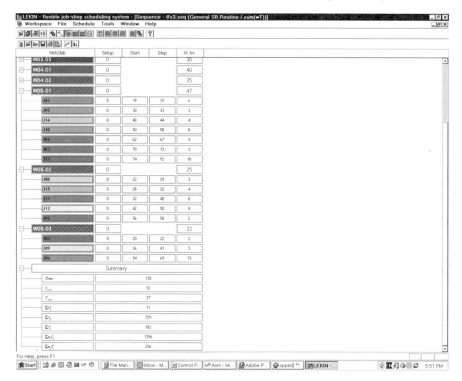

Fig. 13.6. Dispatch-list Interface

capabilities so they can change the sequence in which jobs are processed on a machine or move a job from one machine to another. This sort of interface does not have the disadvantage of the Gantt chart, since the jobs are listed with their job numbers and the scheduler knows exactly where each job is in a sequence. If the scheduler would like to see more attributes of the jobs to be listed (e.g., processing time, due date, completion time under the current schedule, and so on), then more columns can be added next to the job number column, each one with a particular attribute. The disadvantage of the dispatch-list interface is that the scheduler does not have a clear view of the schedule relative to time. The user may not see immediately which jobs are going to be late, which machine is idle most of the time, etc. The dispatch-list interface in Figure 13.6 is also from the LEKIN system.

The third form of user interface is the *capacity buckets* interface. The time axis is partitioned into a number of time slots or buckets. Buckets may correspond to either days, weeks or months. For each machine the processing capacity of a bucket is known. The creation of plans or schedules may in certain environments be accomplished by assigning jobs to machines in given time segments. After such assignments are made, the capacity buckets interface

13.5 User Interfaces and Interactive Optimization

displays for each machine the percentage of the capacity utilized in each time segment. If the decision-maker sees that a machine is overutilized in a given time period, he knows that some jobs in the corresponding bucket have to be rescheduled. The capacity buckets interface contrasts, in a sense, with the Gantt chart interface. A Gantt chart indicates the number of late jobs as well as their respective tardinesses. The number of late jobs and the total amount of tardinesses give an indication of the deficiency in capacity. The Gantt chart is thus a good indicator of the available capacity in the short term (days or weeks) when there are a limited number of jobs (twenty or thirty). Capacity buckets are useful when the scheduler is performing medium or long term planning. The bucket size may be either a week or a month and the total period covered three or four months. Capacity buckets are, of course, a cruder form of information as they do not give an indication of which jobs are completed on time and which ones are completed late.

The fourth form of user interface is the *input-output diagram* or *throughput diagram* interface, which are often of interest when the production is made to stock. These diagrams describe the total amount of orders received, the total amount produced and the total amount shipped, cumulatively over time. The difference, at any point in time, between the first two curves is the total amount of orders waiting for processing and the difference between the second and the third curves equals the total amount of finished goods in inventory. This type of interface specifies neither the number of late jobs nor their respective tardinesses. It does provide the user with information regarding machine utilization and Work-In-Process (WIP).

Clearly, the different user interfaces for the display of information regarding plans or schedules have to be strongly linked with one another. When a user makes changes in either the Gantt chart interface or the dispatch-list interface, the dynamic data may change considerably because of cascading effects or the reoptimization process. Changes made in one interface, of course, have to be shown immediately in the other interfaces as well.

User interfaces that display information regarding the plans or schedules have to be linked to other interfaces as well, e.g., database management interfaces and interfaces of a planning or scheduling engine. For example, a user may modify an existing schedule in the Gantt chart interface by clicking, dragging, and dropping; then he may want to freeze certain jobs in their respective positions. After doing so, he may want to reoptimize the remaining (unfrozen) jobs using an algorithm in the scheduling engine. These algorithms are similar to the algorithms described in Parts II and III for situations where machines are not available during given time periods (because of breakdowns or other reasons). The interfaces that allow the user to manipulate the plans or schedules have to be, therefore, strongly linked with the interfaces for algorithm selection.

User interfaces may also have a separate window that displays the values of all relevant performance measures. If the user has made a change in a plan or schedule the values before and after the change may be displayed.

Typically, performance measures are displayed in plain text format. However, more sophisticated graphical displays may also be used.

Some user interfaces are sophisticated enough to allow the user to split a job into a number of smaller segments and schedule each of these separately. Splitting an operation is equivalent to (possibly multiple) preemptions. The more sophisticated user interfaces also allow different operations of the same job to overlap in time. In practice, this may occur in many settings. For example, a job may start at a downstream machine of a flow shop before it has completed its processing at an upstream machine. This occurs when a job represents a large batch of identical items. Before the entire batch has been completed at the upstream machine, parts of the batch may already have been transported to the next machine and may already have started their processing there.

13.6 Generic Systems vs. Application-Specific Systems

Dozens of software houses have developed systems which they claim can be implemented in many different industrial settings after only minor modifications. It often turns out that the effort involved in customizing such systems is quite substantial. The code developed for the customization of the system may turn out to be more than half the total code of the final version of the system. However, some systems have very sophisticated configurations that allow them to be tailored to different types of industries without much programming effort. These systems are highly modular and have an edge with regard to adjustments to specific requirements. A generic system, if it is highly modular, can be changed to fit a specific environment by adding specific modules, e.g., tailor-made scheduling algorithms. Experts can develop such algorithms and the generic planning and scheduling software supplies standard interfaces or "hooks" that allow the integration of special functions in the package. This concept allows the experts to concentrate on the planning or scheduling problem, while the generic software package supplies the functionalities that are less specific, e.g., user interfaces, data management, standard planning and scheduling algorithms for less complex areas, and so on.

Generic systems may be built either on top of a commercial database system, such as Sybase or Oracle, or a proprietary database system developed specifically for the planning or scheduling system. Generic systems use processing data similar to the data presented in the frameworks described in Chapters 2 and 3. However, the framework in such a system may be somewhat more elaborate than the frameworks presented in Chapters 2 and 3. For example, the database may allow for an alphanumeric order number, that refers to the name of a customer. The order number then relates to several jobs, each one with its own processing time (which often may be referred to as the *quantity* of a batch) and a routing vector that determines the precedence constraints the operations are subject to. The order number has its

13.6 Generic Systems vs. Application-Specific Systems

own due date (committed shipping date), weight (priority factor) and release date (which may be determined by a Material Requirements Planning (MRP) system connected to the planning or scheduling system). The system may include procedures that translate the due date of the order into due dates for the different jobs at the various workcenters. Also, the weights of the different jobs belonging to an order may not be exactly equal to the weight of the order itself. The weights of the different jobs may be a function of the amount of value already added to the product. The weight of the last job pertaining to an order may be larger than the weight of the first job pertaining to that order.

The way the machine or resource environment is represented in the database is also somewhat more elaborate than the way it is described in Chapters 2 and 3. For example, a system typically allows a specification of workcenters and, within each workcenter, a specification of machines.

Most generic planning and scheduling systems have routines for generating a "first" plan or schedule for the user. Of course, such an initial solution rarely satisfies the user. That is why planning and scheduling systems often have elaborate user interfaces that allow the user to manually modify an existing plan or schedule. The automated planning and scheduling capabilities generally consist of a number of different dispatching rules that are basically sorting routines. These rules are usually the same as the priority rules discussed in the previous chapters (SPT, LPT, WSPT, EDD and so on). Some generic systems rely on more elaborate procedures, such as *forward loading* or *backward loading*. Forward loading implies that the jobs are inserted one at the time starting at the beginning of the schedule, that is, at the current time. Backward loading implies that the schedule is generated starting from the back of the schedule, that is, from the due dates, working its way towards the current time (again inserting one job at the time). These insertions, either forward or backward in time, are done according to some priority rule. Some of the more sophisticated automated procedures first identify the bottleneck workcenter(s) or machine(s); they compute time windows during which jobs have to be processed on these machines and then they schedule the jobs on these machines through some algorithmic procedure. After the bottlenecks are scheduled, the procedure schedules the remaining machines through either forward loading or backward loading.

Almost all generic planning and scheduling systems have user interfaces that include Gantt charts and enable the user to manipulate the solutions manually. However, these Gantt chart interfaces are not always perfect. For example, most of them do *not* take into account the cascading and propagation effects referred to in the previous section. They may do some automatic rescheduling on the machine or workcenter where the decision maker has made a change, but they usually do not adapt the schedules on other machines or workcenters to this change. The solutions generated may at times be infeasible. Some systems give the user a warning when, after the modifications, the resulting plan or schedule turns out to be infeasible.

Besides the Gantt chart interface, most systems have at least one other type of interface that either displays the actual plan or schedule or provides important data that is related. The second interface is typically one of those mentioned in the previous section.

Generic systems usually have fairly elaborate report generators, that print out the plan or schedule with alphanumeric characters; such printouts can be done fast and on an inexpensive printer. The printout may then resemble what is displayed, for example, in the dispatch-list interface described in the previous section. It is possible to list the jobs in the order in which they will be processed at a particular machine or workcenter. Besides the job number, other relevant job data may be printed out as well. There are also systems that print out entire Gantt charts. But Gantt charts have the disadvantage mentioned before, namely that it may not be immediately obvious which rectangle or bar corresponds to which job. Usually the bars are too small to append any information to.

Generic systems have a number of advantages over application-specific systems. If the scheduling problem is a fairly standard one and only minor customization of a generic system suffices, then this option is usually less expensive than developing an application-specific system from scratch. An additional advantage is that an established company will maintain the system. On the other hand, most software houses that develop planning or scheduling systems do not provide the source code. This makes the user of the system dependent on the software house even for very minor changes.

In many instances generic systems are simply not suitable and application-specific systems (or modules) have to be developed. There are several good reasons for developing application-specific systems. One reason may be that the planning or scheduling problem is simply so large (because of the number of machines, jobs, or attributes) that a PC-based generic system simply would not be able to handle it. The databases may be very large and the required interface between the shopfloor control system and the planning or scheduling system may be of a kind which a generic system cannot handle. An example of an environment where this is often the case is semiconductor manufacturing.

A second reason to opt for an application-specific system is that the environment may have so many idiosyncrasies that no generic system can be modified in such a way that it can address the problem satisfactorily. The processing environment may have certain restrictions or constraints that are difficult to attach to or build into a generic system. For example, certain machines at a workcenter may have to start with the processing of different jobs at the same time (for one reason or another) or a group of machines may have to sometimes act as a single machine and, at other times, as separate machines. The order portfolio may also have many idiosyncrasies. That is, there may be a fairly common machine environment used in a fairly standard way (that would fit nicely into a generic system), but with too many exceptions on the rules as far as the jobs are concerned. Coding in the special situations represents such a large amount of work that it may be advisable to build a system from scratch.

A third reason for developing an application-specific system is that the user may insist on having the source code and be able to maintain the system within his own organization.

An important advantage of an application-specific system is that manipulating a solution is usually considerably easier than with a generic system.

13.7 Implementation and Maintenance Issues

During the last two decades a large number of planning and scheduling systems have been developed, and many more are under development. These developments have made it clear that a certain proportion of the theoretical research done over the last couple of decades is of very limited use in real world applications. Fortunately, the system development that is going on in industry is currently encouraging theoretical researchers to tackle planning and scheduling problems that are more relevant to the real world. At various academic institutions in Europe, Japan and North America, research is focusing on the development of algorithms as well as on the development of systems; significant efforts are being made in integrating these developments.

Over the last two decades many companies have made large investments in the development and implementation of planning and scheduling systems. However, not all the systems developed or installed appear to be used on a regular basis. Systems, after being implemented, often remain in use only for a limited time; after a while they may be ignored altogether.

In those situations where the systems are in use on a more or less permanent basis, the general feeling is that the operations do run smoother. A system that is in place often does *not* reduce the time the decision-maker spends on planning and scheduling. However, a system usually does enable the user to produce better solutions. Through an interactive Graphics User Interface (GUI) a user is often able to compare different solutions and monitor the various performance measures. There are other reasons for smoother operations besides simply better plans and schedules. A planning or scheduling system imposes a certain "discipline" on the operations. There are now compelling reasons for keeping an accurate database. Plans and schedules are either printed out neatly or visible on monitors. This apparently has an effect on people, encouraging them to actually even do their jobs according to the plan or schedule.

The system designer should be aware of the reasons why some systems have never been implemented or are never used. In some cases, databases are not sufficiently accurate and the team implementing the system does not have the patience or time to improve the database (the persons responsible for the database may be different from the people installing the scheduling system). In other cases, the way in which workers' productivity is measured is not in agreement with the performance criteria the system is based upon. User interfaces may not permit the user of the system to reschedule sufficiently fast in the case of unexpected events. Procedures that enable rescheduling when the main user is absent (for example, if something unexpected happens during third shift) may not be in place. Finally, systems may not be given sufficient time to "settle" or "stabilize" in their environment (this may require many months, if not years).

Even if a system gets implemented and used, the duration during which it remains in use may be limited. Every so often, the organization may change drastically and the system is not flexible enough to provide good plans or schedules for the new environment. Even a change in a manager may derail a system.

In summary, the following points could be taken into consideration when designing, developing and implementing a system.

1. Visualize how the operating environment will evolve over the lifetime of the system before the design process actually starts.
2. Get all the people affected by the system involved in the design process. The development process has to be a team effort and all involved have to approve the design specifications.
3. Determine which part of the system can be handled by off-the-shelf software. Using an appropriate commercial code speeds up the development process considerably.
4. Keep the design of the software modular. This is necessary not only to facilitate the entire programming effort, but also to facilitate changes in the system after its implementation.
5. Make the objectives of the algorithms embedded in the system consistent with the performance measures by which people who must act according to the plans or schedules are being judged.
6. Do not take the data integrity of the database for granted. The system has to be able to deal with faulty or missing data and provide the necessary safeguards.
7. Capitalize on potential side benefits of the system, e.g., spin-off reports for distribution to key people. This enlarges the supporters base of the system.
8. Make provisions to ensure easy rescheduling, not only by the main planner or scheduler but also by others, in case the main user is absent.
9. Keep in mind that the installment of the system requires patience. It may take months or even years before the system runs smoothly. This period should be a period of continuous improvement.
10. Do not underestimate the necessary maintenance of the system after installation. The effort required to *keep* the system in use on a regular basis is considerable.

It appears that in the decade to come, an even larger effort will be made in the design, development and implementation of planning and scheduling

Exercises

systems and that such systems will play an important role in Computer Integrated Manufacturing.

Exercises

13.1. Consider a job shop with machines in parallel at each workcenter (i.e., a flexible job shop). There are hard constraints as well as soft constraints that play a role in the scheduling of the machines. More machines may be installed in the near future. The scheduling process does not have to be done in real time, but can be done overnight. Describe the advantages and disadvantages of an algorithmic approach and of a knowledge-based approach.

13.2. Consider a factory with a single machine with sequence dependent setup times and hard due dates. It does not appear that changes in the environment are imminent in the near future. Scheduling and rescheduling has to be done in real time.

(a) List the advantages and disadvantages of an algorithmic approach and of a knowledge-based approach.

(b) List the advantages and disadvantages of a commercial system and of an application-specific system.

13.3. Design a schedule generation module that is based on a composite dispatching rule for a parallel machine environment with the jobs subject to sequence dependent setup times. Job j has release date r_j and may only be processed on a machine that belongs to a given set M_j. There are three objectives, namely $\sum w_j T_j$, C_{max} and L_{max}. Each objective has its own weight and the weights are time dependent; every time the scheduler uses the system he puts in the relative weights of the various objectives. Design the composite dispatching rule and explain how the scaling parameters depend on the relative weights of the objectives.

13.4. Consider the following three measures of machine congestion over a given time period.

(i) the number of late jobs during the period;
(ii) the average number of jobs waiting in queue during the given period.
(iii) the average time a job has to wait in queue during the period.

How does the selection of congestion measure depend on the objective to be minimized?

13.5. Consider the following scheduling alternatives:

(i) forward loading (starting from the current time);
(ii) backward loading (starting from the due dates);
(iii) scheduling from the bottleneck stage first.

How does the selection of one of the three alternatives depend on the following factors:

(i) degree of uncertainty in the system.
(ii) balanced operations (not one specific stage is a bottleneck).
(iii) due date tightness.

13.6. Consider the ATC rule. The K factor is usually determined as a function of the due date tightness factor θ_1 and the due date range factor θ_2. However, the process usually requires extensive simulation. Design a learning mechanism that refines the function f that maps θ_1 and θ_2 into K during the regular (possibly daily) use of the system's schedule generator.

13.7. Consider an interactive scheduling system with a user interface for schedule manipulation that allows "freezing" of jobs. That is, the scheduler can click on a job and freeze the job in a certain position. The other jobs have to be scheduled around the frozen jobs. Freezing can be done with tolerances, so that in the optimization process of the remaining jobs the frozen jobs can be moved a little bit. This facilitates the scheduling of the unfrozen jobs. Consider a system that allows freezing of jobs with specified tolerances and show that freezing in an environment that does not allow preemptions requires tolerances of at least half the maximum processing time in either direction in order to avoid machine idle times.

13.8. Consider an interactive scheduling system with a user interface that only allows for freezing of jobs with no (zero) tolerances.

(a) Show that in a nonpreemptive environment the machine idle times caused by frozen jobs are always less than the maximum processing time.

(b) Describe how procedures can be designed that minimize in such a scenario machine idle times in conjunction with other objectives, such as the total completion time.

13.9. Consider a user interface of an interactive scheduling system for a bank of parallel machines. Assume that the reoptimization algorithms in the system are designed in such a way that they optimize each machine separately while they keep the current assignment of jobs to machines unchanged. A move of a job (with the mouse) is said to be *reversible* if the move, followed by the reoptimization procedure, followed by the *reverse* move, followed once more by the reoptimization procedure, results in the original schedule.

Suppose a job is moved with the mouse from one machine to another. Show that such a move is reversible if the reoptimization algorithm minimizes the total completion time. Show that the same is true if the reoptimization algorithm minimizes the sum of the weighted tardinesses.

13.10. Consider the same scenario as in the previous exercise. Show that with the type of reoptimization algorithms described in the previous exercise moves that take jobs from one machine and put them on another are *commutative*. That is, the final schedule does not depend on the sequence in which the moves are done, even if all machines are reoptimized after each move.

Comments and References

Many papers and books have been written on the various aspects of production information systems; see, for example, Gaylord (1987), Scheer (1988) and Pimentel (1990).

With regard to the issues concerning the overall development of planning and scheduling systems a relatively large number of papers have been written, often in proceedings of conferences, e.g., Oliff (1988), Karwan and Sweigart (1989), Interrante (1993). See also Kanet and Adelsberger (1987), Kusiak and Chen (1988), Solberg (1989), Adelsberger and Kanet (1991), Pinedo, Samroengraja and Yen (1994), and Pinedo and Yen (1997).

For research focusing specifically on knowledge-based systems, see Smith, Fox and Ow (1986), Shaw and Whinston (1989), Atabakhsh (1991), Noronha and Sarma (1991), Lefrancois, Jobin and Montreuil (1992) and Smith (1992, 1994).

For work on the design and development of user interfaces for planning and scheduling systems, see Kempf (1989) and Woerner and Biefeld (1993).

Chapter 14

Advanced Concepts in Systems Design

14.1 Introduction 345
14.2 Robustness and Reactive Decision Making 346
14.3 Machine Learning Mechanisms 351
14.4 Design of Planning and Scheduling Engines
     and Algorithm Libraries 357
14.5 Reconfigurable Systems 360
14.6 Web-Based Planning and Scheduling Systems .. 362
14.7 Discussion 365

14.1 Introduction

This chapter focuses on a number of issues that have come up in recent years in the design, development, and implementation of planning and scheduling systems. The next section discusses issues concerning uncertainty, robustness and reactive decision making. In practice, plans or schedules often have to be changed because of random events. The more *robust* the original plan or schedule is, the easier the replanning or rescheduling process is. This section focuses on the generation of robust plans and schedules as well as the measurement of their robustness. The third section considers machine learning mechanisms. A system cannot consistently generate good solutions that are to the liking of the user. The decision-maker often has to tweak the plan or schedule generated by the system in order to make it usable. A well-designed system can learn from adjustments made by the user in the past; the mechanism that allows the system to do so is typically referred to as a learning mechanism. The fourth section focuses on the design of planning and scheduling engines. An engine often contains a library of algorithms and routines. One procedure may be more appropriate for one type of instance or data set, while another procedure may be more appropriate for another type of instance. The user should be able to select, for each instance, which procedure to apply. It may even be the case that a user would like to tackle an instance using a combination of various procedures. This fourth section discusses how a planning or scheduling engine should be designed in order to enable the user to adapt and combine algorithms in order to achieve maximum effectiveness. The fifth section focuses on reconfigurable systems. Experience has shown that the development and implementation of systems is very time consuming and costly. In order to reduce the costs, efforts should be made to maintain a high degree of modularity in the design of the system. If the modules are well designed and sufficiently flexible, they can be used over and over again without any major changes. The sixth section focuses on the design aspects of web-based planning and scheduling systems. This section discusses the effects of networking on the design of such systems. The seventh and last section discusses a number of other issues and presents a view on how planning and scheduling systems may evolve in the future.

14.2 Robustness and Reactive Decision Making

In practice, it often happens that soon after a plan or schedule has been generated, an unexpected event happens that forces the decision-maker to make changes. Such an event may, for example, be a machine breakdown or a rush job that suddenly has to be inserted. Many planners and schedulers believe that in practice, most of the time, the decision making process is a *reactive* process. In a reactive process, the planner or scheduler tries to accomplish a number of objectives. He tries to accomodate the original objectives, and also tries to make the new plan or schedule look, as much as possible, like the original one in order to minimize confusion.

The remaining part of this section focuses primarily on reactive decision making in short term scheduling processes. The number of random events that can occur in a short term may, in certain environments, be very high. Rescheduling is in many environments a way of life. One way of doing the rescheduling is to put all the operations not yet started back in the hopper, and generate a new schedule from scratch while taking into account the disruptions that just occurred. The danger is that the new schedule may be completely different from the original schedule, and a big difference may cause some confusion.

If the disruption is minor, e.g., the arrival of just one unexpected job, then a simple change may suffice. For example, the scheduler may insert the unexpected arrival in the current schedule in such a way that the total additional setup is minimized and no other high priority job is delayed. A major disruption, like the breakdown of an important machine, often requires substantial changes in the schedule. If a machine goes down for an extended period of

14.2 Robustness and Reactive Decision Making

time, then the entire workload allocated to that machine over that period has to be transferred to other machines. This may cause extensive delays.

Another way of dealing with the rescheduling process is to somehow anticipate the random events. In order to do so, it is necessary for the original schedule to be robust so that the changes after a disruption are minimal.

Schedule robustness is a concept that is not easy to measure or even define. Suppose the completion time of a job is delayed by δ (because of a machine breakdown or the insertion of a rush job). Let $C''_j(\delta)$ denote the new completion time of job j (i.e., the new time at which job j leaves the system), assuming the sequences of all the operations on all the machines remain the same. Of course, the new completion times of all the jobs are a function of δ. Let Z denote the value of the objective function before the disruption occurred and let $Z'(\delta)$ denote the new value of the objective function. So $Z'(\delta) - Z$ is the difference due to the disruption. One measure of schedule robustness is

$$\frac{Z'(\delta) - Z}{\delta},$$

which is a function of δ. For small values of δ the ratio may be low whereas for larger values of δ it may get progressively worse. It is to be expected that this ratio is increasing convex in δ.

A more accurate measure of robustness can be established when the probabilities of certain events can be estimated in advance. Suppose a perturbation of a random size Δ may occur and the probability the random variable Δ assumes the value δ, i.e., $P(\Delta = \delta)$, can be estimated. If Δ can assume only integer values, then

$$\sum_{\delta=0}^{\infty} \Big(Z'(\delta) - Z\Big) P(\Delta = \delta)$$

is an appropriate measure for the robustness. If the random variable Δ is a continuous random variable with a density function $f(\delta)$, then an appropriate measure is

$$\int_{\delta=0}^{\infty} (Z'(\delta) - Z) f(\delta) d\delta.$$

In practice, it may be difficult to make a probabilistic assessment of random perturbations and one may want to have more practical measures of robustness. For example, one measure could be based on the amount of slack between the completion times of the jobs and their respective due dates. So a possible measure for the robustness of schedule $\mathcal{S}$ is

$$\mathcal{R}(\mathcal{S}) = \frac{\sum_{j=1}^{n} w_j(d_j - C_j)}{\sum w_j d_j}.$$

The larger $\mathcal{R}(\mathcal{S})$, the more robust the schedule. Maximizing this particular measure of robustness is somewhat similar to maximizing the total weighted earliness.

When should a decision-maker opt for a more robust schedule? This may depend on the probability of a disruption as well as on his or her ability to reschedule.

Example 14.2.1 (Measures of Robustness). Consider a single machine and three jobs. The job data are presented in the table below.

jobs	1	2	3
p_j	10	10	10
d_j	10	22	34
w_j	1	100	100

The schedule that minimizes the total weighted tardiness is schedule $1, 2, 3$ with a total weighted tardiness of 0. It is clear that this schedule is not that robust, since two jobs with very large weights are scheduled for completion very close to their respective due dates. Suppose that immediately after schedule $1, 2, 3$ has been fixed a disruption occurs, i.e., at time $0 + \epsilon$, and the machine goes down for $\delta = 10$ time units. The machine can start processing the three jobs at time $t = 10$. If the original job sequence $1, 2, 3$ has to be maintained, then the total weighted tardiness is 1410. The manner in which the total weighted tardiness of sequence $1, 2, 3$ depends on the value of δ is depicted in Figure 14.1.

If the original schedule is $2, 3, 1$, then the total weighted tardiness, with no disruptions, is 20. However, if a disruption does occur at time $0 + \epsilon$, then the impact is considerably less severe than with schedule $1, 2, 3$. If $\delta = 10$, then the total weighted tardiness is 30. The way the total weighted tardiness under sequence $2, 3, 1$ depends on δ is also depicted in Figure 14.1. From Figure 14.1 it is clear that schedule $2, 3, 1$ (even though originally suboptimal) is more robust than schedule $1, 2, 3$.

Under schedule $1, 2, 3$ the robustness is

$$\mathcal{R}(1,2,3) = \frac{\sum_{j=1}^n w_j(d_j - C_j)}{\sum w_j d_j} = \frac{600}{5610} = 0.11,$$

whereas

$$\mathcal{R}(2,3,1) = \frac{2580}{5610} = 0.46.$$

So according to this particular measure of robustness schedule $2, 3, 1$ is considerably more robust.

Suppose that with probability 0.01 a rush job with processing time 10 arrives at time $0 + \epsilon$ and that the decision-maker is not allowed, at the completion of this rush job, to change the original job sequence. If at the outset he had selected schedule $1, 2, 3$, then the total expected weighted tardiness is

$$0 \times 0.9 + 1410 \times 0.1 = 141.$$

14.2 Robustness and Reactive Decision Making

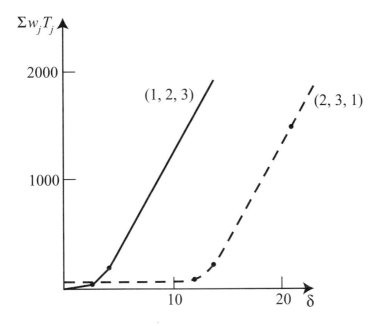

Fig. 14.1. Increase in Objective Value as a Function of Disruption Level

If he had selected schedule $2, 3, 1$, then the total expected weighted tardiness is
$$20 \times 0.9 + 30 \times 0.1 = 21.$$

So, when there is a 10% probability of a disruption it is better to go for the more robust schedule.

Even if a scheduler is allowed to reschedule after a disruption, he still may not choose at time 0 a schedule that is optimal with respect to the original data.

Several other measures of robustness can be defined. For example, assume again that the completion of one job is delayed by δ. However, before computing the effect of the disruption on the objective, each machine sequence is reoptimized separately, i.e., the machine sequences are reoptimized one by one on a machine by machine basis. After this reoptimization the difference in the objective function is computed. The measure of robustness is then a similar ratio as the one defined above. The impact of the disruption is now harder to compute, since different values of δ may result in different schedules. This ratio is, of course, less than the ratio without reoptimization. An even more complicated measure of robustness assumes that after a disruption a reoptimization is done on a more global scale rather than on a machine by machine basis, e.g., under this assumption a disruption may cause an entire job shop to be reoptimized. Other measures of robustness may even allow preemptions in the reoptimization process.

Totally different measures of robustness can be defined based on the capacity utilization of the bottleneck machines (i.e., the percentages of time the machines are utilized) and on the levels of WIP inventory that are kept in front of these machines.

How can one generate robust schedules? One can follow various rules in order to create such schedules, for example,

(i) insert idle times,
(ii) schedule less flexible jobs first,
(iii) do not postpone the processing of any operation unnecessarily, and
(iv) keep always a number of jobs waiting in front of highly utilized machines.

The first rule prescribes the insertion of idle periods on given resources at certain points in time. This is equivalent to scheduling the machines below capacity. The durations of the idle periods as well as their timing within the schedule depend on the expected nature of the disruptions. One could argue that the idle periods in the beginning of the schedule may be kept shorter than the idle periods later in the schedule, since the probability of an event occurring in the beginning may be smaller than later on. In practice, some schedulers follow a rule whereby at any point in time in the current week the machines are utilized up to 90% of capacity, in the next week up to 80% and in the week after that up to 70%. However, one reason for keeping the idle periods in the beginning of the schedule at the same length may be the following: even though the probability of a disruption is small, its relative impact is more severe than that of a disruption that occurs later on in the process.

The second rule suggests that less flexible jobs should have a higher priority than more flexible jobs. If a disruption occurs, then the more flexible jobs remain to be processed. The flexibility of a job is determined, for example, by the number of machines that can do the processing (e.g., the machine eligibility constraints described in Chapter 2). However, the flexibility of a job may also be determined by the setup time structure. Some jobs may require setups that do *not* depend on the sequence. Other jobs may have sequence dependent setup times that are highly variable. The setup times are short only when they follow certain other jobs; otherwise the setup times are very long. Such jobs are clearly less flexible.

The third rule suggests that the processing of a job should not be postponed unnecessarily. From the point of view of inventory holding costs and earliness penalties, it is desirable to start operations as late as possible. From a robustness point of view, it may be desirable to start operations as early as possible. So there is a trade-off between robustness and earliness penalties or inventory holding costs.

The fourth rule tries to make sure that a bottleneck machine never starves because of random events that occur upstream. It makes sense to have always a number of jobs waiting for processing at a bottleneck machine. The reason is the following: if no inventory is kept in front of the bottleneck and the machine feeding the bottleneck suddenly breaks down, then the bottleneck may have to remain idle and may not be able to make up for the lost time later on.

Example 14.2.2 (Starvation Avoidance). Consider a two machine flow shop with 100 identical jobs. Each job has a processing time of 5 time units on machine 1 and of 10 time units on machine 2. Machine 2 is therefore the bottleneck. However, after each job completion on machine 1, machine 1 may have to undergo a maintenance service for a duration of 45 time units during which it cannot do any processing. The probability that such a service is required is 0.01.

The primary objective is the minimization of the makespan and the secondary objective is the average amount of time a job remains in the system, i.e., the time in between the start of a job on machine 1 and its completion on machine 2 (this secondary objective is basically equivalent to the minimization of the Work-In-Process). However, the weight of the primary objective is 1000 times the weight of the secondary objective.

Because of the secondary objective it does not make sense to let machine 1 process the 100 jobs one after another and finish them all by time 500. In an environment in which machine 1 never requires any servicing, the optimal schedule processes the jobs on machine 1 with idle times of 5 time units in between. In an environment in which machine 1 needs servicing with a given probability, it is necessary to have at all times some jobs ready for processing on machine 2. The optimal schedule is to keep consistently 5 jobs waiting for processing on machine 2. If machine 1 has to be serviced, then machine 2 does not lose any time and the makespan does not go up unnecessarily.

This example illustrates the trade-off between capacity utilization and minimization of Work-In-Process.

Robustness and rescheduling have a strong influence on the design of the user interfaces and on the design of the scheduling engine (multi-objective scheduling where one of the performance measures is robustness). Little theoretical research has been done on these issues. This topic may become an important area of research in the near future.

14.3 Machine Learning Mechanisms

In practice, the algorithms embedded in a planning or scheduling system often do not yield plans or schedules that are acceptable to the user. The inadequacy of the algorithms is based on the fact that planning and scheduling problems (which often have multiple objectives) are inherently intractable. It is extremely difficult to develop algorithms that can provide a reasonable and acceptable solution for any instance of a problem in real time.

New research initiatives are focussing on the design and development of learning mechanisms that enable planning and scheduling systems that are in daily use to improve their solution generation capabilities. This process requires a substantial amount of experimental work. A number of machine learning methods have been studied with regard to their applicability to planning and scheduling. These methods can be categorized as follows:

(i) rote learning,
(ii) case-based reasoning,
(iii) induction methods and neural networks,
(iv) classifier systems.

These four classes of learning mechanisms are in what follows described in some more detail.

Rote learning is a form of brute force memorization. The system saves old solutions that gave good results together with the instances on which they were applied. However, there is no mechanism for generalizing these solutions. This form of learning is only useful when the number of possible planning or scheduling instances is limited, i.e., a small number of jobs of very few different types. It is not very effective in a complex environment, when the probability of a similar instance occurring again is very small.

Case-based reasoning attempts to exploit experience gained from similar problems solved in the past. A scheduling problem requires the identification of salient features of past schedules, an interpretation of these features, and a mechanism for determining which case stored in memory is the most useful in the current context. Given the large number of interacting constraints inherent in scheduling, existing case indexing schemes are often inadequate for building the case base and subsequent retrieval, and new ways have to be developed. The following example shows how the performance of a composite dispatching rule can be improved using a crude form of case-based reasoning; the form of case-based reasoning adopted is often referred to as the parameter adjustment method.

Example 14.3.1 (Case-Based Reasoning: Parameter Adjustment). Consider a single machine with n jobs and the total weighted tardiness $\sum w_j T_j$ as the objective to minimize. Moreover, the jobs are subject to sequence dependent setup times s_{jk}. This problem has been considered in Section 5.2 and also in the third section of Appendix C. A fairly effective composite dispatching rule for this scheduling problem is the ATCS rule. When the machine has completed the processing of job l at time t, the ATCS rule calculates the ranking index of job j as

$$I_j(t,l) = \frac{w_j}{p_j} \exp\left(-\frac{\max(d_j - p_j - t, 0)}{K_1 \bar{p}}\right) \exp\left(-\frac{s_{lj}}{K_2 \bar{s}}\right),$$

where $\bar{s}$ is the average setup time of the jobs remaining to be scheduled, K_1 the scaling parameter for the function of the due date of job j and K_2 the

14.3 Machine Learning Mechanisms

scaling parameter for the setup time of job j. As described in Appendix C, the two scaling parameters K_1 and K_2 can be regarded as functions of three factors:

(i) the due date tightness factor θ_1,
(ii) the due date range factor θ_2,
(iii) the setup time severity factor $\theta_3 = \bar{s}/\bar{p}$.

However, it is difficult to find appropriate functions that map the three factors into appropriate values for the scaling parameters K_1 and K_2.

At this point a learning mechanism may be useful. Suppose that in the scheduling system there are functions that map combinations of the three factors θ_1, θ_2 and θ_3 onto two values for K_1 and K_2. These do not have to be algebraic functions; they may be tables of data. When a scheduling instance is considered, the system computes θ_1, θ_2 and θ_3 and looks in the current tables for the appropriate values of K_1 and K_2. (These values for K_1 and K_2 may have to be determined by means of an interpolation). The instance is then solved using the composite dispatching rule with these values for K_1 and K_2. The objective value of the schedule generated is computed as well. However, in that same step, without any human intervention, the system also solves the same scheduling instance using values $K_1 + \delta$, $K_1 - \delta$, $K_2 + \delta$, $K_2 - \delta$ (various combinations). Since the dispatching rule is very fast, this can be done in real time. The performance measures of the schedules generated with the perturbed scaling parameters are also computed. If any of these schedules turns out to be substantially better than the one generated with the original K_1 and K_2, then there may be a reason for changing the mapping from the characteristic factors onto the scaling parameters. This can be done internally by the system without any input from the user of the system.

The learning mechanism described in the example above is an on-line mechanism that operates without supervision. This mechanism is an example of case-based reasoning and can be applied to multi-objective planning and scheduling problems as well, even when a simple index rule does not exist.

The third class of learning mechanisms are of the induction type. The most common form of an induction type learning mechanism is a neural network. A neural net consists of a number of interconnected neurons or units. The connections between units have weights, which represent the strengths of the connections between the units. A multi-layer feedforward net is composed of input units, hidden units and output units (see Figure 14.2). An input vector is processed and propagated through the network starting at the input units and proceeding through the hidden units all the way to the output units. The activation level of input unit i is set to the ith component of the input vector. These values are then propagated to the hidden units via the weighted connections. The activation level of each hidden unit is then computed by summing these weighted values, and by transforming the sum

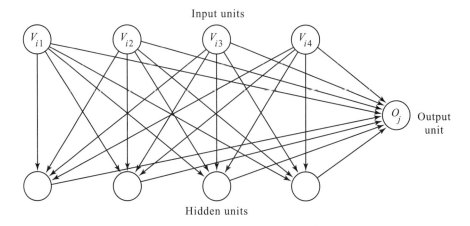

Fig. 14.2. A Four-layer Neural Network

through a function f, that is,

$$a_l = f\left(q_l, \sum_k w_{kl} a_k\right),$$

where a_l is the activation level of unit l, q_l is the bias of unit l, and w_{kl} is the connection weight between nodes k and l. These activation levels are propagated to the output units via the weighted connections between the hidden units and the output units and transformed again by means of the function f above. The neural net's response to a given input vector is composed of the activation levels of the output units that are referred to as the output vector. The dimension of the output vector does not have to be the same as the dimension of the input vector.

The knowledge of the net is stored in the weights and there are well known methods for adjusting the weights on the connections in order to obtain appropriate responses. For each input vector there is a most appropriate output vector. A learning algorithm computes the difference between the output vector of the current net and the most appropriate output vector and suggests incremental adjustments to the weights. One such method is called the backpropagation learning algorithm.

The next example illustrates the application of a neural net to machine scheduling.

Example 14.3.2 (Neural Net for Parallel Machine Scheduling). Consider m non-identical machines in parallel. Machine i has speed v_i and if job j is processed on machine i, then its processing time is $p_{ij} = p_j/v_i$. The jobs have different release dates and due dates and one of the objectives is to minimize the total weighted tardiness (note that the weights in the objective

14.3 Machine Learning Mechanisms

function are not related to the weights in the neural net). The jobs on a machine are subject to sequence dependent setup times s_{jk}. For each machine there is already a partial schedule in place which consists of jobs already assigned; more jobs are released as time goes on. At each new release it has to be decided to which machine the job should be assigned. The neural net has to support this decision-making process.

Typically, the encoding of the data in the form of input vectors is crucial to the problem solving process. In the parallel machines application, each input pattern represents a description of the attributes of a sequence on one of the machines. The values of the attributes of a sequence are determined as follows. Each new job is first positioned where its processing time is the shortest (including the setup time immediately preceding it and the setup time immediately following it). After this insertion the following attributes are computed with regard to that machine:

(i) the increase in the total weighted completion times of all the jobs already scheduled on the machine;
(ii) the number of additional jobs that are late on the machine;
(iii) the average additional latenesses of jobs already scheduled on the machine;
(iv) the current number of jobs on the machine.

However, the importance of each individual attribute is relative. For example, knowing that the number of jobs on a machine is five does not mean much without knowing the number of jobs on the other machines. Let N_{il} be attribute l with respect to machine i. Two transformations have to be applied to N_{il}.

Translation: The N_{il} value of attribute l under a given sequence on machine i is transformed as follows.

$$N'_{il} = N_{il} - \min(N_{1l}, \ldots, N_{ml}) \qquad i = 1, \ldots, m, \quad l = 1, \ldots, k.$$

In this way, $N'_{i^*l} = 0$ for the best machine, machine i^*, with respect to attribute l and the value N'_{il} corresponds to the difference with the best value.

Normalization: The N'_{il} value is transformed by normalizing over the maximum value in the context.

$$N''_{il} = \frac{N'_{il}}{\max(N'_{1l}, \ldots, N'_{ml})} \qquad i = 1, \ldots, m, \quad l = 1, \ldots, k.$$

Clearly,
$$0 \leq N''_{il} \leq 1.$$

These transformations make the comparisons of the input patterns corresponding to the different machines significantly easier. For example, if the value of attribute l is important in the decision making process, then a machine with the l-th attribute close to zero is more likely to be selected.

A neural net architecture to deal with this problem can be of the structure described in Figure 14.2. This is a three layer network with four input nodes (equal to the number of attributes k), four hidden nodes and one output node. Each input node is connected to all the hidden nodes as well as to the output node. The four hidden nodes are connected to the output node as well.

During the training phase of this network an extensive job release process has to be considered, say 1000 jobs. Each job release generates m input patterns (m being the number of machines) which have to be fed into the network. During this training process the *desired* output of the net is set equal to 1 when the machine associated with the given input is selected by the expert for the new release and equal to 0 otherwise. For each input vector there is a desired output and the learning algorithm has to compute the error or difference between the current neural net output and the desired output in order to make incremental changes in the connection weights. A well-known learning algorithm for the adjustment of connection weights is the backpropagation learning algorithm; this algorithm requires the choice of a so-called learning rate and a momentum term. At the conclusion of the training phase the connection weights are fixed.

Using the network after the completion of the training phase requires that every time a job is released the m input patterns are fed into the net; the machine associated with the output closest to 1 is then selected for the given job.

In contrast to the learning mechanism described in Example 14.3.1 the mechanism described in Example 14.3.2 requires off-line learning with supervision, i.e., training.

The fourth class of learning mechanisms consists of the classifier systems. A common form of classifier system can be implemented via a genetic algorithm (see Appendix C). However, a chromosome (or string) in such a algorithm now does not represent a schedule, but rather a list of rules (e.g., priority rules) that are to be used in the successive steps (or iterations) of an algorithmic framework designed to generate schedules for the problem at hand. For example, consider a framework for generating job shop schedules that is similar to Algorithm B.4.2. However, this algorithm can be modified by replacing its Step 3 with a priority rule that selects an operation from set Ω'. A schedule for the job shop can now be generated by doing nm successive iterations of this algorithm where nm is the total number of operations. Every time a new operation has to be scheduled, a given priority rule is used to select the operation from the current set Ω'. The information that specifies which priority rule should be used in each iteration can be stored in a string of length nm for a genetic algorithm. The fitness of such a string is the value of the objective function obtained when all the (local) rules are applied successively in the given framework. This representation of a solution (by specifying rules), however, requires relatively intricate cross-over operators in order to get feasible off-spring. The genetic algorithm is thus used to search over the space of rules and not over the space of actual schedules. The genetic algorithm serves this way as a meta-strategy that controls the use of priority rules.

14.4 Design of Planning and Scheduling Engines and Algorithm Libraries

A planning or scheduling engine in a system often contains a library of algorithmic procedures. Such a library may include basic dispatching rules, composite dispatching rules, shifting bottleneck techniques, local search techniques, branch-and-bound procedures, beam search techniques, mathematical programming routines, and so on. For a specific instance of a problem one procedure may be more suitable than another. The appropriateness of a procedure may depend on the amount of CPU time available or the length of time the user is willing to wait for a solution.

The user of such a planning or scheduling system may want to have a certain flexibility in the usage of the various types of procedures in the library. The desired flexibility may simply imply an ability to determine which procedure to apply to the given instance of the problem, or it may imply more elaborate ways of manipulating a number of procedures. A planning or scheduling engine may have modules that allow the user

(i) to analyze the data of an instance and determine algorithmic parameters,
(ii) to set up algorithms in parallel,
(iii) to set up algorithms in series,
(iv) to integrate algorithms.

For example, an algorithm library allows a user to do the following: he may have statistical procedures at hand which he can apply to the data set in order to generate some statistics, such as average processing time, range of processing times, due date tightness, setup time severity, and so on. Based on these statistics the user can select a procedure and specify appropriate levels for its parameters (e.g., scaling parameters, lengths of tabu-lists, beam widths, number of iterations, and so on).

If a user has more than one computer or processor at his disposal, he may want to apply different procedures concurrently (i.e., in parallel), since he may not know in advance which one is the most suitable for the instance under consideration. The different procedures function then completely independently from one another.

A user may also want to concatenate procedures i.e., set various procedures up in series. That is, he or she would set the procedures up in such a way that the output of one serves as an input to another, e.g., the outcome of a dispatching rule serves as the initial solution for a local search procedure. The transfer of data from one procedure to the next is usually relatively simple. For example, it may be just a schedule, which, in the case of a single machine, is a permutation of the jobs. In a parallel machine environment or a job shop environment it may be a collection of sequences, one for each machine.

Example 14.4.1 (Concatenation of Procedures). Consider a scheduling engine that allows a user to feed the outcome of a composite dispatching rule into a local search procedure. This means that the output of the first stage, i.e., the dispatching rule, is a complete schedule. The schedule is feasible and the starting times and completion times of all the operations have been determined. The output data of this procedure (and the input data for the next procedure) may contain the following information:

(i) the sequence of operations on each machine;
(ii) the start time and completion time of each operation;
(iii) the values of specific objective functions.

The output data does not have to contain all the data listed above; for example, it may include only the sequence of operations on each machine. The second procedure, i.e., the local search procedure, may have a routine that can compute the start and completion times of all the operations given the structure of the problem and the sequences of the operations.

Another level of flexibility allows the user not only to set up procedures in parallel or in series, but to integrate the procedures in a more complex manner. When different procedures are integrated within one framework, they do not work independently from one another; the effectiveness of one procedure may depend on the input or feedback received from another. Consider a branch-and-bound procedure or a beam search procedure for a scheduling problem. At each node in the search tree, one has to obtain either a lower bound or an estimate for the total penalty that will be incurred by the jobs that have not yet been scheduled. A lower bound can often be obtained by assuming that the remaining jobs can be scheduled while allowing preemptions. A preemptive version of a problem is often easier to solve than its nonpreemptive counterpart and the optimal solution of the preemptive problem provides a lower bound for the optimal solution of the nonpreemptive version.

Another example of an integration of procedures arises in decomposition techniques. A machine-based decomposition procedure is typically a heuristic designed for a complicated scheduling problem with many subproblems. A framework for a procedure that is applicable to the main problem can be constructed as in Chapter 5. However, the user may want to be able to specify, knowing the particular problem or instance, which procedure to apply on the subproblem.

If procedures have to be integrated, then one often has to work within a general framework (sometimes also referred to as a control structure) in which one or more specific types of subproblems have to be solved many times. The user may want to have the ability to specify certain parameters within this framework. For example, if the framework is a search tree for a beam search, then the user would like to be able to specify the beam width as well as the filter width. The subproblem that has to be solved at each node of the search tree has to yield (with little computational effort) a good estimate for the contribution to the objective by those jobs that have not yet been scheduled.

The transfer of data between procedures within an integrated framework may be complicated. It may be the case that data concerning a subset of

14.4 Design of Planning and Scheduling Engines and Algorithm Libraries 359

jobs or a subset of operations has to be transferred. It may also be the case that the machines are not available at all times. The positions of the jobs already scheduled on the various machines may be fixed, implying that the procedure that has to schedule the remaining jobs must know at what times the machines are still available. If there are sequence dependent setup times, then the procedure also has to know which job was the last one processed on each machine, in order to compute the sequence dependent setup time for the next job.

Example 14.4.2 (Integration of Procedures in a Branching Scheme).
Consider a branch-and-bound approach for a single machine problem with the total weighted tardiness objective and jobs that are subject to sequence dependent setup times. Moreover, the machine may not be available for certain periods of time due to maintenance. The jobs are scheduled in a forward manner, i.e., a partial schedule consists of a sequence of jobs that starts at time zero. At each node of the branching tree a bound has to be established for the total weighted tardiness of the jobs still to be scheduled. If a procedure is called to generate a lower bound for all schedules that are descendants from any particular node, then the following input data has to be provided:

(i) the set of jobs already scheduled and the set of jobs still to be scheduled;
(ii) the time periods that the machine remains available;
(iii) the last job in the current partial schedule (in order to determine the sequence dependent setup time).

The output data of the procedure may contain a sequence of operations as well as a lower bound. The required output may be just the lower bound; the actual sequence may not be of interest.

If there are no setup times then a schedule can also be generated in a backward manner (since the value of the makespan is then known in advance).

Example 14.4.3 (Integration of Procedures in a Decomposition Scheme). Consider a shifting bottleneck framework for a flexible job shop with at each workcenter a number of machines in parallel.

At each iteration a subset of the workcenters has already been scheduled and an additional workcenter must be scheduled. The sequences of the operations at the workcenters already scheduled imply that the operations of the workcenter to be scheduled in the subproblem is subject to delayed precedence constraints. When the procedure for the subproblem is called, a certain amount of data has to be transferred. These data may include:

(i) the release date and due date of each operation;
(ii) the precedence constraints between the various operations;
(iii) the necessary delays that go along with the precedence constraints.

The output data consists of the sequence of the operations as well as their start times and completion times. It also contains the values of given performance measures.

It is clear that the type of information and the structure of the information is more complicated than in a simple concatenation of procedures.

These forms of integration of procedures have led to the development of so-called description languages for planning and scheduling. A description language is a high level language that enables a planner or a scheduler to write the code for a complex integrated algorithm using only a limited number of concise statements or commands. Each statement in a description language involves the application of a relatively powerful procedure. For example, a statement may carry the instruction to apply a tabu-search procedure to a given set of jobs in a given machine environment. The input to such a statement consists of the set of jobs, the machine environment, the processing restrictions and constraints, the length of the tabu-list, an initial schedule, and the total number of iterations. The output consists of the best schedule generated by the tabu-search procedure. Other statements can be used for setting up procedures in parallel or concatenate procedures.

14.5 Reconfigurable Systems

The last two decades have witnessed the development of a large number of planning and scheduling systems in industry and in academia. Some of these systems are application-specific, others are generic. In implementations application-specific systems tend to do somewhat better than generic systems that are customized. However, application-specific systems are often hard to modify and adapt to changing environments. Generic systems are usually somewhat better designed and more modular. Nevertheless, any customization of such systems typically requires a significant investment.

Considering the experience of the last two decades, it appears useful to provide guidelines that facilitate and standardize the design and the development of planning and scheduling systems. Efforts have to be made to provide guidelines as well as tools for systems development. The most recent designs tend to be object-oriented.

There are many advantages in following an object-oriented design approach for the development of a planning or scheduling system. First, the design is modular, which makes maintenance and modification of the system relatively easy. Second, large segments of the code are reusable. This implies that two systems that are inherently different still may share a significant amount of code. Third, the designer thinks in terms of the behavior of objects, not in lower level detail. In other words, the object-oriented design approach can speed up the design process and separate the design process from its implementation.

Object oriented systems are usually designed around two basic entities, namely *objects* and *methods*. Objects refer to various types of entities or concepts. The most obvious ones are jobs and machines or activities and resources.

14.5 Reconfigurable Systems

However, a plan or a schedule is also an object and so are user-interface components, such as buttons, menus and canvases. There are two basic relationships between object types, namely the *is-a* relationship and the *has-a* relationship. According to an is-a relationship one object type is a special case of another object type. According to a has-a relationship an object type may consist of several other object types. Objects usually carry along static information, referred to as attributes, and dynamic information, referred to as the state of the object. An object may have several attributes that are descriptors associated with the object. An object may be in any one of a number of states. For example, a machine may be *busy*, *idle*, or *broken down*. A change in the state of an object is referred to as an *event*.

A method is implemented in a system through one or more operators. Operators are used to manipulate the attributes corresponding to objects and may result in changes of object states, i.e., events. On the other hand, events may trigger operators as well. The sequence of states of the different objects can be described by a state-transition or event diagram. Such an event diagram may represent the links between operators and events. An operator may be regarded as the manner in which a method is implemented in the software. Any given operator may be part of several methods. Some methods may be very basic and can be used for simple manipulations of objects, e.g., a pairwise interchange of two jobs in a schedule. Others may be very sophisticated, such as an intricate heuristic that can be applied to a given set of jobs (objects) in a given machine environment (also objects). The application of a method to an object usually triggers an *event*.

The application of a method to an object may cause information to be transmitted from one object to another. Such a transmission of information is usually referred to as a message. Messages represent information (or content) that are transmitted from one object (for example, a schedule) via a method to another object (for example, a user interface display). A message may consist of simple attributes or of an entire object. Messages are transmitted when events occur (caused by the application of methods to objects). Messages have been referred to in the literature also as memos. The transmission of messages from one object to another can be described by a transition event diagram, and requires the specification of protocols.

A planning or scheduling system may be object-oriented in its conceptual design and/or in its development. A system is object-oriented in its conceptual design if the design of the system is object-oriented throughout. This implies that every concept used and every functionality of the system is either an object or a method of an object (whether it is in the data or knowledge base, the algorithm library, the planning or scheduling engine or the user interfaces). Even the largest modules within the system are objects, including the algorithm library and the user interface modules. A system is object-oriented in its development if only the more detailed design aspects are object-oriented and the code is based on a programming language with object-oriented extensions such as C++.

Many planning and scheduling systems developed in the past have object-oriented aspects and tend to be object-oriented in their development. A number of these systems also have conceptual design aspects that are object-oriented. Some rely on inference engines for the generation of feasible plans or schedules and others are constraint based relying on constraint propagation algorithms and search. These systems usually do not have engines that perform very sophisticated optimization.

Not that many systems implemented in the past have been designed according to an object-oriented philosophy throughout. Some aspects that are typically not object-oriented are:

  (i) the design of planning and scheduling engines,
 (ii) the design of the user interfaces and
(iii) the specification of the precedence, routing and layout constraints.

Few existing engines have extensive libraries of algorithms at their disposal which are easily reconfigurable and that would benefit from a modular object-oriented design (an object-oriented design would require a detailed specification of operators and methods). Since most planning and scheduling environments would benefit from highly interactive optimization, schedule generators have to be strongly linked to interfaces that allow schedulers to manipulate schedules manually. Still, object-oriented design has not had yet a major impact on the design of user interfaces for scheduling systems. The precedence constraints, the routing constraints, and the machine layout constraints are often represented by rules in a knowledge base and an inference engine must generate a plan or schedule that satisfies the rules. However, these constraints can be modeled conceptually easily using graph and tree objects that then can be used by an object oriented planning or scheduling engine.

14.6 Web-Based Planning and Scheduling Systems

With the ongoing development in information technology, conventional single-user stand-alone systems have become available in networks and on the Internet. Basically there are three types of web-based systems:

  (i) information access systems,
 (ii) information coordination systems,
(iii) information processing systems.

In information access systems, information can be retrieved and shared through the Internet, through EDI or through other electronic systems. The server acts as an information repository and distribution center, such as a homepage on the Internet.

In information coordination systems, information can be generated as well as retrieved by many users (clients). The information flows go in many directions and the server can synchronize and manage the information, such as in project management and in electronic markets.

14.6 Web-Based Planning and Scheduling Systems

In information processing systems the servers can process the information and return the results of this processing to the clients. In this case, the servers function as application programs that are transparent to the users.

Web-based planning and scheduling systems are information processing systems that are very similar to the interactive planning or scheduling systems described in previous sections, except that a web-based planning or scheduling system is usually a strongly distributed system. Because of the client-server architecture of the Internet, all the important components of a planning or scheduling system, i.e., its database, its engine, and its user interface, may have to be adapted. The remaining part of this section focuses on some of the typical design features of web-based planning and scheduling systems.

The advantages of having servers that make planning and scheduling systems available on the web are the following. First, the input-output interfaces (used for the graphical displays) can be supported by local hosts rather than by servers at remote sites. Second, the server as well as the local clients can handle the data storage and manipulation. This may alleviate the workload at the server sites and give local users the capability and flexibility to manage the database. Third, multiple servers can collaborate on the solution of large-scale and complicated planning and scheduling problems. A single server can provide a partial solution and the entire problem can be solved using distributed computational resources.

In order to retain all the functions inherent in an interactive planning or scheduling system, the main components of a system have to be restructured in order to comply with the client-server architecture and to achieve the advantages listed above. This restructuring affects the design of the database, the engine as well as the user interface.

The design of the database has the following characteristics: The process manager as well as the planning or scheduling manager reside at the servers. However, some data can be kept at the client for display or further processing. Both the Gantt chart and the dispatch lists are representations of the solution generated by the engine. The local client can cache the results for fast display and further processing, such as editing. Similarly, both the server and the client can process the information. Figure 14.3 exhibits the information flow between the server and local clients. A client may have a general purpose database management system (such as Sybase or Excel) or an application-specific planning or scheduling database for data storage and manipulation.

The design of the planning or scheduling engine has the following characteristics: A local client can select for the problem that he has to deal with an algorithm from a library that resides at a remote server. Often, there is no algorithm specifically designed for his particular planning or scheduling problem and he may want to create a composite procedure using some of the algorithms that are available in the library. The server or client algorithm generator may function as a workplace for users to create new procedures. Figure 14.4 shows how a new composite procedure can result in a new algorithm that then can be included in both the server and the client libraries.

364 14 Advanced Concepts in Systems Design

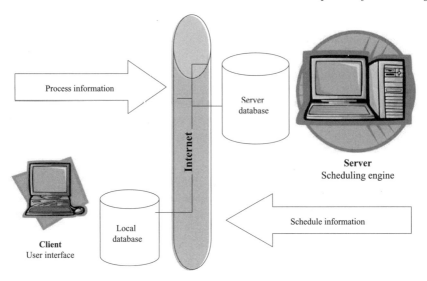

Fig. 14.3. Information Flow Between Server and Client

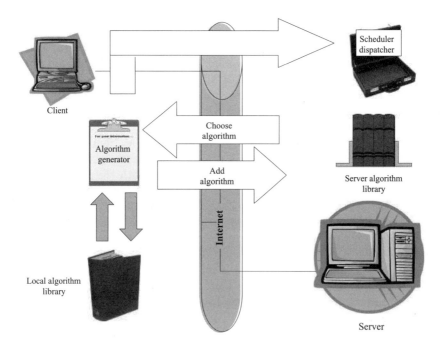

Fig. 14.4. Process of Constructing New Methods

This local workplace can speed up the process of constructing intermediate and final composite methods and extend the server and client libraries at the same time.

The Internet also has an effect on the design of the user interfaces. Using existing Internet support, such as HTML (HyperText Markup Language), Java, Java script, Perl and CGI (Common Gateway Interface) functions, the graphical user interfaces of planning and scheduling systems can be implemented as library functions at the server sites. Through the use of appropriate browsers (such as Netscape), users can enter data or view schedules with a dynamic hypertext interface. Moreover, a user can also develop interface displays that link server interface functions to other applications. In Figure 14.3 it is shown how display functions can be supported either by remote servers or by local clients.

Thus it is clear that servers can be designed in such a way that they can help local clients solve their planning or scheduling problems. The local clients can manipulate data and construct new planning and scheduling methods. The servers function as regular interactive planning and scheduling systems except that now they can be used in a multi-user environment on the Internet.

Web-based planning and scheduling systems can be used in several ways. One way is based on personal customization and another way is based on unionization. Personal customization implies that a system can be customized to satisfy an individual user's needs. Different users may have different requirements, since each has his own way of using information, applying planning and scheduling procedures, and solving problems. Personalized systems can provide shortcuts and improve system performance. Unionization means that a web-based planning or scheduling system can be used in a distributed environment. A distributed system can exchange information efficiently and collaborate effectively in solving hard planning and scheduling problems.

With the development of Internet technology and client-server architectures, new tools can be incorporated in planning and scheduling systems for solving large-scale and complicated problems. It appears that web-based systems may well lead to viable personalized interactive planning and scheduling systems.

14.7 Discussion

Many teams in industry and academia are currently developing planning and scheduling systems. The database (or object base) management systems are usually off-the-shelf, developed by companies that specialize in these systems, e.g., Oracle. These commercial databases are typically not specifically geared for planning and scheduling applications; they are of a more generic nature.

Dozens of software development and consulting companies specialize in planning and scheduling applications. They may specialize even in certain niches, e.g., planning and scheduling applications in the process industries or in the microelectronics industries. Each of these companies has its own systems with elaborate user interfaces and its own way of doing interactive optimization.

Research and development in planning and scheduling algorithms and in learning mechanisms will most likely only take place in academia or in large industrial research centers. This type of research needs extensive experimentation; software houses often do not have the time for such developments.

In the future, the Internet may allow for the following types of interaction between software companies and universities that develop systems on the one side and companies that need planning and scheduling services (customers) on the other side. A customer may use a system that is available on the web, enter its data and run the system. The system gives the customer the values of the performance measures of the solution generated. However, the customer cannot yet see the plan or schedule. If the performance measures of the solution are to the liking of the customer, then he may decide to purchase the solution from the company that owns the system.

Exercises

14.1. One way of constructing robust schedules is by inserting idle times. Describe all the factors that influence the timing, the frequency and the duration of the idle periods.

14.2. Consider all the nonpreemptive schedules on a single machine with n jobs. Define a measure for the "distance" (or the "difference") between two schedules.
(a) Apply the measure when the two schedules consist of the same set of jobs.
(b) Apply the measure when one set of jobs has one more job than the other set.

14.3. Consider the same set of jobs as in Example 14.2.1. Assume that there is a probability p that the machine needs servicing beginning at time 2. The servicing takes 10 time units.
(a) Assume that neither preemption nor resequencing is allowed (i.e., after the servicing has been completed, the machine has to continue processing the job it was processing before the servicing). Determine the optimal sequence(s) as a function of p.
(b) Assume preemption is not allowed but resequencing is allowed. That is, after the first job has been completed the scheduler may decide not to start the job he originally scheduled to go second. Determine the optimal sequence(s) as a function of p.
(c) Assume preemption as well as resequencing are allowed. Determine the optimal sequence(s) as a function of p.

Exercises

14.4. Consider two machines in parallel that operate at the same speed and two jobs. The processing times of each one of the two jobs is equal to one time unit. At each point in time each machine has a probability 0.5 of breaking down for one time unit. Job 1 can only be processed on machine 1 whereas job 2 can be processed on either one of the two machines. Compute the expected makespan under the Least Flexible Job first (LFJ) rule and under the Most Flexible Job first (MFJ) rule.

14.5. Consider a single machine scheduling problem with the jobs being subject to sequence dependent setup times. Define a measure of job flexibility that is based on the setup time structure.

14.6. Consider the following instance of a single machine with sequence dependent setup times. The objective to be minimized is the makespan. There are 6 jobs. The sequence dependent setup times are specified in the table below.

k	0	1	2	3	4	5	6
s_{0k}	-	1	$1+\epsilon$	K	$1+\epsilon$	$1+\epsilon$	K
s_{1k}	K	-	1	$1+\epsilon$	K	$1+\epsilon$	$1+\epsilon$
s_{2k}	$1+\epsilon$	K	-	1	$1+\epsilon$	K	$1+\epsilon$
s_{3k}	$1+\epsilon$	$1+\epsilon$	K	-	1	$1+\epsilon$	K
s_{4k}	K	$1+\epsilon$	$1+\epsilon$	K	-	1	$1+\epsilon$
s_{5k}	$1+\epsilon$	K	$1+\epsilon$	$1+\epsilon$	K	-	1
s_{6k}	1	$1+\epsilon$	K	$1+\epsilon$	$1+\epsilon$	K	-

Assume K to be very large. Define as the neighbourhood of a schedule all schedules that can be obtained through an adjacent pairwise interchange.

(a) Find the optimal sequence.

(b) Determine the makespans of all schedules that are neighbors of the optimal schedule.

(c) Find a schedule, with a makespan less than K, of which all neighbors have the same makespan. (The optimal sequence may be described as a "brittle" sequence, while the last sequence may be described as a more "robust" sequence.)

14.7. Consider a flow shop with limited intermediate storages that is subject to a cyclic schedule as described in Section 6.2. Machine i now has at the completion of each operation a probability p_i that it goes down for an amount of time x_i.

a) Define a measure for the congestion level of a machine.

b) Suppose that originally there are no buffers between machines. Now a total of k buffer spaces can be inserted between the m machines and the allocation has to be done in such a way that the schedules are as robust as possible. How does the allocation of the buffer spaces depend on the congestion levels at the various machines?

14.8. Explain why rote learning is an extreme form of case-based reasoning.

14.9. Describe how a branch-and-bound approach can be applied to a scheduling problem with m identical machines in parallel, the jobs subject to sequence dependent setup times and the total weighted tardiness as objective. That is, generalize the discussion in Example 14.4.2 to parallel machines.

14.10. Consider Example 14.4.3 and Exercise 14.9. Integrate the ideas presented in an algorithm for the flexible job shop problem.

14.11. Consider a scheduling description language that includes statements that can call different scheduling procedures for a scheduling problem with m identical machines in parallel, the total weighted tardiness objective and the n jobs released at different points in time. Write the specifications for the input and the output data for three statements that correspond to three procedures of your choice. Develop also a statement for setting the procedures up in parallel and a statement for setting the procedures up in series. Specify for each one of these last two statements the appropriate input and output data.

14.12. Suppose a scheduling description language is used for coding the shifting bottleneck procedure. Describe the type of statements that are required for such a code.

Comments and References

There is an extensive literature on planning and scheduling under uncertainty (i.e., PERT, stochastic scheduling). However, the literature on PERT and on stochastic scheduling, in general, does not address the issue of robustness per se. But robustness concepts have received some special attention in the literature; see, for example, the work by Leon and Wu (1994), Leon, Wu and Storer (1994), Mehta and Uzsoy (1999), Wu, Storer and Chang (1991), Wu, Byeon and Storer (1999). For an overview of research in reactive planning and scheduling, see the excellent survey by Smith (1992) and the framework presented by Vieira, Herrmann and Lin (2003). For more detailed work on reactive scheduling and rescheduling in job shops, see Bierwirth and Mattfeld (1999) and Sabuncuoglu and Bayiz (2000). For an industrial application of reactive scheduling, see Elkamel and Mohindra (1999).

Research on learning mechanisms in planning and scheduling systems started in the late eighties; see, for example, Shaw (1988), Shaw and Whinston (1989), Yih (1990), Shaw, Park, and Raman (1992), and Park, Raman and Shaw (1997). The parametric adjustment method for the ATCS rule in Example 14.3.1 is due to Chao and Pinedo (1992). An excellent overview of learning mechanisms for scheduling systems is presented in Aytug, Bhattacharyya, Koehler and Snowdon (1994). The book by Pesch (1994) focuses on learning in scheduling through genetic algorithms (classifier systems).

A fair amount of development work has been done recently on the design of adaptable planning and scheduling engines. Akkiraju, Keskinocak, Murthy and Wu

(1998, 2001) discuss the design of an agent-based approach for a scheduling system developed at IBM. Feldman (1999) describes in detail how algorithms can be linked and integrated and Webster (2000) presents two frameworks for adaptable scheduling algorithms.

The design, development and implementation of modular or reconfigurable planning and scheduling systems is often based on objects and methods. For objects and methods, see Booch(1994), Martin (1993), and Yourdon(1994). For modular design with regard to databases and knowledge bases, see, for example, Collinot, LePape and Pinoteau (1988), Fox and Smith (1984), Smith (1992), and Smith, Muscettola, Matthys, Ow and Potvin (1990)). For an interesting design of a scheduling engine, see Sauer (1993). A system design proposed by Smith and Lassila (1994) extends the modular philosophy for planning and scheduling systems farther than any previous system. This is also the case with the approach by Yen (1995) and Pinedo and Yen (1997).

The paper by Yen (1997) contains the material on web-based planning and scheduling systems that is presented in Section 14.6.

Chapter 15

What Lies Ahead?

15.1 Introduction 371
15.2 Planning and Scheduling in Manufacturing 372
15.3 Planning and Scheduling in Services 373
15.4 Solution Methods............................ 375
15.5 Systems Development........................ 377
15.6 Discussion 378

15.1 Introduction

With so many different types of applications, it is not surprising that there is such a great variety of planning and scheduling models. Moreover, the numerous solution methods provide a host of procedures for the myriad of problems. Any given application typically requires its own type of planning and scheduling engine as well as customized user interfaces. The overall architecture of a system may therefore be very application-specific. The decision support systems that have been designed for planning and scheduling in the various industries tend to be quite different from one another.

Over the last decade there has been a tendency to build larger systems that have more capabilities and that are better integrated within the ERP system of the enterprise. Especially in the manufacturing world there has been a tendency to design and develop integrated systems with multiple functionalities. Especially in supply chain management the systems (and their underlying models) have become more and more elaborate. The dimensions according to which such a system can be measured include the number of facilities in the network as well as the various time horizons over which the system must optimize. Integration may occur in scope, space and time.

In service organizations the systems have become more and more integrated as well. In airlines, fleet scheduling systems and crew scheduling systems have become more and more integrated as well as interactive. In call centers, personnel scheduling systems interact with operator assignment and call routing systems.

The difficulties in the modeling, the design, and the development of large integrated systems lie typically in the coordination and integration of smaller modules with a narrower scope. There are many forms of integration and many types of interfaces between modules. For example, within a system a medium term planning module may have to interact with a short term scheduling module; a short term scheduling module may have to interact with reactive scheduling procedures. The models that have been discussed in the open literature and that are covered in this book tend to be narrow in scope; integration affects the modeling as well as the solution methods to be used. The difficulties often tend to lie on the interfaces between the different modeling paradigms.

Because of these forms of integration, the various modules in a system (that are designed to deal with different types of problems) must exchange data with one another. Since the solution method in one module often attempts to perform some form of local optimization in an iterative manner, it may be the case that the various modules must exchange data with one another regularly in order to solve their problems. Because the solution methods in the different modules of a system often attempt to optimize at different levels (with regard to the horizon, the level of detail, etc.), it may be the case that in the transfer of data between the modules the data have to undergo some form of transformation (e.g., aggregation).

15.2 Planning and Scheduling in Manufacturing

Many different planning and scheduling models in manufacturing have been analyzed in the literature in detail. Part II of this book focuses on some of the more important model categories, namely

(i) project planning and scheduling,
(ii) machine scheduling and job shop scheduling,
(iii) scheduling of flexible assembly systems,
(iv) economic lot scheduling, and
(v) planning and scheduling in supply chains.

Clearly, Part II does not cover all the planning and scheduling models in manufacturing that have been considered in the literature. Some of the more narrow areas in manufacturing with very special scheduling problems have not been covered in this book. Examples of such niche areas are:

(i) crane scheduling, and
(ii) scheduling of robotic cells.

In both these areas a fairly large number of papers have appeared. However, the structure of these problems tend to be very special and it is not clear of how much use they are to other planning and scheduling areas.

Some other classes of models have received little or no attention in the literature and are therefore not discussed in this book either. For example, not much work has been done on models that combine continuous aspects of scheduling problems with discrete aspects. In the process industries it is often the case that planning and scheduling involves Make-To-Order (MTO) production as well as Make-To-Stock (MTS), i.e., it may be the case that a certain part of the production is MTO and the remaining is MTS. The MTO part consists of a set of orders with committed shipping dates (due dates), whereas the MTS part is concerned with setup costs, inventory carrying costs, and stockout probabilities. The MTO part has strong discrete aspects whereas the MTS part has strong continuous aspects. Even though this hybrid type of scheduling problem is very common in various process industries, very little research has been done in this area. Such models may gain more research attention in the near future.

Another class of models that has not received much attention in the literature are models that deal with the concept of *product assembly* or *production synchronization*, i.e., parts and raw material have to arrive at a point of assembly at more or less the same time, just before the assembly is about to start. If two production processes have to be synchronized in such a way that the completion of one item in one process has to occur more or less at the same time as the completion of the corresponding item in the other process, then the associated scheduling problems tend to be hard.

Some classes of models which already have been considered in the literature may in the future be generalized and extended in new directions. First, models considered in the literature often focus on a single objective. In practice, it may very well be the case that several objectives have to be considered at the same time and that the user would like to see a parametric analysis in order to evaluate the trade-offs. Second, models that have been analyzed under the assumption that the solution procedure would be used in one particular manner (e.g., allowing overnight number crunching) may have to be studied again under the assumption that the procedure would be used in a different manner (e.g., in real time or in a distributed fashion). Third, models which assume that a plan or schedule can be created entirely from scratch (without any initial conditions or without a partial schedule already in place) may have to be generalized to allow for the existence of initial conditions or a partial schedule.

15.3 Planning and Scheduling in Services

A number of important application areas in services have been covered in this book, namely

(i) project planning and scheduling,

(ii) scheduling of meetings, exams, and so on,
(iii) scheduling of entertainment and sport events,
(iv) transportation scheduling, and
(v) workforce scheduling.

All five areas have received a significant amount of attention from the academic community, developers (software houses), and users. The five areas listed have very different characteristics. The first, second, and fifth area are somewhat generic and are important in many different industries; the third and fourth application area are very industry-specific.

Systems in the first and second application area seem to be easier to design and implement than systems in the other three areas. Project planning, scheduling of meetings and exams, as well as rostering and timetabling tend to be somewhat less difficult to develop and do not require a significant amount of interfacing with other enterprise-wide systems.

The third application area, tournament scheduling, typically does not involve any real time scheduling. Tournament schedules are created in advance and there are usually no constraints on the computer time. The algorithms may be complicated but the system is usually a stand-alone system.

Systems implementations in the fourth and fifth application area tend to be more challenging. In these application areas it is often the case that planning and scheduling systems have to be linked to other decision support systems. It may involve, for example,

(i) integration of fleet scheduling and crew scheduling in transportation,
(ii) integration of fleet scheduling and yield management in transportation, and
(iii) integration of personnel scheduling and queue management in call centers.

Even though the systems have become more and more integrated, the different problem areas still tend to be analyzed separately. That is, not much work has been done on hybrid models that encompass different types of problem areas and have a more global objective.

Just like in manufacturing, there are various niche areas in services that have received some research attention in the past and that have not been included in this book. These niche areas include:

(i) equipment maintenance and order scheduling;
(ii) operating room scheduling in hospitals;
(iii) scheduling of check processing in banks.

In an equipment maintenance model, it is typically assumed that there are m resources in parallel (e.g., m different types of repair men) and each resource is capable of doing a certain type of task. Each customer that comes in requires a number of different services (e.g., repairs), that can be done concurrently and independently by the various resources. A customer can depart only after all requested services have been completed. Each customer is given in advance a time when all his services are expected to be completed; this time serves as a due date. One objective is to minimize the total weighted tardiness.

There are several applications of planning and scheduling in health care, e.g., nurse scheduling (see Chapter 12). Another important application of planning and scheduling in health care concerns the scheduling of operating rooms in hospitals. Scheduling of operating rooms tends to be difficult because of its stochastic nature. The durations of the operations can only be estimated in advance, but the actual times have a certain variability. An additional difficulty is that a sequence of planned surgeries can be interrupted at times by an emergency surgery. An objective function may be, for example, the minimization of the expected waiting times of the patients.

Future research in planning and scheduling in services may also focus on generalizations and extensions of existing models, similar to the extensions mentioned with regard to the manufacturing models. It may involve models with multiple objectives. It may involve models which assume that the problems have to be solved in a different mode. For example, a system may have to function in a reactive mode, i.e., when something unexpected occurs, the system has to reschedule all the activities in real time. Some research has already been done in this direction: airlines rely heavily on crew recovery programs in case flight schedules have been thrown out of whack because of, say, weather conditions.

15.4 Solution Methods

The different solution methods (described in Appendixes A, B, C, and D) have various important characteristics. First, one characteristic is the quality of the solution; that is, how close to optimal is the solution generated? A second characteristic is the computation time involved; that is, does the technique work in a real time environment or does it need a significant amount of computer time? A third characteristic concerns the development time and ease of maintenance, e.g., how easy is it to adapt the code to a slightly different problem?

There are many solution methods available, including exact optimization techniques (e.g., integer programming), heuristics (e.g., decomposition techniques), and constraint programming techniques. Moreover, these basic techniques can in many ways be combined with one another in the form of hybrid techniques.

Over the last decade a significant amount of research and development work has been done in the application of integer programming techniques to specific scheduling problems. The techniques include branch-and-cut, branch-and-price (also referred to as column generation), and branch-cut-and-price. The main goal is always to improve the performance of the integer programming techniques; a significant amount of progress has already been made.

An enormous amount of work has focused on heuristic methods, including decomposition techniques and local search techniques. Some of this effort has been directed to real time applications; other work has been based on the assumption of an unlimited amount of computer time.

During the last decade a fair amount of effort has focused on the development and implementation of constraint programming applications. Several software firms, including ILOG and Dash Optimization, have developed elaborate software packages for constraint programming applications.

The appropriateness of a technique depends, of course, also on the type of application (i.e., the structure of the problem) and the manner in which the technique is supposed to be used (e.g., in a reactive mode). Both mathematical programming and constraint programming have their advantages and disadvantages; however, their strenghts and weaknesses are often complementary. If the data set is fuzzy and subject to frequent changes, then it may not make sense to implement a very expensive and time consuming optimization technique. In such a case a heuristic may be more appropriate. It is often advantageous to do first an analysis concerning the accuracy and robustness of the data set.

The next stage in the development of solution methods will focus on the development of hybrid techniques that combine constraint programming with optimization approaches and heuristics. For example, Dash Optimization recently developed the Xpress-CP constraint programming tool. This programming tool can be used in conjunction with the Xpress optimizer tool Xpress-MP in the Xpress-Mosel language. Xpress-CP combines the mathematical optimization software Xpress-MP and the constraint programming software CHIP in a hybrid optimization framework. An advantage of Xpress-CP is that the mathematical programming and constraint programming methods are embedded in the same software environment and the problem can be formulated as one model. Such an architecture facilitates the development of a more sophisticated solution technique that uses mathematical programming and contraint programming methodologies in an integrated manner. The model can be written in the Xpress-Mosel language; the mathematical programming and constraint programming solvers can be invoked easily without a need for intricate programming or any particular expertise in the methodologies. Built on Mosel's Native Interface technology, the Mosel module Xpress-CP provides a high level of abstraction through a large collection of types and constraint structures that support several functions and procedures. In addition, the CP solver contains high level primitives to guide during the search procedure and can follow various established strategies and heuristics. This, together with its flexible design, makes Xpress-CP a useful tool for modeling and solving complex scheduling problems.

Many planning and scheduling problems have multiple objectives and the weights of the various objectives are either subjective or vary over time. Users of planning and scheduling systems may at times be interested in doing some form of parametric analysis and determining the trade-offs. This form of parametric optimization may require special solution techniques as well. Some of these techniques may be generic, whereas others may be application-specific.

Large planning and scheduling problems are often difficult to analyze and may benefit from distributed processing. If a problem is hard, it may be advantageous to decompose (i.e., partition) the problem into smaller subproblems which then are solved separately on different servers. Certain types of solution techniques may be more amenable to distributed processing than other types of solution techniques. Finding the proper approaches to decompose problems and to aggregate the results of the various subproblems may lead to challenging research in the future. An interesting example of such a framework is the so-called A-Team architecture of IBM, which was developed at IBM's T.J. Watson Research Center.

15.5 Systems Development

An enormous amount of systems development is going on in manufacturing as well as in services. There are many ways in which each module within a system can be designed (whether it is the database, the planning or scheduling engine, or the user interface). In the design of a system, the data transfer and the information exchange play an extremely important role.

Supply chain management needs on a global level systems for long term, medium term, and short term optimization; on a local level it needs systems for short term optimization as well. All these systems have to exchange data with one another. The output of one system (e.g., a plan or schedule) may be an input for another. However, the data usually cannot be exchanged without some appropriate transformation (either some form of aggregation or some form of disaggregation). The optimal forms of communication between decision support systems require a certain amount of research.

Data transfer and information exchange have been undergoing major transformations recently. Because of the technological advances that have taken place over the last decade with the emergence of the Internet, significant changes are taking place with regard to the design and the implementation of decision support systems in supply chain management. The Internet has a significant impact on various aspects of planning and scheduling in supply chains. In many scenarios planning and scheduling functions are becoming more web-based. It enhances the level of communication between the various stages as well as with suppliers and customers; it has the potential of reducing bull-whip effects. It allows for

(i) more accurate, more accessible, and more timely information;
(ii) the establishment of e-Hubs and Vendor Managed Inventories (VMI);
(iii) auctions and B2B communication for the procurement of commodity items.

Having more accurate and more timely information has a significant positive effect on short term and/or reactive scheduling, but not a major effect on long term planning. An important issue is how to aggregate data and determine the form and the timing of the transfer of the information obtained. Information with regard to disruptions in the supply or in the demand are conveyed now more rapidly upstream as well as downstream. The Material Requirements Planning system may be able to react faster and marketing may be able to adjust prices quicker.

In order to improve inventory management, companies have been designing e-Hubs for their Vendor Managed Inventory (VMI) systems which provide vendors access to current information concerning inventory levels of products they have to supply. An e-Hub has the effect that release dates, which are important for short term scheduling processes, are known more accurately. So there is less variability and schedules are more reliable.

With multiple suppliers of items that are somewhat commodity the Internet provides a capability for auctions and bidding (B2B). Assembly operations in particular can make use of B2B. B2B tends to be applicable to commodity items of which quality specifications are easily established and for which there are multiple vendors.

In service industries many elaborate systems have been designed and integration has become an important aspect of system design as well. A number of software houses have been considering how to integrate personnel scheduling with call routing mechanisms. Such systems have also become web-based. Appointment systems in general tend to be online.

The development of special languages and tools will facilitate the development and maintenance of large integrated systems. For example, the availability of tools such as Xpress-MP from Dash Optimization and the Rave language from Carmen Systems have had a very positive effect on systems development.

In the future, generic as well as application-specific systems will follow more and more an object-oriented design. At the same time, there may be new developments in description languages for planning and scheduling. Such developments will lead to more flexibility in the design as well as in the use of planning and scheduling systems. It will also be easier to reconfigure a system when there are changes in the environment.

15.6 Discussion

The main factors that govern the design features of a system include the environment in which the system has to be installed and the manner in which it is supposed to be used. For example, the manner in which the system uses the Internet and the amount of distributed processing that is being done, are important factors in the selection of the solution approach (e.g., mathematical programming, constraint programming, etc.). Once a decision has been made with regard to the approach, then the model formulation, the solution technique, and the system design cannot be separated from one another. All these aspects are closely linked and have a major influence on one another.

A significant amount of research and development is being done in every direction and at all levels. More elaborate and sophisticated models are being formulated and analyzed and more effective (hybrid) algorithms are being designed and more efficient optimization modules are being developed. Since there is such a great variety in planning and scheduling problems, system reconfigurability (from the modeling point of view as well as from the solution approach point of view) is of the utmost importance. Such requirements are stimulating the research in object oriented design as well as in description languages for planning and scheduling.

Comments and References

Kjenstad (1998) studied coordination issues in the planning and scheduling of supply chains. Pinedo, Seshadri and Shantikumar (1999) analyzed the interactions between personnel scheduling in call centers and other call center functions.

There is an extensive literature on the scheduling of robotic cells; see, for example, the survey papers by Crama, Kats, Van de Klundert and Levner (2000) and Dawande, Geismar, Sethi and Sriskandarajah (2004). For some interesting papers on crane and hoist scheduling, see Daganzo (1989), Yih (1994), Ge and Yih (1995), Armstrong, Gu and Lei (1996), Lee, Lei and Pinedo (1997), and Liu, Jiang and Zhou (2002).

Sung and Yoon (1998), Leung, Li and Pinedo (2004, 2005) have focused on equipment repair and order scheduling. There is a broad literature on scheduling in health care, in particular on operating room scheduling; see, for example, Blake and Carter (1997, 2002), Blake and Donald (2002), and Cayirli and Veral (2004).

Rachlin, Goodwin, Murthy, Akkiraju, Wu, Kumaran and Das (2001) describe IBM's A-Team architecture. Murthy, Akkiraju, Goodwin, Keskinocak, Rachlin, Wu, Kumaran, Yeh, Fuhrer, Aggarwal, Sturzenbecker, Jayaraman and Daigle (1999) discuss a cooperative multi-objective decision support for the paper industry. Akkiraju, Keskinocak, Murthy and Wu (2001) describe an agent-based approach to multi-machine scheduling.

Keskinocak, Goodwin, Wu, Akkiraju and Murthy (2001) and Sadeh, Hildum and Kjenstad (2003) focus on decision support systems for managing an electronic supply chain. Yen (1997) considered interactive scheduling agents on the Internet. A significant amount of research is currently being done on how the Internet can be used to solve very large optimization problems; see, for example, Fourer (1998) and Fourer and Goux (2001).

Yen (1995) considers the use of scheduling discription languages in the development of scheduling systems.

Appendices

A	Mathematical Programming: Formulations and Applications	383
B	Exact Optimization Methods	395
C	Heuristic Methods	413
D	Constraint Programming Methods	437
E	Selected Scheduling Systems	447
F	The Lekin System User's Guide	451

References 459

Notation 489

Name Index 493

Subject Index 501

Appendix A

Mathematical Programming: Formulations and Applications

A.1 Introduction 383
A.2 Linear Programming Formulations 383
A.3 Nonlinear Programming Formulations 386
A.4 Integer Programming Formulations 388
A.5 Set Partitioning, Set Covering, and Set Packing . 390
A.6 Disjunctive Programming Formulations 391

A.1 Introduction

In this appendix we give an overview of the types of problems that can be formulated as mathematical programs. The applications discussed concern only planning and scheduling problems. In order to understand the examples the reader should be familiar with the notation and terminologies introduced in Chapters 2 and 3.

This appendix is aimed at people who are already familiar with elementary operations research techniques. It makes an attempt to put various notions and problem definitions in perspective. Relatively little will be said about the standard solution techniques for solving these problems.

A.2 Linear Programming Formulations

The most basic mathematical program is the *Linear Program (LP)*. An LP refers to an optimization problem in which the objective and the constraints are linear in the variables to be determined. An LP can be expressed as follows:

minimize $\quad c_1 x_1 + c_2 x_2 + \cdots + c_n x_n$

subject to

$$a_{11}x_1 + a_{12}x_2 + \cdots + a_{1n}x_n \leq b_1$$
$$a_{21}x_1 + a_{22}x_2 + \cdots + a_{2n}x_n \leq b_2$$
$$\vdots$$
$$a_{m1}x_1 + a_{m2}x_2 + \cdots + a_{mn}x_n \leq b_m$$
$$x_j \geq 0 \qquad \text{for } j = 1, \ldots, n.$$

The objective is the minimization of costs. The $c_1, \ldots, c_n$ vector is referred to as the cost vector. The variables $x_1, \ldots, x_n$ have to be determined so that the objective function $c_1 x_1 + \cdots + c_n x_n$ is minimized. The column vector $a_{1j}, \ldots, a_{mj}$ is referred to as activity vector j. The value of the variable x_j refers to the level at which this activity j is performed. The $b_1, \ldots, b_m$ is referred to as the resource vector. The fact that in linear programming n denotes the number of activities and in scheduling theory n refers to the number of jobs is a mere coincidence; that in linear programming m denotes the number of resources and in scheduling theory m refers to the number of machines is a coincidence as well. The representation above can be written in matrix form:

$$\text{minimize} \quad \bar{c}\bar{x}$$

subject to

$$\mathbf{A}\bar{x} \leq \bar{b}$$
$$\bar{x} \geq 0.$$

There are several algorithms or classes of algorithms for solving an LP. The two most important ones are

(i) the simplex methods and
(ii) the interior point methods.

Although simplex methods work very well in practice, it is not known if there is any version of the simplex method that solves the LP problem in polynomial time. The best known example of an interior point method is the *Karmarkar* algorithm, which is known to solve the LP problem in polynomial time. Many textbooks cover these subjects in great detail.

A special case of the linear program is the so-called *transportation* problem. In the transportation problem the matrix $\mathbf{A}$ has a special form. The matrix has mn columns and $m+n$ rows and takes the form

$$\mathbf{A} = \begin{bmatrix} \bar{1} & 0 & \cdots & 0 \\ 0 & \bar{1} & \cdots & 0 \\ \vdots & \vdots & \ddots & \vdots \\ 0 & 0 & \cdots & \bar{1} \\ \mathbf{I} & \mathbf{I} & \cdots & \mathbf{I} \end{bmatrix}$$

A.2 Linear Programming Formulations

where the $\bar{1}$ is a row vector with n 1's and the $\mathbf{I}$ is an $n \times n$ identity matrix. All but two entries in each column (activity) of this $\mathbf{A}$ matrix are zero; the two nonzero entries are equal to 1. This matrix is associated with the following problem. Consider a situation in which items have to be shipped from m sources to n destinations. A column (activity) in the $\mathbf{A}$ matrix represents a route from a given source to a given destination. The cost associated with this column (activity) is the cost of transporting one item from the given source to the given destination. The first m entries in the $b_1, \ldots, b_{m+n}$ vector represent the supplies at the m sources, while the last n entries of the $b_1, \ldots, b_{m+n}$ vector represent the demands at the n different destinations. Usually it is assumed that the sum of the demands equals the sum of the supplies and the problem is to transport all the items from the sources to the demand points and minimize the total cost incurred. (When the sum of the supplies is less than the sum of the demands there is no feasible solution and when the sum of the supplies is larger than the sum of the demands an artificial destination can be created where the surplus is sent to at zero cost).

The matrix $\mathbf{A}$ of the transportation problem is an example of a matrix with the so-called *total unimodularity property*. A matrix has the total unimodularity property if the determinant of every square submatrix within the matrix has a value -1, 0 or 1. It can be easily verified that this is the case with the matrix of the transportation problem. This total unimodularity property has an important consequence: if the values of the supplies and demands are all integers, then there is an optimal solution $x_1, \ldots, x_n$, that is a vector of integers and the simplex method will find such a solution.

The transportation problem is important in planning and scheduling theory for a number of reasons. First, there are planning and scheduling problems that can be formulated as transportation problems. Second, transportation problems are often used to obtain bounds in branch-and-bound procedures for NP-hard planning and scheduling problems.

The following example focuses on a scheduling problem that can be formulated as a transportation problem.

Example A.2.1 (Parallel Machines and the Transportation Problem). Consider m machines in parallel. The speed of machine i is v_i. There are n identical jobs that all require the same amount of processing, say 1 unit. If job j is processed on machine i, its processing time is $1/v_i$. Preemptions are not allowed. If job j is completed at C_j a penalty $h_j(C_j)$ is incurred. Let the variable x_{ijk} be equal to 1 if job j is scheduled as the kth job on machine i and 0 otherwise. So the variable x_{ijk} is associated with an activity. The cost of operating this activity at unit level is

$$c_{ijk} = h_j(C_j) = h_j(k/v_i).$$

Assume that there are a total of $n \times m$ positions (a maximum of n jobs can be assigned to any machine). Clearly, not all positions will be filled. The n jobs are equivalent to the n sources in the transportation problem and the

$n \times m$ positions are the destinations. The problem can be formulated easily as an LP.

$$\text{minimize} \sum_{i=1}^{m} \sum_{j=1}^{n} \sum_{k=1}^{n} c_{ijk} x_{ijk}$$

subject to

$$\sum_{i=1}^{m} \sum_{k=1}^{n} x_{ijk} = 1 \qquad \text{for } j = 1, \ldots, n$$

$$\sum_{j=1}^{n} x_{ijk} \leq 1 \qquad \text{for } i = 1, \ldots, m, \quad k = 1, \ldots, n$$

$$x_{ijk} \geq 0 \qquad \text{for } i = 1, \ldots, m, \quad j = 1, \ldots, n, \quad k = 1, \ldots, n$$

The first set of constraints ensures that job j is assigned to one and only one position. The second set of constraints ensures that each position i, k has at most one job assigned to it. Actually from the LP formulation it is not immediately clear that the optimal values of the variables x_{ijk} have to be either 0 or 1. From the constraints it may appear at first sight that an optimal solution of the LP formulation may result in x_{ijk} values between 0 and 1. Because of the total unimodularity property, the constraints do not specifically have to require that the variables be either 0 or 1.

A special case of the transportation problem is the *assignment* problem. A transportation problem is referred to as an assignment problem when $n = m$ (the number of sources is equal to the number of destinations) and at each source there is a supply of exactly one item and at each destination there is a demand of exactly one item. The assignment problem is also important in scheduling theory. Single machine problems with the n jobs having identical processing times often can be formulated as assignment problems.

Example A.2.2 (Single Machine and the Assignment Problem). Consider a special case of the problem discussed in *Example A.2.1*, namely the single machine version. Let the processing time of each job be equal to 1. The objective is again $\sum h_j(C_j)$. So there are n jobs and n positions, and the assignment of job j to position k has cost $h_j(k)$ associated with it.

A.3 Nonlinear Programming Formulations

A *Nonlinear Program (NLP)* is a generalization of a linear program that allows the objective function and the constraints to be nonlinear in $x_1, \ldots, x_n$.

Under certain convexity assumptions on the objective function and the constraints there are necessary and sufficiency conditions for a solution to be

A.3 Nonlinear Programming Formulations

optimal. These conditions are in the literature referred to as the Kuhn-Tucker conditions. So if the objective function and constraints satisfy the required convexity assumptions, then the optimality of a solution can be verified easily via these conditions.

There are a number of methods for solving nonlinear programming problems. These methods tend to be different from the methods used for linear programming problems. (However, they may use linear programming methods as subroutines in their overall framework.) The most commonly used methods for nonlinear programming are:

(i) gradient methods and
(ii) penalty and barrier function methods.

In this book nonlinear programs appear in Sections 4.5 and 7.4 (the allocation of resources in project scheduling with nonlinear costs and the lot scheduling problems with arbitrary schedules). In both applications the objective functions are nonlinear and the constraints linear. The problem described in Section 4.5 can be solved either by a gradient method or by a penalty function method. The problem in Section 7.4, which has a single inequality constraint, can be solved in a different manner. If the constraint is not tight, then the problem is an unconstrained problem and is easy. If the constraint is tight, then the problem is a nonlinear progamming problem with an equality constraint and this type of problem can be solved relatively easily as follows.

Consider a nonlinear programming problem with multiple equality constraints. Such a problem can be transformed into an unconstrained problem using so-called Lagrangian multipliers. For example, consider the nonlinear programming problem

$$\text{minimize} \quad g(x_1, \ldots, x_n)$$

subject to

$$f_1(x_1, \ldots, x_n) = 0$$
$$\vdots$$
$$f_m(x_1, \ldots, x_n) = 0$$

This problem can be transformed into the following unconstrained optimization problem with the objective function

$$\text{minimize} \quad g(x_1, \ldots, x_n) + \lambda_1 f_1(x_1, \ldots, x_n) + \cdots + \lambda_m f_m(x_1, \ldots, x_n)$$

If the original objective function $g(x_1, \ldots, x_n)$ satisfies certain convexity conditions then the optimal solution can be obtained by taking the partial derivative of the unconstrained problem with respect to x_j and set that equal to zero. This yields n equations with $n + m$ unknowns $(x_1, \ldots, x_n, \lambda_1, \ldots, \lambda_m)$.

These n equations together with the original set of m constraints result in a system of $n+m$ equations with $n+m$ unknowns. This technique is the one used in Section 7.4.

A.4 Integer Programming Formulations

An *Integer Program (IP)* is a linear program with the additional requirement that the variables $x_1, \ldots, x_n$ have to be integers. If only a subset of the variables are required to be integer and the remaining ones are allowed to be real, the problem is referred to as a *Mixed Integer Program (MIP)*. In contrast with the LP, an efficient (polynomial time) algorithm for the IP or MIP does *not* exist.

Many scheduling problems can be formulated as integer programs. In this section we give two examples of integer programming formulations. The first example describes an integer programming formulation for the nonpreemptive single machine problem with the total weighted completion time $\sum w_j C_j$ as objective. Even though this problem is quite easy and can be solved by the simple Weighted Shortest Processing Time first (WSPT) priority rule (see Section 5.2 and Appendix C), the problem still serves as a useful example. The formulation is a generic one and can be used for scheduling problems with multiple machines as well.

Example A.4.1 (A Single Machine and Integer Programming). Consider the nonpreemptive single machine scheduling problem with the jobs subject to precedence constraints and the total weighted completion time objective. Let x_{jk} denote a $0-1$ decision variable which assumes the value 1 if job j precedes job k in the sequence and 0 otherwise. The values x_{jj} have to be 0 for all j. The completion time of job j is then equal to $\sum_{k=1}^{n} p_k x_{kj} + p_j$. The integer programming formulation of the problem without precedence constraints thus becomes

$$\text{minimize} \sum_{j=1}^{n}\sum_{k=1}^{n} w_j p_k x_{kj} + \sum_{j=1}^{n} w_j p_j$$

subject to

$$x_{kj} + x_{jk} = 1 \qquad \text{for } j, k = 1, \ldots, n, \ j \neq k,$$
$$x_{kj} + x_{lk} + x_{jl} \geq 1 \qquad \text{for } j, k, l = 1, \ldots, n, \ j \neq k, \ j \neq l, \ k \neq l,$$
$$x_{jk} \in \{0,1\} \qquad \text{for } j, k = 1, \ldots, n,$$
$$x_{jj} = 0 \qquad \text{for } j = 1, \ldots, n.$$

We can replace the third set of constraints with a combination of (i) a set of linear constraints that require all x_j to be nonnegative, (ii) a set of linear constraints requiring all x_j to be less than or equal to 1 and (iii) a set of

A.4 Integer Programming Formulations

constraints requiring all x_j to be integer. Constraints requiring certain precedences between the jobs can be added easily by specifying the corresponding x_{jk} values.

Often, there is more than one integer programming formulation of the same problem. In Exercise A.1 a different integer programming formulation has to be developed for the same problem.

In what follows an integer programming formulation is presented for the job shop problem with the makespan objective described in Chapter 5. This integer programming formulation is similar to the one presented in Chapter 4 for the workforce constrained project scheduling problem.

Example A.4.2 (A Job Shop and Integer Programming). Consider a job shop as described in Chapter 5. There are n jobs and m machines. Job j has to be processed on machine i for a duration of p_{ij}. The route of each one of the jobs is given and recirculation is allowed. If operation (i,j) of job j on machine i has to be completed before operation (h,j) on machine h is allowed to start, then this routing constraint may be regarded as a precedence constraint $(i,j) \to (h,j)$. Let the arc set A denote the set of all such routing (precedence) constraints. Let x_{ijt} denote a $0-1$ decision variable that is equal to 1 if operation (i,j) is completed exactly at time t and 0 otherwise. Let H denote an upper bound for the makespan $C_{\max}$; such an upper bound can be obtained fairly easily (take, for example, the sum of all the processing times of all jobs over all machines). So the completion time of operation (i,j) can be expressed as

$$C_{ij} = \sum_{t=1}^{H} t\, x_{ijt}.$$

In order to formulate the integer program, assume that the makespan $C_{\max}$ is also a decision variable. (However, the makespan decision variable is clearly not a $0-1$ variable.) The integer program that minimizes the makespan can now be formulated as follows.

minimize $C_{\max}$

subject to

$$\sum_{t=1}^{H} t\, x_{ijt} - C_{\max} \leq 0 \qquad \text{for } i=1,\ldots,m;\ j=1,\ldots,n$$

$$\sum_{t=1}^{H} t\, x_{ijt} + p_{ij} - \sum_{t=1}^{H} t\, x_{hjt} \leq 0 \qquad \text{for } (i,j) \to (h,j) \in A.$$

$$x_{ijt} \in \{0,1\} \quad \text{for} \quad i=1,\ldots,m;\ j=1,\ldots,n;\ t=1,\ldots,H.$$

In Section A.6 of this appendix a so-called disjunctive programming formulation is given for this problem.

There are several methods for solving Integer Programs. These methods are discussed in Appendix B.

A.5 Set Partitioning, Set Covering, and Set Packing

There are several special types of Integer Programming formulations that have numerous planning and scheduling applications in practice. This section focuses on three such formulations, namely Set Partitioning, Set Covering, and Set Packing.

The integer programming formulation of the Set Partitioning problem has the following structure.

minimize $\quad c_1x_1+c_2x_2+\cdots+c_nx_n$

subject to

$$a_{11}x_1 + a_{12}x_2 + \cdots + a_{1n}x_n = 1$$
$$a_{21}x_1 + a_{22}x_2 + \cdots + a_{2n}x_n = 1$$
$$\vdots$$
$$a_{m1}x_1 + a_{m2}x_2 + \cdots + a_{mn}x_n = 1$$
$$x_j \in \{0,1\} \qquad \text{for } j=1,\ldots,n.$$

All a_{ij} values are either 0 or 1.

When the equal signs (=) in the constraints are replaced by greater than or equal ($\geq$), the problem is referred to as the Set Covering problem, and when the equal signs are replaced by less than or equal ($\leq$), then the problem is referred to as the Set Packing problem. In practice, the objective of the Set Packing problem is typically a profit maximization objective.

The mathematical model underlying the Set Partitioning problem can be described as follows: Assume m different elements and n different subsets of these m elements. Each subset contains one or more elements. If $a_{ij}=1$, then element i is part of subset j, and if $a_{ij}=0$, then element i is not part of subset j. The objective is to find a collection of subsets such that each element is part of exactly one subset. The objective is to find that collection of subsets that have a minimum cost. In the Set Covering problem, each element has to be part of at least one subset. In the Set Packing problem each subset yields a certain profit π_j and the total profit has to be maximized in such a way that each element is part of at most one subset.

An example of the Set Partitioning problem is the crew scheduling problem considered in Chapter 12. Each element is a flight leg and each subset is a round trip. The objective is to cover each flight leg exactly once and minimize the total cost of all the round trips. Each column in the **A** matrix is a round trip and each row is a flight leg that must be covered exactly once by one round trip. If flight leg i is part of round trip j, then a_{ij} is 1, otherwise a_{ij} is 0. Let x_j denote a $0-1$ decision variable that assumes the value 1 if round trip j is selected and 0 otherwise. The optimization problem is then to select, at minimum cost, a set of round trips that satisfies the constraints. The constraints in this problem are often called the partitioning equations. For a feasible solution $(x_1, \ldots, x_n)$, the variables that are equal to 1 are referred to as the *partition*.

Another example of a Set Partitioning problem is the aircraft routing and scheduling problem considered in Chapter 11. It is clear that in the aviation industry the constraint that each flight leg should be covered *exactly* once by a round trip is important. However, in the trucking industry, it may be possible to have one leg covered by several round trips. The constraints may therefore be relaxed and the problem then assumes a Set Covering structure.

An example of a Set Packing problem is the tanker scheduling problem described in Chapter 11. The reason why the tanker scheduling problem is a packing problem and not a partitioning problem is based on the fact that not every cargo has to be transported by a company owned tanker. If it is advantageous to assign a cargo to an outside charter, then that is allowed.

A.6 Disjunctive Programming Formulations

There is a large class of mathematical programs in which the constraints can be divided into a set of *conjunctive* constraints and one or more sets of *disjunctive* constraints. A set of constraints is called conjunctive if each one of the constraints has to be satisfied. A set of constraints is called disjunctive if at least one of the constraints has to be satisfied but not necessarily all.

In the standard linear program all constraints are conjunctive. The mixed integer program described in Example A.4.1 in essence contains pairs of disjunctive constraints. The fact that the integer variable x_{jk} has to be either 0 or 1 can be enforced by a pair of disjunctive linear constraints: either $x_{jk} = 0$ or $x_{jk} = 1$. This implies that the single machine problem with precedence constraints and the total weighted completion time objective can be formulated as a disjunctive program as well.

Example A.6.1 (A Disjunctive Program for a Single Machine). Before expressing the nonpreemptive single machine problem with the jobs subject to precedence constraints and the total weighted completion time objective in the form of a disjunctive program, it is of interest to represent the problem by a disjunctive graph model. Let N denote the set of nodes which correspond to the n jobs. Between any pair of nodes (jobs) j and k in this graph exactly one of the following three conditions has to hold:

(i) job j precedes job k,
(ii) job k precedes job j and
(iii) jobs j and k are independent with respect to one another.

The set of directed arcs A represent the precedence relationships between the jobs. These arcs are the so-called conjunctive arcs. Let set I contain all the pairs of jobs that are independent of one another. Each pair of jobs $(j, k) \in I$

are now connected with one another by two arcs going in opposite directions. These arcs are referred to as disjunctive arcs. The problem is to select from each pair of disjunctive arcs between two independent jobs j and k one arc which indicates which one of the two jobs goes first. The selection of disjunctive arcs has to be such that these arcs together with the conjunctive arcs do not contain a cycle. The selected disjunctive arcs together with the conjunctive arcs determine a schedule for the n jobs.

Let the variable x_j in the disjunctive program formulation denote the completion time of job j. The set A denotes the set of precedence constraints $j \to k$ which require job j to be processed before job k.

$$\text{minimize} \sum_{j=1}^{n} w_j x_j$$

subject to

$$x_k - x_j \geq p_k \quad \text{for all } j \to k \in A,$$
$$x_j \geq p_j \quad \text{for } j = 1, \ldots, n,$$
$$x_k - x_j \geq p_k \quad \text{or} \quad x_j - x_k \geq p_j \quad \text{for all } (j,k) \in I.$$

The first and second set of constraints are conjunctive constraints. The third set is a set of disjunctive constraints.

Example A.6.2 (A Disjunctive Program for a Job Shop). To present a disjunctive program for a job shop as described in Chapter 5, let the variable y_{ij} denote the starting time of operation (i,j). Recall that set N denotes the set of all operations (i,j), and set A the set of all routing constraints $(i,j) \to (k,j)$ which require job j to be processed on machine i before it is processed on machine k. The following mathematical program minimizes the makespan.

$$\text{minimize } C_{\max}$$

subject to

$$y_{kj} - y_{ij} \geq p_{ij} \quad \text{for all } (i,j) \to (k,j) \in A$$
$$C_{\max} - y_{ij} \geq p_{ij} \quad \text{for all } (i,j) \in N$$
$$y_{ij} - y_{il} \geq p_{il} \quad \text{or} \quad y_{il} - y_{ij} \geq p_{ij} \quad \text{for all } (i,l) \text{ and } (i,j), \; i = 1, \ldots, m$$
$$y_{ij} \geq 0 \quad \text{for all } (i,j) \in N$$

In this formulation, the first set of constraints ensure that operation (k,j) cannot start before operation (i,j) is completed. The third set of constraints are the disjunctive constraints; they ensure that some ordering exists among operations of different jobs that have to be processed on the same machine. Because of these constraints this formulation is referred to as a disjunctive programming formulation.

Example A.6.3 (A Disjunctive Program for a Job Shop Continued).

Consider the following example with four machines and three jobs. The route, i.e., the machine sequence, as well as the processing times are given in the table below.

jobs	machine sequence	processing times
1	1, 2, 3	$p_{11} = 10$, $p_{21} = 8$, $p_{31} = 4$
2	2, 1, 4, 3	$p_{22} = 8$, $p_{12} = 3$, $p_{42} = 5$, $p_{32} = 6$
3	1, 2, 4	$p_{13} = 4$, $p_{23} = 7$, $p_{43} = 3$

The objective consists of the single variable $C_{\max}$. The first set of constraints consists of seven constraints: two for job 1, three for job 2 and two for job 3. For example, one of these is

$$y_{21} - y_{11} \geq 10 \ (= p_{11}).$$

The second set consists of ten constraints, one for each operation. An example is

$$C_{\max} - y_{11} \geq 10 \ (= p_{11}).$$

The set of disjunctive constraints contains eight constraints: three each for machines 1 and 2 and one each for machines 3 and 4 (there are three operations to be performed on machines 1 and 2 and two operations on machines 3 and 4). An example of a disjunctive constraint is

$$y_{11} - y_{12} \geq 3 \ (= p_{12}) \quad \text{or} \quad y_{12} - y_{11} \geq 10 \ (= p_{11}).$$

The last set includes ten nonnegativity constraints, one for each starting time.

That a scheduling problem can be formulated as a disjunctive program does not imply that there is a standard solution procedure available that will work satisfactorily. Minimizing the makespan in a job shop is a very hard problem and solution procedures are either based on enumeration or on heuristics. To obtain optimal solutions branch-and-bound methods are required.

The same techniques that can be applied to integer programs can also be applied to disjunctive programs. The application of branch-and-bound to a disjunctive program is straightforward. First the LP relaxation of the disjunctive program has to be solved (i.e., the LP obtained after deleting the set of disjunctive constraints). If the optimal solution of the LP by chance satisfies all disjunctive constraints, then the solution is optimal for the disjunctive program as well. However, if one of the disjunctive constraints is violated, say the constraint

$$x_k - x_j \geq p_k \quad \text{or} \quad x_j - x_k \geq p_j,$$

then two additional LP's are generated. One has the extra constraint $x_k - x_j \geq p_k$ and the other has the extra constraint $x_j - x_k \geq p_j$. The procedure is in all other respects similar to a standard branch-and-bound procedure for integer programming as described in Appendix B.

Exercises

A.1. Consider the problem described in Example A.4.1. In order to formulate a different integer program for the same problem introduce the 0-1 decision variable x_{jt}. This variable is 1 if job j starts exactly at time t and 0 otherwise.

(a) Show that the objective of the integer program is

$$\text{minimize} \sum_{j=1}^{n} \sum_{t=0}^{l} w_j(t+p_j)x_{jt}$$

(b) Formulate the constraint sets for this integer program.

A.2. Formulate the instance of Example A.6.3 as an integer program of the type described in Example A.4.2.

Comments and References

Many books have been written on linear programming, integer programming and combinatorial optimization. Examples of some relatively recent ones are Papadimitriou and Steiglitz (1982), Parker and Rardin (1988), Padberg (1995), Wolsey (1998), Nemhauser and Wolsey (1999), and Schrijver (2003). For excellent surveys of mathematical programming formulations of machine scheduling problems, see Blazewicz, Dror and Weglarz (1991) and Van De Velde (1991).

Appendix B

Exact Optimization Methods

B.1 Introduction 395
B.2 Dynamic Programming 396
B.3 Optimization Methods for Integer Programs.... 400
B.4 Examples of Branch-and-Bound Applications ... 402

B.1 Introduction

There are planning and scheduling problems that are inherently easy; they can be formulated as linear programs that are readily solvable through the use of existing efficient algorithms. Other easy problems can be solved by different algorithms that are also efficient. These efficient algorithms are usually referred to as polynomial time algorithms. That a problem can be solved by an efficient, polynomial time, algorithm implies that very large instances of that problem, with hundreds or even thousands of jobs, still can be solved in a relatively short time on a computer.

However, there are many more planning and scheduling problems that are intrinsically very hard. These problems are referred to as *NP-hard*. They are typically combinatorial problems that cannot be formulated as linear programs and there are no simple rules or algorithms that yield optimal solutions in a limited amount of computer time. Often, it may be possible to have nice and elegant integer programming or disjunctive programming formulations for such problems, but still solving these to optimality may require an enormous amount of computer time.

There are various classes of methods that are useful for obtaining optimal solutions for such NP-Hard problems. One class of methods is referred to as *Dynamic Programming*. Dynamic programming is one of the more widely used techniques for dealing with combinatorial optimization problems. It is a procedure that is based on a divide and conquer approach. Dynamic Programming can be applied to problems that are solvable in polynomial time, as well as problems that are NP-Hard.

If a planning and scheduling problem can be formulated as an Integer Program, then various other techniques can be applied. The best known methods for solving Integer Programs are:

(i) branch-and-bound methods,
(ii) cutting plane (polyhedral) methods,
(iii) hybrid methods.

The first class of methods, branch-and-bound, is one of the most popular class of techniques used for Integer Programming. The branching refers to a partitioning of the solution space; each part of the solution space is then considered separately. The bounding refers to the development of lower bounds for parts of the solution space (assuming the objective has to be minimized). If a lower bound on the objective in one part of the solution space is larger than an integer solution already found in a different part of the solution space, the corresponding part of the former solution space can be disregarded.

The second class of methods, cutting plane methods, focuses on the linear program relaxation of the integer program. These methods generate *additional* linear constraints, i.e., cutting planes, that have to be satisfied for the variables to be integer. These additional inequalities constrain the feasible set more than the original set of linear inequalities without cutting off integer solutions. Solving the LP relaxation of the IP with the additional constraints then yields a new solution, that may be integer. If the solution is integer, the procedure stops as the solution obtained is optimal for the original IP. If the variables are not integer, more inequalities are generated.

Hybrid methods typically combine ideas from various different approaches. For example, the cutting plane method has become popular in recent years through its use in combination with branch-and-bound. When branch-and-bound is used in conjunction with cutting plane techniques it is referred to as branch-and-cut.

B.2 Dynamic Programming

Dynamic programming is basically a complete enumeration scheme that attempts, via a divide and conquer approach, to minimize the amount of computation to be done. The approach solves a series of subproblems until it finds a solution for the original problem. It determines the optimal solution for each subproblem and its contribution to the objective function. At each iteration it determines the optimal solution for a subproblem, which is larger than all previously solved subproblems. It finds a solution for the current subproblem by utilizing all the information obtained earlier in the solutions of all the previous subproblems.

B.2 Dynamic Programming

Dynamic programming is characterized by three types of equations, namely

(i) initial conditions;
(ii) a recursive relation and
(iii) an optimal value function.

In scheduling a choice can be made between forward dynamic programming and backward dynamic programming. The following example illustrates the use of forward dynamic programming.

Example B.2.1 (Forward Dynamic Programming Formulation). Consider a single machine and n jobs. If job j is completed at time C_j, then a cost $h_j(C_j)$ is incurred. The problem is to sequence the jobs in such a way that the objective

$$\sum_{j=1}^{n} h_j(C_j)$$

is minimized. This problem is a very important problem in scheduling theory as it includes many important objective functions as special cases. The objective is, for example, a generalization of the $\sum w_j T_j$ objective; the problem is therefore NP-hard. Let J denote a subset of the n jobs and assume that all the jobs in set J are processed before any one of the jobs not in set J. Let

$$V(J) = \sum_{j \in J} h_j(C_j),$$

provided the set of jobs J is processed first. The dynamic progamming formulation of the problem is based on the following initial conditions, recursive relation and optimal value function.

Initial Conditions:

$$V(\{j\}) = h_j(p_j), \qquad j = 1, \ldots, n$$

Recursive Relation:

$$V(J) = \min_{j \in J} \left(V(J - \{j\}) + h_j\left(\sum_{k \in J} p_k\right) \right)$$

Optimal Value Function:

$$V(\{1, \ldots, n\})$$

The idea behind this dynamic programming procedure is relatively straightforward. At each iteration the optimal sequence for a subset of the jobs (say a subset J that contains l jobs) is determined, assuming this subset goes first. This is done for *every* subset of size l. There are $n!/(l!(n-l)!)$ such subsets. For each subset the contribution of the l scheduled jobs to the objective function is computed. Through the recursive relation this is expanded to every subset which contains $l+1$ jobs. Each one of the $l+1$ jobs is considered as a candidate to go first. When using the recursive relation the actual sequence of the l jobs of the smaller subset does not have to be taken into consideration; only the contribution of the l jobs to the objective has to be known. After the value $V(\{1,\ldots,n\})$ has been determined the optimal sequence is obtained through a simple backtracking procedure.

The computational complexity of this problem can be determined as follows. The value of $V(J)$ has to be determined for all subsets that contain l jobs. There are $n!/l!(n-l)!$ subsets. So the total number of evaluations that have to be done are

$$\sum_{l=1}^{n} \frac{n!}{l!(n-l)!} = O(2^n).$$

Example B.2.2 (Application of Forward Dynamic Programming).
Consider the problem described in the previous example with the following jobs.

jobs	1	2	3
p_j	4	3	6
$h_j(C_j)$	$C_1 + C_1^2$	$3 + C_2^3$	$8C_3$

So $V(\{1\}) = 20$, $V(\{2\}) = 30$ and $V(\{3\}) = 48$. The second iteration of the procedure considers all sets containing two jobs. Applying the recursive relation yields

$$V(\{1,2\}) = \min\Big(V(\{1\}) + h_2(p_1 + p_2), V(\{2\}) + h_1(p_2 + p_1)\Big)$$
$$= \min(20 + 346,\ 30 + 56) = 86$$

So if jobs 1 and 2 precede job 3, then job 2 has to go first and job 1 has to go second. In the same way it can be determined that $V(\{1,3\}) = 100$ with job 1 going first and job 3 going second and that $V(\{2,3\}) = 102$ with job 2 going first and job 3 going second. The last iteration of the procedure considers set $\{1,2,3\}$.

$$V(\{1,2,3\}) = \min\Big(V(\{1,2\}) + h_3(p_1 + p_2 + p_3), V(\{2,3\}) + h_1(p_1 + p_2 + p_3),$$
$$V(\{1,3\}) + h_2(p_1 + p_2 + p_3)\Big).$$

So
$$V(\{1,2,3\}) = \min\Big(86 + 104,\ 102 + 182,\ 100 + 2200\Big) = 190.$$

It follows that jobs 1 and 2 have to go first and job 3 last. The optimal sequence is $2, 1, 3$ with objective value 190.

B.2 Dynamic Programming

In the following example the same problem is handled through a backward dynamic programming procedure. In scheduling problems the backward version typically can be used only for problems with a makespan that is schedule independent (e.g., single machine problems without sequence dependent setups, multiple machine problems with jobs that have identical processing times).

The use of backwards dynamic programming is nevertheless important as it is somewhat similar to the dynamic programming procedure discussed in the next section for stochastic scheduling problems.

Example B.2.3 (Backward Dynamic Programming Formulation).
Consider again a single machine scheduling problem with $\sum h_j(C_j)$ as objective and no preemptions. It is clear that the makespan $C_{\max}$ is schedule independent and that the last job is completed at $C_{\max}$ which is equal to the sum of the n processing times.

Again, J denotes a subset of the n jobs and it is assumed that J is processed first. Let J^C denote the complement of J. So set J^C is processed last. Let $V(J)$ denote the minimum contribution of the set J^C to the objective function. In other words, $V(J)$ represent the minimum additional cost to complete all *remaining* jobs after all jobs in set J have been completed.

The backward dynamic programming procedure is now characterized by the following initial conditions, recursive relation and optimal value function.

Initial Conditions:

$$V(\{1,\ldots,j-1,j+1,\ldots,n\}) = h_j(C_{\max}) \qquad j = 1,\ldots,n$$

Recursive Relation:

$$V(J) = \min_{j \in J^C} \left(V(J \cup \{j\}) + h_j\left(\sum_{k \in J \cup \{j\}} p_k \right) \right)$$

Optimal Value Function:

$$V(\emptyset)$$

Again, the procedure is relatively straightforward. At each iteration, the optimal sequence for a subset of the n jobs, say a subset J^C of size l, is determined, assuming this subset goes *last*. This is done for every subset of size l. Through the recursive relation this is expanded for every subset of size $l+1$. The optimal sequence is obtained when the subset comprises all jobs. Note that, as in Example B.2.1, subset J goes first; however, in Example B.2.1 set J denotes the set of jobs already scheduled while in this example set J denotes the set of jobs still to be scheduled.

B.3 Optimization Methods for Integer Programs

Integer programs are often solved via branch-and-bound. A very basic branch-and-bound procedure can be described as follows. Suppose one solves the LP relaxation of an IP (that is, the IP without the integrality constraints). If the solution of the LP relaxation happens to be integer, say $\bar{x}^0$, then this solution is optimal for the original integer program as well. If $\bar{x}^0$ is not integer, then the value of the optimal solution of the LP relaxation, $\bar{c}\bar{x}^0$, still serves as a lower bound for the value of the optimal solution for the original integer program.

If one of the variables in $\bar{x}^0$, is not integer, say $x_j = r$, then the branch-and-bound procedure proceeds as follows. The integer programming problem is split into two subproblems by adding two mutually exclusive and exhaustive constraints. In one subproblem, say SP(1), the original integer program is modified by adding the additional constraint

$$x_j \leq \lfloor r \rfloor,$$

where $\lfloor r \rfloor$ denotes the largest integer smaller than r, while in the other subproblem, say SP(2), the original integer program is modified by adding the additional constraint

$$x_j \geq \lceil r \rceil$$

where $\lceil r \rceil$ denotes the smallest integer larger than r. It is clear that the optimal solution of the original integer program has to lie in the feasible region of one of these two subproblems.

The branch-and-bound procedure now considers the LP relaxation of one of the subproblems, say SP(1), and solves it. If the solution is integer, then this branch of the tree does not have to be explored further, as this solution is the optimal solution of the original integer programming version of SP(1). If the solution is not integer, SP(1) has to be split into two subproblems, say SP(1,1) and SP(1,2) through the addition of mutually exclusive and exhaustive constraints.

Proceeding in this manner a tree is constructed. From every node that corresponds to a noninteger solution a branching occurs to two other nodes, and so on. The bounding process is straightforward. If a solution at a node is noninteger, then this value provides a lower bound for all the solutions in its offspring. The branch-and-bound procedure stops when all nodes of the tree either have an integer solution or a noninteger solution that is higher than an integer solution at another node. The node with the best integer solution provides an optimal solution for the original integer program.

An enormous amount of research and experimentation has been done on branch-and-bound techniques. For example, the branching technique as well as the bounding technique described above are relatively simple. Several more sophisticated ways of applying branch-and-bound have proven to be very useful in practice, namely:

(i) Lagrangean relaxation,

B.3 Optimization Methods for Integer Programs

(ii) branch-and-cut, and

(iii) branch-and-price (often also referred to as column generation).

Lagrangean relaxation is a sophisticated technique for establishing lower bounds in a branch-and-bound procedure. It generates lower bounds that are substantially better (higher) than the LP relaxation bounds described above and better bounds typically cut down the overall computation time substantially. Lagrangean relaxation, instead of dropping the integrality constraints, relaxes some of the main constraints. However, the relaxed constraints are not totally dropped. Instead, they are dualized or weighted in the objective function with suitable Lagrangean multipliers to discourage violations.

Branch-and-cut is a class of methods that are based on a combination of branch-and-bound with cutting plane techniques. Branch-and-cut uses in each subproblem of the branching tree a cutting plane algorithm to generate a lower bound. That is, a cutting plane algorithm is applied to the problem formulation that includes the additional constraints introduced at that node.

A branch-and-cut method solves a sequence of linear programming relaxations of the integer programming problem to be solved; the cutting plane method improves the relaxation of the problem to more closely approximate the integer programming problem, and the branch-and-bound proceeds by the usual divide and conquer approach to solve the problem. A pure branch-and-bound approach can be sped up considerably by the employment of a cutting plane scheme, either just at the top of the tree, or at every node of the tree. Lately, these techniques have found applications in crew scheduling and truck dispatching problems.

Branch-and-price is often used to solve integer programs that have a huge number of variables (columns). A branch-and-price algorithm always works with a restricted problem in a sense that only a subset of the variables is taken into account; the variables outside the subset are fixed at 0 and the corresponding columns are disregarded. From the theory of Linear Programming it is known that after solving this restricted problem to optimality, each variable that is included has a negative *potential savings* or, equivalently, a positive so-called *reduced cost*. If each variable that is not included in the restricted problem also has a negative potential savings, then an optimal solution for the original problem is found. However, if there are variables with a positive potential savings, then one or more of these variables should be included in the restricted problem. The main idea behind column generation is that the occurrence of variables with positive potential savings is not verified by enumerating all variables, but rather by solving an optimization problem. This optimization problem is called the pricing problem and is defined as the problem of finding the variable with maximum potential savings (or minimum reduced cost). To apply column generation effectively it is important to find a good method for solving the pricing problem. A branch-and-bound algorithm in which the lower bounds are computed by solving LP relaxations through column generation is called a branch-cut-and-price algorithm. Branch-and-

B.4 Examples of Branch-and-Bound Applications

In this section we illustrate the use of branch-and-bound with two examples. First, consider a single machine and n jobs. Job j has release date r_j and due date d_j. The objective is to minimize the maximum lateness without preemptions. This problem is NP-hard. It is important because it appears frequently as a subproblem in heuristic procedures for flow shop and job shop scheduling (see, for example, the shifting bottleneck procedure described in Chapter 5). It has received therefore considerable attention, that has resulted in a number of reasonably efficient branch-and-bound procedures.

A branch-and-bound procedure for this problem can be designed as follows. The branching process may be based on the fact that schedules are developed starting from the beginning of the schedule. There is a single node at level 0 which is at the top of the tree. At this node none of the jobs have been put into any position in the sequence. There are n branches going down to n nodes at level 1. Each node at this level corresponds to a partial solution with a specific job in the first position of the schedule. So, at each of these nodes there are still $n-1$ jobs whose positions in the schedule are yet to be determined. There are $n-1$ arcs emanating from each node at level 1 to level 2. There are therefore $n \times (n-1)$ nodes at level 2. At each node at level 2, the two jobs in the first two positions are specified; at level k, the jobs in the first k positions are fixed.

Actually, it is often not necessary to consider every remaining job as a candidate for the next position. If at a node at level $k-1$ jobs $j_1, \ldots, j_{k-1}$ are assigned to the first $k-1$ positions, job c has to be considered as a candidate for position k only if

$$r_c < \min_{l \in J} \Big(\max(t, r_l) + p_l \Big),$$

where J denotes the set of jobs not yet scheduled and t denotes the time the machine completes job j_{k-1} and is free to start the next job. If job c does not satisfy this inequality, i.e., if

$$r_c \geq \min_{l \in J} \Big(\max(t, r_l) + p_l \Big),$$

then it makes sense to put the job that minimizes the R.H.S. in position k and job c in position $k+1$. This would not affect the completion time of job c in any way. So, in this case, job c does not have to be considered for position k.

There are several bounding schemes for generating lower bounds for nodes in the search tree. An easy lower bound for a node at level $k-1$ can be

B.4 Examples of Branch-and-Bound Applications

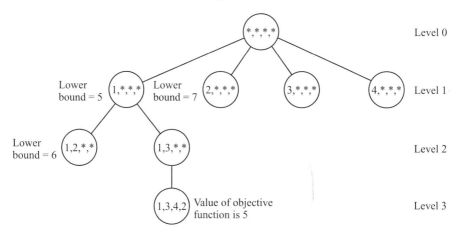

Fig. B.1. Branch and Bound Tree

established by scheduling the remaining jobs J according to the *preemptive EDD* rule. The preemptive EDD rule is known to be optimal for this particular problem when preemptions are allowed and thus provides a lower bound for the nonpreemptive problem. If a preemptive EDD rule yields a nonpreemptive schedule, then all nodes with a higher lower bound can be disregarded.

Example B.4.1 (Single Machine Scheduling). Consider a single machine with the following 4 jobs.

jobs	1	2	3	4
p_j	4	2	6	5
r_j	0	1	3	5
d_j	8	12	11	10

The objective is to minimize the maximum lateness $L_{\max}$. At level 1 of the search tree there are four nodes: $(1, *, *, *)$, $(2, *, *, *)$, $(3, *, *, *)$ and $(4, *, *, *)$. It is easy to see that nodes $(3, *, *, *)$ and $(4, *, *, *)$ may be disregarded immediately. Job 3 is released at time 3; if job 2 would start its processing at time 1, job 3 still can start at time 3. Job 4 is released at time 5; if job 1 would start its processing at time 0, job 4 still can start at time 5 (see Figure B.1).

Computing a lower bound for node $(1, *, *, *)$ according to the preemptive EDD rule yields a schedule where job 3 is processed during the time interval [4,5], job 4 during the time interval [5,10], job 3 (again) during interval [10,15] and job 2 during interval [15,17]. The $L_{\max}$ of this schedule, which provides the lower bound for node $(1, *, *, *)$, is 5. In a similar way a lower bound can be obtained for node $(2, *, *, *)$. The value of this lower bound is 7.

Consider node $(1, 2, *, *)$ at level 2. The lower bound for this node is 6 and is determined by the (nonpreemptive) schedule $1, 2, 4, 3$. Proceed with node

$(1, 3, *, *)$ at level 2. The lower bound is 5 and determined by the (nonpreemptive) schedule $1, 3, 4, 2$. As the lower bound for node $(1, *, *, *)$ is 5 and the lower bound for node $(2, *, *, *)$ is larger than 5 it follows that schedule $1, 3, 4, 2$ has to be optimal.

The more general problem with jobs subject to precedence constraints can be handled in a similar way. Since the problem without precedence constraints is a special case of the problem with precedence constraints, the latter is, from a complexity point of view, at least as hard. However, from an enumeration point of view, the problem with precedence constraints is actually somewhat easier than the one without precedence constraints; schedules that violate precedence constraints can be ruled out immediately. This reduces the number of schedules that have to be considered.

This nonpreemptive single machine scheduling problem with release dates, due dates, precedence constraints, and the maximum lateness objective, is a very important scheduling problem. This is because job shop scheduling problems are often decomposed in such a way that in the solution process many single machine problems of this type have to be solved (see the Shifting Bottleneck heuristic in Chapter 5).

The second example of a branch-and-bound application concerns the job shop described in Chapter 5. The branching as well as the bounding procedures that are applicable to the job shop problem are typically of a special design. In order to describe one of the branching procedures we limit ourselves to a specific class of schedules.

Definition 1 (Active Schedule). *A feasible schedule is called active if it cannot be altered in any way that some operation is completed earlier and no other operation is completed later.*

A schedule being active implies that when a job arrives at a machine, this job is processed as early as possible according to the prescribed sequence. In order to get a better understanding of active schedules, consider the non-active schedule depicted in Figure B.2. In this non-active schedule there is an idle period in between the processing of two operations and this idle period is long enough to accomodate the operation of a job that is waiting for processing at that time. This schedule is clearly non-active. This implies that an active schedule cannot have any idle period in which the operation of a waiting job could fit.

From the definition it follows that an active schedule has the property that it is impossible to reduce the makespan without increasing the starting time of some operation. Of course, there are many different active schedules. It can be shown that there exists an active schedule that is optimal among all possible schedules.

A branching scheme that is often used is based on the generation of all active schedules. All such schedules can be generated by a simple algorithm. In this algorithm Ω denotes the set of all operations all of whose predecessors

B.4 Examples of Branch-and-Bound Applications   405

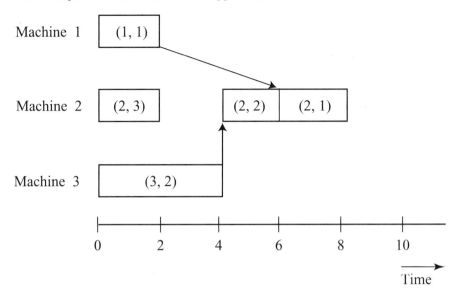

Fig. B.2. A Nonactive Schedule

have already been scheduled (i.e., the set of all schedulable operations) and r_{ij} the earliest possible starting time of operation (i, j) in Ω. Set Ω' is a subset of set Ω.

Algorithm B.4.2 (Generation of all Active Schedules).

Step 1. *(Initial Conditions)*

  Let Ω contain the first operation of each job;

  Let $r_{ij} = 0$, for all $(i, j) \in \Omega$.

Step 2. *(Machine Selection)*

  Compute for the current partial schedule

  $$t(\Omega) = \min_{(i,j) \in \Omega} \{r_{ij} + p_{ij}\}$$

  and let i^* denote the machine on which the minimum is achieved.

Step 3. *(Branching)*

  Let Ω' denote the set of all operations (i^*, j) on machine i^* such that

  $$r_{i^*j} < t(\Omega)$$

  For each operation in Ω' consider an (extended) partial schedule with that operation as the next one on machine i^*.

  For each such (extended) partial schedule delete the operation from Ω, include its immediate follower in Ω and return to Step 2.

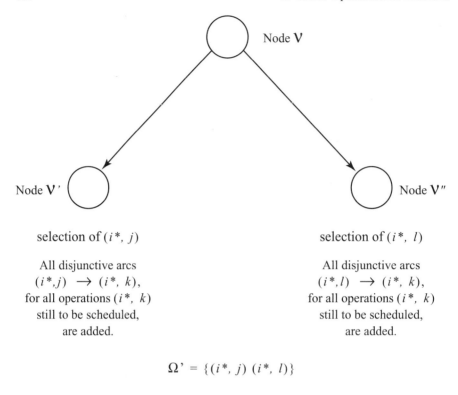

Fig. B.3. Branching Tree for Branch and Bound Approach

Algorithm B.4.3 is the basis for the branching process. Step 3 performs the branching from the node that is characterized by the given partial schedule; the number of branches is equal to the number of operations in Ω'. With this algorithm one can generate the entire tree and the nodes at the very bottom of the tree correspond to all the active schedules.

So a node $\mathcal{V}$ in the tree corresponds to a partial schedule and the partial schedule is characterized by a selection of disjunctive arcs that corresponds to the order in which all the predecessors of a given set Ω have been scheduled. A branch out of node $\mathcal{V}$ corresponds to the selection of an operation $(i^*, j) \in \Omega'$ as the next to go on machine i^*. The disjunctive arcs $(i^*, j) \rightarrow (i^*, k)$ then have to be added to machine i^* for all operations (i^*, k) still to be scheduled on machine i^*. This implies that the newly created node at the lower level, say node $\mathcal{V}'$, which corresponds to a partial schedule with only one additional operation scheduled, contains a number of additional selected disjunctive arcs (see Figure B.3). Let D' denote the set of disjunctive arcs selected at the newly created node. Refer to the graph with all the conjunctive arcs and set D' as graph $G(D')$. The number of branches sprouting from node $\mathcal{V}$ is equal to the number of operations in Ω'.

B.4 Examples of Branch-and-Bound Applications

To find a lower bound for the makespan at node $\mathcal{V}'$, consider graph $G(D')$. The length of the critical path in this graph already results in a lower bound for the makespan at node $\mathcal{V}'$. Call this lower bound $LB(\mathcal{V}')$. Better (higher) lower bounds can be obtained for this node as follows.

Consider machine i and assume that all *other* machines are allowed to process, at any point in time, multiple operations simultaneously (since not all disjunctive arcs are selected yet in $G(D')$, it may be the case that, at some points in time, multiple operations require processing on a machine simultaneously). However, we force machine i to process its operations one after another. First, compute the earliest possible starting times r_{ij} of all the operations (i,j) on machine i; that is, determine in graph $G(D')$ the length of the longest path from the source to node (i,j). Second, for each operation (i,j) on machine i, compute the minimum amount of time needed between the completion of operation (i,j) and the lower bound $LB(\mathcal{V}')$, by determining the longest path from node (i,j) to the sink in $G(D')$. This amount of time, together with the lower bound on the makespan, translates then into a due date d_{ij} for operation (i,j), i.e., d_{ij} is equal to $LB(\mathcal{V}')$ minus the length of the longest path from node (i,j) to the sink, plus p_{ij}. Consider now the problem of sequencing the operations on machine i as a single machine problem with jobs released at different points in time, no preemptions allowed, and the maximum lateness as the objective to minimize (this is a problem considered in Section 5.2). Even though this problem is NP hard, there are relatively fast algorithms available that result in good solutions. The optimal sequence obtained for this problem implies a selection of disjunctive arcs that can be added (temporarily) to D'. This then may lead to a longer overall critical path in the graph, a longer makespan and a better (higher) lower bound for node $\mathcal{V}'$. At node $\mathcal{V}'$ this can be done for each of the m machines separately. The longest makespan obtained this way can be used as a lower bound at node $\mathcal{V}'$. Of course, the temporary disjunctive arcs inserted to obtain the lower bound are deleted as soon as the best lower bound is determined.

Even though it appears somewhat of a burden to solve m NP-hard scheduling problems to obtain one lower bound for another NP-hard problem, it turns out that this type of bounding procedure performs quite well in practice.

Example B.4.3 (Application of Branch-and-Bound). Consider the instance described in Example 5.2.1. The initial graph contains only conjunctive arcs and is depicted in Figure B.4.a. The makespan corresponding to this graph is 22. Applying the branch-and-bound procedure to this instance results in the following branch-and-bound tree.

Level 1: Applying Algorithm B.4.3 yields

$$\Omega = \{(1,1),\ (2,2),\ (1,3)\},$$
$$t(\Omega) = \min\,(0+10, 0+8, 0+4) = 4,$$
$$i^* = 1,$$
$$\Omega' = \{(1,1),(1,3)\}.$$

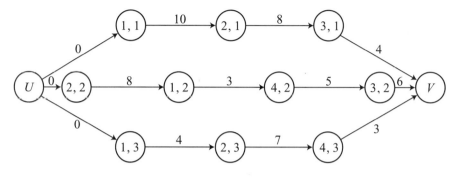

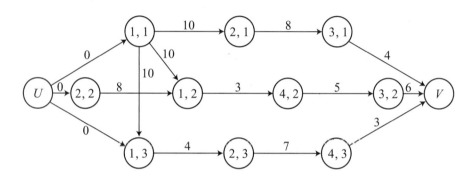

Fig. B.4. Precedence Graphs at Level 1 in Example B.4.4

So there are two nodes of interest at level 1, one corresponding to operation $(1,1)$ being processed first on machine 1 and the other to operation $(1,3)$ being processed first on machine 1.

If operation $(1,1)$ is scheduled first, then the two disjunctive arcs depicted in Figure B.4.b are added to the graph. The node is characterized by the two disjunctive arcs

$$(1,1) \rightarrow (1,2),$$
$$(1,1) \rightarrow (1,3).$$

The addition of these two disjunctive arcs immediately increases the lower bound on the makespan to 24. In order to improve this lower bound one can generate for machine 1 an instance of the maximum lateness problem with release dates and due dates. The release date of job j in this single machine problem is determined by the longest path from the source to node $(1,j)$ in Figure B.4.b. The due date of job j is computed by finding the longest path from node $(1,j)$ to the sink, subtracting p_{1j} from the length of this longest path, and subtracting the resulting value from 24. These computations lead to the following single machine problem for machine 1.

B.4 Examples of Branch-and-Bound Applications

jobs	1	2	3
p_{1j}	10	3	4
r_{1j}	0	10	10
d_{1j}	10	13	14

The sequence that minimizes $L_{\max}$ in this single machine problem is $1, 2, 3$ with $L_{\max} = 3$. This implies that a lower bound for the makespan at the corresponding node is $24 + 3 = 27$. An instance of the minimization of the maximum lateness problem corresponding to machine 2 can be generated in the same way. The release dates and due dates also follow from Figure B.4 (assuming a makespan of 24), and are as follows.

jobs	1	2	3
p_{2j}	8	8	7
r_{2j}	10	0	14
d_{2j}	20	10	21

The optimal sequence is $2, 1, 3$ with $L_{\max} = 4$. This yields a better lower bound for the makespan at the node corresponding to operation $(1, 1)$ scheduled first, i.e., $24 + 4 = 28$. Analyzing machines 3 and 4 in the same way does not yield a better lower bound.

The second node at Level 1 corresponds to operation $(1, 3)$ being scheduled first. If $(1, 3)$ is scheduled to go first, two different disjunctive arcs are added to the original graph, yielding a lower bound of 26. The associated instance of the maximum lateness problem for machine 1 has an optimal sequence $3, 1, 2$ with $L_{\max} = 2$. This implies that the lower bound for the makespan at this node, corresponding to operation $(1, 3)$ scheduled first, is also equal to 28. Analyzing machines 2, 3 and 4 does not result in a better lower bound.

The next step is to branch from node (1,1) at Level 1 and generate the nodes at the next level.

Level 2: Applying Algorithm B.4.3 now yields

$$\Omega = \{(2,2), (2,1), (1,3)\},$$
$$t(\Omega) = \min(0+8, 10+8, 10+4) = 8,$$
$$i^* = 2,$$
$$\Omega' = \{(2,2)\}.$$

There is one node of interest at this part of Level 2, the node corresponding to operation $(2, 2)$ being processed first on machine 2 (see Figure B.5). Two disjunctive arcs are added to the graph, namely $(2, 2) \to (2, 1)$ and $(2, 2) \to (2, 3)$. So this node is characterized by a total of four disjunctive arcs:

$$(1,1) \to (1,2),$$

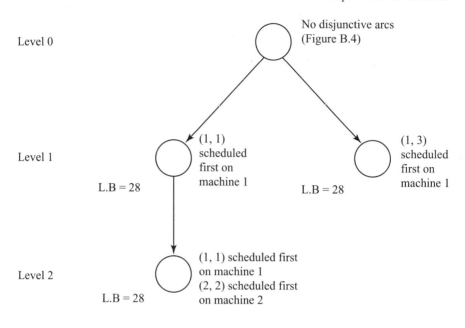

Fig. B.5. Branching in Example B.4.4

$$(1,1) \to (1,3),$$
$$(2,2) \to (2,1),$$
$$(2,2) \to (2,3).$$

This leads to an instance of the maximum lateness problem for machine 1 with the following release dates and due dates (assuming a makespan of 28).

jobs	1	2	3
p_{1j}	10	3	4
r_{1j}	0	10	10
d_{1j}	14	17	18

The optimal job sequence is $1, 3, 2$ and $L_{\max} = 0$. This implies that the lower bound for the makespan at the corresponding node is $28 + 0 = 28$. Analyzing machines 2, 3 and 4 in the same way does not increase the lower bound.

Continuing the branch-and-bound procedure results in the following job sequences for the four machines.

machine	job sequence
1	$1, 3, 2$ (or $1, 2, 3$)
2	$2, 1, 3$
3	$1, 2$
4	$2, 3$

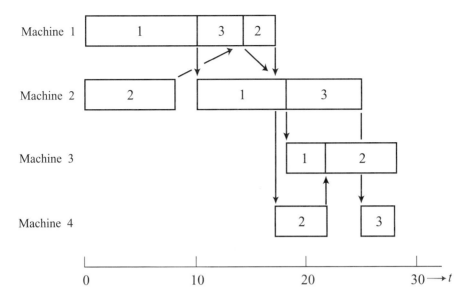

Fig. B.6. Optimal Schedule in Example B.4.4

The makespan under this optimal schedule is 28 (see Figure B.6).

The approach described above is based on complete enumeration and is guaranteed to lead to an optimal schedule. However, with a large number of machines and a large number of jobs the computation time is prohibitive. Already with 20 machines and 20 jobs it will be very hard to find the optimal schedule.

Exercises

B.1. Solve the instance considered in Example B.2.2 using the Backward Dynamic Programming formulation presented in Example B.2.3.

B.2. Consider the instance of the job shop problem described in Exercise 5.5. Solve the problem using the branch-and-bound procedure described in Section B.4.

Comments and References

There are many excellent textbooks covering dynamic programming; see, for example, Denardo (1982) and Bertsekas (1987).

Many researchers have developed branch-and-bound approaches for scheduling problems; see, for example, Fisher (1976), Fisher (1981), Sarin, Ahn, and Bishop

(1988), and Luh and Hoitomt (1993). Branch-and-bound techniques have often been applied to minimize the makespan in job shops; see, for example, Lomnicki (1965), Brown and Lomnicki (1966), McMahon and Florian (1975), Barker and McMahon (1985), Carlier and Pinson (1989), Applegate and Cook (1991), and Hoitomt, Luh and Pattipati (1993). Some of the branching schemes of these branch-and-bound approaches are based on the generation of active schedules. The concept of an active schedule was first introduced by Giffler and Thompson (1960). For an overview of branch-and-bound techniques applied to job shop scheduling, see Pinson (1995).

Hoffman and Padberg (1993) applied branch-and-cut to airline crew scheduling and Bixby and Lee (1998) applied branch-and-cut to truck scheduling.

Appendix C

Heuristic Methods

C.1 Introduction 413
C.2 Basic Dispatching Rules...................... 414
C.3 Composite Dispatching Rules................. 417
C.4 Beam Search 421
C.5 Local Search: Simulated Annealing and
  Tabu-Search 424
C.6 Local Search: Genetic Algorithms............. 431
C.7 Discussion 433

C.1 Introduction

Some planning and scheduling problems can be formulated as linear programs and are therefore inherently easy; they can be solved readily through the use of existing efficient algorithms. Other easy planning and scheduling problems can be solved through different algorithms that are also efficient; these algorithms are referred to as polynomial time algorithms. That a problem can be solved by an efficient, polynomial time, algorithm implies that very large instances of that problem, with hundreds or even thousands of jobs, still can be solved in a relatively short time on a computer.

However, there are many more scheduling problems that are intrinsically very hard, i.e., NP-hard. They cannot be formulated as linear programs and there are no simple rules or algorithms that yield optimal solutions in a limited amount of computer time. It may be possible to formulate these problems as integer or disjunctive programs, but solving these to optimality may require an enormous amount of computer time. Since in practice that amount of computer time is not always available, one is usually satisfied with an "acceptable" feasible solution that presumably is not far from optimal.

This appendix describes a number of general purpose techniques that have proven useful in industrial scheduling systems. Most of these methods typically do not guarantee an *optimal* solution; their goal is to find a reasonably good solution in a relatively short time. Although many such techniques exist, only a few representative ones are discussed here in detail.

The next section presents a number of simple and basic dispatching rules. The third section describes a method of combining the basic dispatching rules into so-called hybrid or composite rules. The fourth section describes a derivative of branch-and-bound named beam search. This method aims at eliminating search alternatives in an intelligent way in order not to have to examine the entire solution space. On the one hand it requires less computer time than branch-and-bound, but on the other hand it does not guarantee an optimal solution. The following two sections present procedures based on local search techniques, namely simulated annealing, tabu-search and genetic algorithms. These are fairly generic techniques that can be applied to different planning and scheduling problems with only minor customization. The last section discusses how several empirical methods can be combined into a single framework.

There are many other empirical procedures. The selection described in this chapter is not meant to be exhaustive, but rather aims at providing a flavor of the thinking behind such heuristics.

C.2 Basic Dispatching Rules

A dispatching rule is a rule that prioritizes all the jobs that are waiting for processing on a machine. The prioritization scheme may take into account the jobs' attributes, the machines' attributes as well as the current time. Whenever a machine has been freed a dispatching rule selects, among the jobs waiting, the job with the highest priority. Research in dispatching rules has been active for several decades and many different rules have been developed and studied in the literature.

Dispatching rules can be classified in various ways. For example, a distinction can be made between *static* and *dynamic* rules. Static rules are not time dependent. They are just a function of the job and/or machine data. Dynamic rules, on the other hand, are time dependent. An example of a dynamic rule is the *Minimum Slack (MS) first* which orders jobs according to their remaining slack. This remaining slack is defined as $\max(d_j - p_j - t, 0)$, where d_j is the due date, p_j the processing time, and t the current time. This implies that at some point in time job j may have a higher priority than job k, while at a later point in time they may have the same priority.

A second way of classifying dispatching rules is according to the information they are based upon. A *local* rule uses only information pertaining to either the queue where the job is waiting or the machine (or workcenter) where the job is queued. Most of the rules described in this section can be

C.2 Basic Dispatching Rules

used as local rules. A *global* rule may use information pertaining to other machines, such as the processing time of the job on the next machine on its route or the current queue length at that machine. Of course, there are many basic dispatching rules. In what follows we give only a limited sample.

The *Service in Random Order (SIRO)* rule. According to this priority rule, whenever a machine is freed, the next job is selected at random from those waiting for processing. No attempt is made to optimize any objective.

The *Earliest Release Date first (ERD)* rule. This rule is equivalent to the well-known First-Come-First-Served rule. This rule in a sense minimizes the variation in the waiting times of the jobs at a machine.

The *Earliest Due Date first (EDD)* rule. Whenever a machine is freed the job with the earliest due date is selected to be processed next. This rule tends to minimize the maximum lateness among the jobs waiting for processing. Actually, in a single machine setting, with n jobs available at time zero, the EDD rule does minimize the maximum lateness.

The *Minimum Slack first (MS)* rule is a variation of the EDD rule. If a machine is freed at time t the remaining slack of each job at that time, defined as $\max(d_j - p_j - t, 0)$, is computed. The job with the minimum slack is scheduled next. This rule tends to minimize due date related objectives.

The *Weighted Shortest Processing Time first (WSPT)* rule. Whenever a machine is freed the job with the highest ratio of weight (w_j) over processing time (p_j) is scheduled next, i.e., the jobs are ordered in decreasing order of w_j/p_j. This rule tends to minimize the weighted sum of the completion times, i.e., $\sum w_j C_j$. In a single machine setting, with n jobs available at time zero, the WSPT rule does minimize the $\sum w_j C_j$. When all the weights are equal the WSPT rule reduces to the *Shortest Processing Time first (SPT)* rule.

The *Longest Processing Time first (LPT)* rule. This rule orders the jobs in decreasing order of processing times. When there are machines in parallel, this rule tends to balance the workload over the machines. The reasoning behind this is the following. It is advantageous to keep jobs with short processing times for later, because these jobs are useful at the end for balancing the workload. After the assignment of jobs to machines has been determined, the jobs on any given machine can be resequenced without affecting the workload balance.

The *Shortest Setup Time first (SST)* rule. Whenever a machine is freed, this rule selects for processing the job with the shortest setup time.

The *Least Flexible Job first (LFJ)* rule. This rule is used when there are a number of non-identical machines in parallel and the jobs are subject to machine eligibility constraints. Job j can only be processed on a specific subset of the m machines, say M_j. Whenever a machine is freed the job that can be processed on the smallest number of other machines is selected, i.e., the job with the fewest processing alternatives.

The *Critical Path (CP)* rule. The Critical Path rule is used with operations subject to precedence constraints. It selects as the next job the one at the head of the longest string of processing times in the precedence constraints graph.

The *Largest Number of Successors (LNS)* rule. This rule may also be used when the jobs are subject to precedence constraints. It selects as the next job the one that has the largest number of jobs following it.

The *Shortest Queue at the Next Operation (SQNO)* rule is used in job shops. Whenever a machine is freed the job with the shortest queue at the next machine on its route is selected for processing. The length of the queue at the next machine can be measured in a number of ways. It may be simply the number of jobs waiting in queue or it may be the total amount of work waiting in queue.

Table C.1 gives a summary of these dispatching rules. A number of these rules result in optimal schedules in certain machine environments and are reasonable heuristics in others. All of these rules have variations that can be applied in more complicated settings.

The basic dispatching rules described above are of limited use. When a complex objective has to be minimized none of the basic dispatching rules may perform effectively. In the next section we describe a framework for combining basic dispatching rules that can make the resulting rules significantly more effective.

Table C.1. Summary of Dispatching Rules

	RULE	DATA	OBJECTIVES
Rules Dependent on Release Dates and Due Dates	ERD EDD MS	r_j d_j d_j	Variance in Throughput Times Maximum Lateness Maximum Lateness
Rules Dependent on Processing Times	LPT SPT WSPT CP LNS	p_j p_j p_j, w_j $p_j, prec$ $p_j, prec$	Load Balancing over Parallel Machines Sum of Completion Times, WIP Weighted Sum of Completion Times, WIP Makespan Makespan
Miscellaneous	SIRO SST LFJ SQNO	— s_{jk} M_j —	Ease of Implementation Makespan and Throughput Makespan and Throughput Machine Idleness

C.3 Composite Dispatching Rules

Basic dispatching rules are useful when one attempts to find a reasonably good schedule with regard to a single objective such as the makespan, the sum of the completion times or the maximum lateness.

However, in practice objectives are often very complicated. A realistic objective may be a combination of several objectives and it may also be a function of time or a function of the set of jobs waiting for processing. Sorting the jobs on the basis of one or two parameters may not lead to acceptable schedules. More elaborate dispatching rules that consider a number of parameters can address more complicated objective functions. Some of these more elaborate rules are merely a combination of a few of the basic dispatching rules listed above, and are referred to as *composite* dispatching rules.

To explain the structure of these composite dispatching rules, we first introduce a general framework. We then describe two of the more widely used composite rules.

A composite dispatching rule is a ranking expression that combines a number of basic dispatching rules. A basic rule is, as described in the previous section, a function of attributes of jobs and/or machines. An attribute is any property associated with either a job or a machine; it may be constant or time dependent. Examples of job attributes are weight, processing time and due date. Examples of machine attributes are speed, the number of jobs waiting for processing and the total amount of processing that is waiting in queue. The extent to which a given attribute affects the overall priority of a job is determined by the basic rule that uses it and a scaling parameter. Each basic rule in the composite dispatching rule has its own scaling parameter that is chosen to properly scale the contribution of the basic rule to the total ranking expression. The scaling parameters may be fixed by the designer of the rule, or variable and a function of the particular job set to be scheduled. If they depend on the particular job set to be scheduled, they require the computation of some job set statistics that characterize the particular scheduling instance as accurately as possible (for example, whether or not the due dates of the jobs are tight). These statistics, which are also called *factors*, usually do not depend on the schedule and can be computed easily from the given job and machine attributes.

The functions that map the statistics into the scaling parameters have to be determined by the designer of the rule. Experience may offer a reasonable guide, but extensive computer simulation is often required. These functions are usually determined only once, before the rule is released for regular use.

Each time the composite dispatching rule is used for generating a schedule, the necessary statistics are computed. Based on the values of these statistics, the values of the scaling parameters are set by the predetermined functions. After the scaling parameters have been fixed the dispatching rule is applied to the job set.

One example of a composite dispatching rule is a rule for a single machine and n jobs (all available at time zero) with the sum of the weighted tardinesses ($\sum w_j T_j$) as objective. This problem is inherently very hard and no efficient algorithm is known for this problem. As branch-and-bound methods are prohibitively time consuming even with only 30 jobs, it is important to have a heuristic that provides a reasonably good schedule with a reasonable computational effort. Some heuristics come immediately to mind; namely, WSPT (which is optimal when all release dates and due dates are zero) and EDD or MS (which are optimal when all due dates are sufficiently loose and spread out). It is natural to seek a heuristic or priority rule that combines the characteristics of these dispatching rules. The *Apparent Tardiness Cost (ATC)* heuristic is a composite dispatching rule that combines WSPT and MS. (Recall that the MS rule selects at time t the job with the minimum slack $\max(d_j - p_j - t, 0)$.) Under the ATC rule, jobs are scheduled one at a time; that is, every time the machine is freed a ranking index is computed for each remaining job. The job with the highest ranking index is then selected to be processed next. This index is a function of the time t when the machine becomes free, and the p_j, w_j and d_j of the remaining jobs. It is defined as

$$I_j(t) = \frac{w_j}{p_j} \exp\left(-\frac{\max(d_j - p_j - t, 0)}{K\bar{p}}\right),$$

where K is the scaling parameter, which can be determined empirically, and $\bar{p}$ is the average of the processing times of the remaining jobs. If K is very large the ATC rule reduces to WSPT. If K is very small and there are no overdue jobs, the rule reduces to MS. If K is very small and there are overdue jobs, the rule reduces to WSPT applied to the overdue jobs.

In order to obtain good schedules, the value of K (also called the look-ahead parameter) must be appropriate for the particular instance of the problem. This can be done by first performing a statistical analysis of the particular scheduling instance under consideration. There are several statistics that can be used to help characterize scheduling instances. The *due date tightness* factor θ_1 is defined as

$$\theta_1 = 1 - \frac{\bar{d}}{C_{\max}},$$

where $\bar{d}$ is the average of the due dates. Values of θ_1 close to 1 indicate that the due dates are tight and values close to 0 indicate that the due dates are loose. The due date range factor θ_2 is defined as

$$\theta_2 = \frac{d_{\max} - d_{\min}}{C_{\max}}.$$

A high value of θ_2 indicates a wide range of due dates, while a low value indicates a narrow range of due dates. A significant amount of experimental research has been done to establish the relationships between the scaling parameter K and the factors θ_1 and θ_2.

C.3 Composite Dispatching Rules

Thus, when one wishes to minimize $\sum w_j T_j$ on a single machine or in a more complicated machine environment (parallel machines, flexible flow shops), one first characterizes the particular problem instance through these two statistics. One then determines the value of K as a function of these characterizing factors as well as the particular machine environment. After fixing K, one applies the rule.

Several generalizations of the ATC rule have been developed in order to take release dates and sequence dependent setup times into account. Such a generalization, the *Apparent Tardiness Cost with Setups (ATCS)* rule, is designed for the following problem: There are n jobs and a single machine. The jobs are subject to sequence dependent setups s_{jk}. The objective is once again to minimize the sum of the weighted tardinesses. The priority of any job j depends on the job just completed when the machine is freed. ATCS combines WSPT, MS and SST in a single ranking index. The rule calculates the index of job j upon the completion of job l at time t as

$$I_j(t,l) = \frac{w_j}{p_j} \exp\left(-\frac{\max(d_j - p_j - t, 0)}{K_1 \bar{p}}\right) \exp\left(-\frac{s_{lj}}{K_2 \bar{s}}\right),$$

where $\bar{s}$ is the average of the setup times of the jobs remaining to be scheduled, K_1 the due date related scaling parameter and K_2 the setup time related scaling parameter. Note that the scaling parameters are dimensionless quantities, making them independent of the units used to express the various quantities.

The two scaling parameters, K_1 and K_2, can be regarded as functions of three factors:

(i) the due date tightness factor θ_1,
(ii) the due date range factor θ_2,
(iii) the setup time severity factor $\theta_3 = \bar{s}/\bar{p}$.

These statistics are not as easy to determine as in the previous case. Even with a single machine the makespan is now schedule dependent because of the setup times. Before computing the θ_1 and θ_2 factors the makespan has to be estimated. A simple estimate for the makespan on a single machine can be

$$\hat{C}_{\max} = \sum_{j=1}^{n} p_j + n\bar{s}.$$

This quantity is likely to overestimate the makespan as the final schedule will take advantage of setup times that are lower than average. The definitions of θ_1 and θ_2 have to be modified by replacing the makespan with its estimate.

An experimental study of the ATCS rule, although inconclusive, has suggested some guidelines for the selection of the two parameters K_1 and K_2. The following rules can be used for selecting values for K_1 and K_2:

$$K_1 = 4.5 + \theta_2, \qquad \text{for } \theta_2 \leq 0.5$$
$$K_1 = 6 - 2\theta_2, \qquad \text{for } \theta_2 \geq 0.5.$$
$$K_2 = \theta_1/(2\sqrt{\theta_3}).$$

Example C.3.1 (The ATCS Rule). Consider the following instance of the single machine, total weighted tardiness problem with four jobs subject to sequence dependent setups.

jobs	1	2	3	4
p_j	13	9	13	10
d_j	12	37	21	22
w_j	2	4	2	5

The setup times s_{0j} of the first job in the sequence are presented in the table below.

jobs	1	2	3	4
s_{0j}	1	1	3	4

The sequence dependent setup times of the jobs following the first job are:

jobs	1	2	3	4
s_{1j}	–	4	1	3
s_{2j}	0	–	1	0
s_{3j}	1	2	–	3
s_{4j}	4	3	1	–

The average processing time $\bar{p}$ is approximately 11 while the average setup time $\bar{s}$ is approximately 2. An estimate for the makespan is

$$\hat{C}_{max} = \sum_{j=1}^{n} p_j + n\bar{s} = 45 + 4 \times 2 = 53.$$

The due date range factor $\theta_2 = 25/53 \approx 0.47$, the due date tightness factor $\theta_1 = 1 - 23/53 \approx 0.57$ and the setup time severity factor is $\theta_3 = 2/11 \approx 0.18$. According to the suggested rules, the value of K_1 should be 5 and that of K_2 should be 0.7. To determine which job goes first, we have to compute $I_j(0,0)$, for $j = 1, \ldots, 4$.

$$I_1(0,0) = \frac{2}{13} \exp\left(-\frac{(12-13)^+}{55}\right) \exp\left(-\frac{1}{1.4}\right) \approx 0.15 \times 1 \times 0.51 = 0.075,$$

$$I_2(0,0) = \frac{4}{9} \exp\left(-\frac{(37-9)^+}{55}\right) \exp\left(-\frac{1}{1.4}\right) \approx 0.44 \times 0.6 \times 0.47 = 0.131,$$

$$I_3(0,0) = \frac{2}{13} \exp\left(-\frac{(21-13)^+}{55}\right) \exp\left(-\frac{3}{1.4}\right) \approx 0.15 \times 0.86 \times 0.103 = 0.016,$$

$$I_4(0,0) = \frac{5}{10} \exp\left(-\frac{(22-10)^+}{55}\right) \exp\left(-\frac{4}{1.4}\right) \approx 0.50 \times 0.80 \times 0.06 = 0.024.$$

Job 2 has the highest priority in spite of the fact that its due date is the latest. As its setup time is 1, its completion time is 10. So at the second iteration $I_1(10, 2)$, $I_3(10, 2)$ and $I_4(10, 2)$ have to be computed. To simplify the computations the values of $K_1\bar{p}$ and $K_2\bar{s}$ can be kept the same. Continuing the application of the ATCS rule results in the sequence $2, 4, 3, 1$ with the total weighted tardiness equal to 98. Complete enumeration shows that this sequence is optimal. Note that this sequence always selects, whenever the machine is freed, one of the jobs with the smallest setup time.

The ATCS rule can easily be applied to scheduling problems with parallel machines, i.e., m identical machines in parallel, n jobs subject to sequence dependent setups and the total weighted tardiness as objective. Of course, the look-ahead parameters K_1 and K_2 have to be determined as a function of θ_1, θ_2, θ_3, and m.

C.4 Beam Search

Enumerative branch-and-bound methods are currently the most widely used methods for obtaining optimal solutions to NP-hard scheduling problems. The disadvantage of branch-and-bound is that it can be extremely time consuming, since the number of nodes one has to consider is often very large. Beam search is a derivative of branch-and-bound, which tries to eliminate branches in an intelligent way so that not all branches have to be examined. It thus requires less computer time, but can no longer guarantee an optimal solution.

Consider again a single machine problem with n jobs. Assume that for each node at level k the jobs for the first k positions have been fixed. Recall, from the branch-and-bound discussion in Section B.4, that there is a single node at level 0 with n branches emanating to n nodes at level 1. Each node at level 1 branches out into $n-1$ nodes at level 2, resulting in a total of $n(n-1)$ nodes at level 2. At level k there are $n!/(n-k)!$ nodes and at the bottom level, level n, there are $n!$ nodes.

With the branch-and-bound method one attempts to eliminate a node by determining a lower bound on the objective values of all schedules that belong to the offspring of that node. If the lower bound is higher than the objective value of a known schedule, the node is eliminated and its offspring disregarded. If one could obtain a reasonably good schedule through some clever heuristic before starting the branch-and-bound procedure, it may be possible to eliminate many nodes. Dominance rules (see Section B.4) may also reduce the number of nodes to be investigated. However, even after these eliminations branch-and-bound usually still has too many nodes to evaluate. The main advantage of branch-and-bound is that, after evaluating all nodes, the final solution is known with certainty to be optimal.

With beam search only the most promising nodes at level k are selected as nodes to branch from. The remaining nodes at that level are discarded *permanently*. The number of nodes retained is called the *beam width* of the search. Clearly, the process that determines which nodes to retain is a crucial component of this method. Evaluating each node carefully, in order to obtain an accurate estimate for the potential of its offspring, may be time consuming. There is a trade-off here: a crude prediction is quick, but may result in discarding good solutions, while a more thorough evaluation may be prohibitively time consuming. At this point a two-stage approach is useful. For all the nodes generated at level k, we first do a crude evaluation. Based on the outcomes of the crude evaluations we select a number of nodes for a thorough evaluation, while discarding (filtering out) the remaining nodes. The number of nodes selected for a thorough evaluation is referred to as the *filter width*. After a thorough evaluation of all nodes that passed the filter, we choose a subset of these nodes. The number of nodes in this subset is equal to the beam width which therefore must be smaller than the filter width. From this subset the branches to the next level are generated.

A simple example of a crude prediction is the following. The contribution of the partial schedule to the objective and the due date tightness (or some other statistic) of the jobs remaining to be scheduled are computed. Based on these values the nodes at a given level can be compared with one another and an overall assessment is made.

Every time a node has to undergo a thorough evaluation, all the jobs that have not yet been scheduled are scheduled according to a composite dispatching rule. Such a schedule can still be generated reasonably fast as it only requires sorting. The result of such a schedule is an indication of the promise of the node. If a large number of jobs is involved, nodes may be filtered out by first examining a partial schedule obtained by scheduling only a subset of the remaining jobs with a dispatching rule. The value of this extended partial schedule may be used to decide whether or not to discard a node. A node that is retained may be analyzed more thoroughly by having all its remaining jobs scheduled with the composite dispatching rule. The value of this schedule's objective then represents an upper bound on the best schedule among the offspring of that node. The following example illustrates a simplified version of beam search.

Example C.4.1 (Beam search). Consider the following instance of the single machine total weighted tardiness problem with all jobs available at time zero and no sequence dependent setups (i.e., a special case of the problem described in Section C.3).

jobs	1	2	3	4
p_j	10	10	13	4
d_j	4	2	1	12
w_j	14	12	1	12

As the number of jobs is rather small only one type of prediction is made for the nodes at any particular level and no filtering mechanism is used. The

C.4 Beam Search

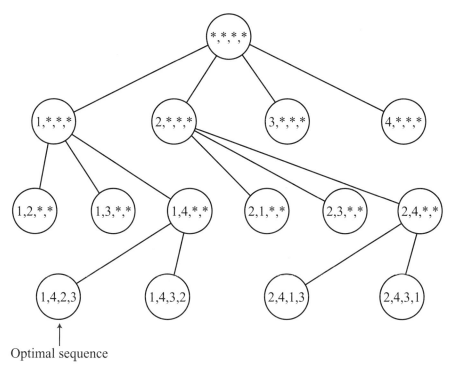

Fig. C.1. Beam Search Applied to Single-machine Weighted Tardiness Problem

beam width is chosen to be 2; so at each level only two nodes are retained. The prediction at a node is made by sequencing the unscheduled jobs using the ATC rule. As the due date range factor $\theta_2 = 11/37$ and the due date tightness factor $\theta_1 = 32/37$, the look-ahead parameter is chosen to be 5.

A search tree is constructed assuming the sequence is developed starting from $t = 0$. Again, at the kth level of the tree, jobs are put into the kth position. At level 1 of the tree there are four nodes: $(1, *, *, *)$, $(2, *, *, *)$, $(3, *, *, *)$ and $(4, *, *, *)$, see Figure C.1. Applying the ATC rule to the three remaining jobs at each one of the four nodes results in four sequences: (1,4,2,3), (2,4,1,3), (3,4,1,2) and (4,1,2,3) with objective values 408, 436, 771 and 440. As the beam width is 2, only the first two nodes are retained.

Each of these two nodes leads to three nodes at level 2. Node $(1, *, *, *)$ leads to nodes $(1, 2, *, *)$, $(1, 3, *, *)$ and $(1, 4, *, *)$ and node $(2, *, *, *)$ leads to nodes $(2, 1, *, *)$, $(2, 3, *, *)$ and $(2, 4, *, *)$. Applying the ATC rule to the remaining two jobs in each of the 6 nodes at level 2 results in nodes $(1, 4, *, *)$ and $(2, 4, *, *)$ being retained; the remaining four nodes are discarded.

The two nodes at level 2 branch out to four nodes at level 3 (the last level), namely nodes (1,4,2,3), (1,4,3,2), (2,4,1,3) and (2,4,3,1). Of these four, sequence (1,4,2,3) is the best with a total weighted tardiness of 408. Solving the problem with branch-and-bound shows that this sequence is actually optimal.

C.5 Local Search: Simulated Annealing and Tabu-Search

So far, all the algorithms and heuristics described in this chapter have been of the constructive type. They start without a schedule and gradually construct a schedule by adding one job at a time.

In this section and the next we describe algorithms of the improvement type. Algorithms of the improvement type are conceptually completely different from algorithms of the constructive type. They start out with a complete schedule, which may be selected arbitrarily, and then try to obtain a better schedule by manipulating the current schedule. An important class of improvement type algorithms are the local search procedures. A local search procedure does not guarantee an optimal solution. It usually attempts to find a better schedule than the current one in the *neighbourhood* of the current one. Two schedules are *neighbors*, if one can be obtained through a well defined modification of the other. At each iteration, a local search procedure performs a search within the neighbourhood and evaluates the various neighbouring solutions. The procedure either accepts or rejects a candidate solution as the next schedule to move to, based on a given acceptance-rejection criterion.

One can compare the various local search procedures on the following four design criteria:

(i) The schedule representation needed for the procedure.
(ii) The neighbourhood design.
(iii) The search process within the neighbourhood.
(iv) The acceptance-rejection criterion.

The representation of a schedule may at times be nontrivial. A nonpreemptive single machine schedule can be specified by a simple permutation of the n jobs. A nonpreemptive job shop schedule can be specified by m consecutive strings, each one representing a permutation of n operations on a specific machine. Based on this information, the starting times and completion times of all operations can be computed. However, when preemptions are allowed the format of the schedule representation becomes significantly more complicated.

The design of the neighbourhood is a very important aspect of a local search procedure. For a single machine a neighbourhood of a particular schedule may be simply defined as all schedules that can be obtained by performing a single adjacent pairwise interchange. This implies that there are $n-1$ schedules in the neighbourhood of the original schedule. A larger neighbourhood for a single machine schedule may be defined by taking an arbitrary job in the

C.5 Local Search: Simulated Annealing and Tabu-Search

schedule and inserting it in another position in the schedule. Clearly, each job can be inserted in $n-1$ other positions. The entire neighbourhood contains less than $n(n-1)$ neighbors as some of these neighbors are identical. The neighbourhood of a schedule in a more complicated machine environment is usually more complex.

An interesting example is a neighbourhood designed for the job shop problem with the makespan objective. In order to describe this neighbourhood, we have to use the concept of a critical path. A critical path in a job shop schedule consists of a set of operations of which the first one starts out at time $t = 0$ and the last one finishes at time $t = C_{\max}$. The completion time of each operation on a critical path is equal to the starting time of the next operation on that path; two successive operations either belong to the same job or are processed on the same machine (see Figure C.2). A schedule may have multiple critical paths that may overlap. Finding the critical path(s) in a given schedule for a job shop problem with the makespan as objective is relatively straightforward. It is clear that in order to reduce the makespan, changes have to be made in the sequence(s) of the operations on the critical path(s). A simple neighbourhood of an existing schedule can be designed as follows: the set of schedules whose corresponding sequences of operations on the machines can be obtained by interchanging a pair of adjacent operations on the critical path of the current schedule. Note that in order to interchange a pair of operations on the critical path, the operations must be on the same machine and belong to different jobs. If there is a single critical path, then the number of neighbors within the neighbourhood is at most the number of operations on the critical path minus 1.

Experiments have shown that this type of neighbourhood for the job shop problem is too simple to be effective. The number of neighbouring schedules that are better than the existing schedule tends to be very limited. More sophisticated neighbourhoods have been designed that perform better. One of these is referred to as the *One Step Look-Back Adjacent Interchange*.

Example C.5.1 (Neighbourhood of a job shop schedule). A neighbor of a current schedule is obtained by first performing an adjacent pairwise interchange between two operations (i, j) and (i, k) on the critical path. After the interchange operation (i, k) is processed before operation (i, j) on machine i. Consider job k to which operation (i, k) belongs and refer to that operation of job k that immediately precedes operation (i, k) as operation (h, k) (it is processed on machine h). On machine h, interchange operation (h, k) and the operation preceding (h, k) on machine h, say operation (h, l) (see Figure C.3). From the figure it is clear that, even if the first interchange (between (i, j) and (i, k)) does not result in an improvement, the second interchange between (h, k) and (h, l) may lead to an overall improvement.

Actually, this design can be made more elaborate by backtracking more than one step. These types of interchanges are referred to as *Multi-Step Look-Back Interchanges*.

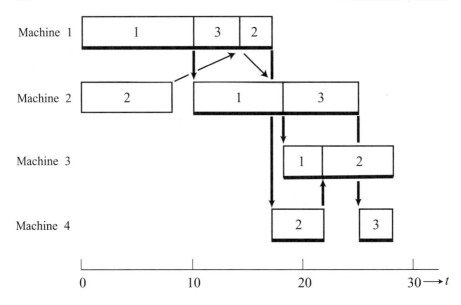

Fig. C.2. Optimal Schedule With Critical Paths

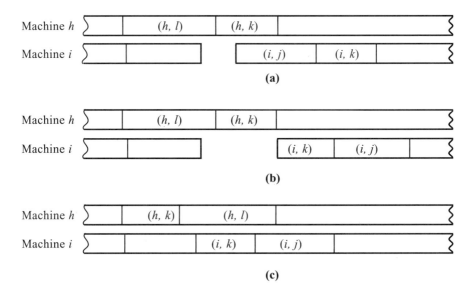

Fig. C.3. One-step look-back interchange for (a) current schedule; (b) schedule after interchange of (i,j) and (i,k); (c) schedule after interchange of (h,l) and (h,k).

C.5 Local Search: Simulated Annealing and Tabu-Search

In exactly the same way *One Step Look-Ahead Interchange* and *Multi-Step Look-Ahead Interchanges* can be constructed.

The search process within a neighbourhood can be done in a number of ways. A simple way is to select schedules in the neighbourhood at random, evaluate these schedules and decide which one to accept. However, it may pay to do a more organized search and select first schedules that appear promising. One may want to consider swapping those jobs that affect the objective the most. For example, when the total weighted tardiness has to be minimized, one may want to move jobs that are very tardy more towards the beginning of the schedule.

The acceptance-rejection criterion is usually the design aspect that distinguishes a local search procedure the most. The difference between the two procedures discussed in the remaining part of this section, simulated annealing and tabu-search, lies mainly in their acceptance-rejection criteria. In simulated annealing the acceptance-rejection criterion is based on a probabilistic process while in tabu-search it is based on a deterministic process.

We first discuss simulated annealing. This procedure is a search process that has its origin in the fields of material science and physics. It was first developed as a simulation model for describing the physical annealing process of condensed matter.

The simulated annealing procedure goes through a number of iterations. At iteration k of the procedure, there is a current schedule $\mathcal{S}_k$ as well as a best schedule found so far, $\mathcal{S}_0$. For a single machine problem these schedules are sequences (permutations) of the jobs. Let $G(\mathcal{S}_k)$ and $G(\mathcal{S}_0)$ denote the corresponding values of the objective function. Note that $G(\mathcal{S}_k) \geq G(\mathcal{S}_0)$. The value of the best schedule obtained so far, $G(\mathcal{S}_0)$, is often referred to as the aspiration criterion. The algorithm, in its search for an optimal schedule, moves from one schedule to another. At iteration k, a search for a new schedule is conducted within the neighbourhood of $\mathcal{S}_k$. First, a so-called *candidate* schedule, say $\mathcal{S}_c$, is selected from the neighbourhood. This selection of a candidate schedule can be done at random or in an organized, possibly sequential, way. If $G(\mathcal{S}_c) < G(\mathcal{S}_k)$, a move is made, setting $\mathcal{S}_{k+1} = \mathcal{S}_c$. If $G(\mathcal{S}_c) < G(\mathcal{S}_0)$, then $\mathcal{S}_0$ is set equal to $\mathcal{S}_c$. However, if If $G(\mathcal{S}_c) \geq G(\mathcal{S}_k)$, a move is made to $\mathcal{S}_c$ with probability

$$P(\mathcal{S}_k, \mathcal{S}_c) = \exp\left(\frac{G(\mathcal{S}_k) - G(\mathcal{S}_c)}{\beta_k}\right);$$

with probability $1 - P(\mathcal{S}_k, \mathcal{S}_c)$ schedule $\mathcal{S}_c$ is rejected in favor of the current schedule, setting $\mathcal{S}_{k+1} = \mathcal{S}_k$. Schedule $\mathcal{S}_0$ does not change when it is better than schedule $\mathcal{S}_c$. The

$$\beta_1 \geq \beta_2 \geq \beta_3 \geq \cdots > 0$$

are control parameters referred to as cooling parameters or temperatures (in analogy with the annealing process mentioned above). Often β_k is chosen to be a^k for some a between 0 and 1.

From the above description of the simulated annealing procedure it is clear that moves to worse solutions are allowed. The reason for allowing these moves is to give the procedure the opportunity to move away from a local minimum and find a better solution later on. Since β_k decreases with k, the acceptance probability for a non-improving move is lower in later iterations of the search process. The definition of the acceptance probability also ensures that if a neighbor is significantly worse, its acceptance probability is very low and the move is unlikely to be made.

Several stopping criteria are used for this procedure. One way is to let the procedure run for a prespecified number of iterations. Another is to let the procedure run until no improvement has been obtained for a given number of iterations.

The method can be summarized as follows:

Algorithm C.5.2 (Simulated Annealing).

Step 1.

Set $k = 1$ and select β_1.

Select an initial sequence $\mathcal{S}_1$ using some heuristic.

Set $\mathcal{S}_0 = \mathcal{S}_1$.

Step 2.

Select a candidate schedule $\mathcal{S}_c$ from the neighbourhood of $\mathcal{S}_k$.

If $G(\mathcal{S}_0) < G(\mathcal{S}_c) < G(\mathcal{S}_k)$, set $\mathcal{S}_{k+1} = \mathcal{S}_c$ and go to Step 3.

If $G(\mathcal{S}_c) < G(\mathcal{S}_0)$, set $\mathcal{S}_0 = \mathcal{S}_{k+1} = \mathcal{S}_c$ and go to Step 3.

If $G(\mathcal{S}_c) > G(\mathcal{S}_k)$ generate a random number U_k from a Uniform(0,1) distribution;

If $U_k \leq P(\mathcal{S}_k, \mathcal{S}_c)$ set $\mathcal{S}_{k+1} = \mathcal{S}_c$ otherwise set $\mathcal{S}_{k+1} = \mathcal{S}_k$ and go to Step 3.

Step 3.

Select $\beta_{k+1} \leq \beta_k$.

Increment k by 1.

If $k = N$ then STOP, otherwise go to Step 2.

The effectiveness of simulated annealing depends on the design of the neighbourhood and on how the search is conducted within this neighbourhood. If the neighbourhood is designed in a way that facilitates moves to better solutions and moves out of local minima, then the procedure will perform well. The search within a neighbourhood can be done randomly or in a more organized way. For example, the contribution of each job to the objective function can be computed and the job with the highest impact on the objective can be selected as a candidate for an interchange.

Over the last two decades simulated annealing has been applied to many scheduling problems, in academia as well as in industry, with considerable success.

In the remaining part of this section we describe the tabu-search procedure. Tabu-search is in many ways similar to simulated annealing in that it also moves from one schedule to another with the next schedule being possibly worse than the one before. For each schedule, a neighbourhood is defined as in simulated annealing. The search for a neighbor within the neighbourhood as a potential candidate to move to is again a design issue. As in simulated annealing, this can be done randomly or in an organized way. The basic difference between tabu-search and simulated annealing lies in the mechanism used for approving a candidate schedule. In tabu-search the mechanism is not probabilistic but rather of a deterministic nature. At any stage of the process a tabu-list of mutations, which the procedure is *not* allowed to make, is kept. Mutations on the tabu-list may be, for example, pairs of jobs that may not be interchanged. The tabu-list has a fixed number of entries (usually between 5 and 9), which depends upon the application. Every time a move is made by a mutation in the current schedule, the *reverse* mutation is entered at the top of the tabu-list; all other entries are pushed down one position and the bottom entry is deleted. The reverse mutation is put on the tabu-list to avoid returning to a local minimum that has been visited before. Actually, at times a reverse mutation that is tabu could have led to a new schedule, not visited before, with an objective value lower than any one obtained before. This may happen when the mutation is close to the bottom of the tabu-list and a number of moves have already been made since the mutation was put on the list. Thus, if the number of entries in the tabu-list is too small cycling may occur; if it is too large the search may be unduly constrained. The method can be summarized as follows:

Algorithm C.5.3 (Tabu-Search).

Step 1.

Set $k = 1$.

Select an initial sequence S_1 using some heuristic.

Set $S_0 = S_1$

Step 2.

Select a candidate schedule S_c from the neighbourhood of S_k.

If the move $S_k \to S_c$ is prohibited by a mutation on the tabu-list, set $S_{k+1} = S_k$ and go to Step 3.

If the move $S_k \to S_c$ is not prohibited by any mutation on the tabu-list, set $S_{k+1} = S_c$ and enter reverse mutation at the top of the tabu-list; push all other entries in the tabu-list one position down; delete the entry at the bottom of the tabu-list.

If $G(\mathcal{S}_c) < G(\mathcal{S}_0)$, set $\mathcal{S}_0 = \mathcal{S}_c$;

Go to Step 3.

Step 3.

Increment k by 1.

If $k = N$ then STOP, otherwise go to Step 2.

The following example illustrates the method.

Example C.5.4 (Tabu-Search). Consider the following instance of the single machine total weighted tardiness problem with four jobs.

jobs	1	2	3	4
p_j	10	10	13	4
d_j	4	2	1	12
w_j	14	12	1	12

The neighbourhood of a schedule is defined as all schedules that can be obtained through adjacent pairwise interchanges. The tabu-list is a list of pairs of jobs (j, k) that were swapped within the last two moves and cannot be swapped again. Initially, the tabu-list is empty.

As a first schedule the sequence $\mathcal{S}_1 = 2, 1, 4, 3$ is chosen, with objective function value

$$\sum w_j T_j(2, 1, 4, 3) = 500.$$

The aspiration criterion is therefore 500. There are three schedules in the neighbourhood of $\mathcal{S}_1$, namely $1, 2, 4, 3$; $2, 4, 1, 3$ and $2, 1, 3, 4$. The respective values of the objective function are 480, 436 and 652. Selection of the best non-tabu sequence results in $\mathcal{S}_2 = 2, 4, 1, 3$. The aspiration criterion is changed to 436. The tabu-list is updated and contains now the pair (1,4). The values of the objective functions of the neighbors of $\mathcal{S}_2$ are:

sequence	4,2,1,3	2,1,4,3	2,4,3,1
$\sum w_j T_j$	460	500	608

Note that the second move is tabu. However, the first move is better than the second anyhow. The first move results in a schedule that is worse than the best one so far. The best one is the current one, which therefore is a local minimum. Nevertheless, $\mathcal{S}_3 = 4, 2, 1, 3$ and the tabu-list is updated and now contains $\{(2, 4), (1, 4)\}$. Neighbors of $\mathcal{S}_3$ with the corresponding values of the objective functions are:

sequence	2,4,1,3	4,1,2,3	4,2,3,1
$\sum w_j T_j$	436	440	632

Now, although the best move is to $2,4,1,3$ ($\mathcal{S}_2$), this move is tabu. Therefore $\mathcal{S}_4$ is chosen to be $4,1,2,3$. Updating the tabu-list results in $\{(1,2),(2,4)\}$ and the pair (1,4) drops from the tabu-list as the length of the list is kept to 2. Neighbors of $\mathcal{S}_4$ and their corresponding objective function values are:

sequence	1,4,2,3	4,2,1,3	4,1,3,2
$\sum w_j T_j$	408	460	586

The schedule $4,2,1,3$ is tabu, but the best move is to the schedule $1,4,2,3$. So $\mathcal{S}_5 = 1,4,2,3$. The corresponding value of the objective is better than the aspiration criterion. So the aspiration criterion becomes 408. The tabu-list is updated by adding (1,4) and dropping (2,4). Actually, $\mathcal{S}_5$ is a global minimum, but tabu-search, being unaware of this fact, continues.

The information carried along in tabu-search consists of the tabu-list as well as the best solution obtained so far during the search process. Recently, more powerful versions of tabu-search have been proposed; these versions retain more information. One version uses a so-called tabu-tree. In this tree each node represents a solution or schedule. While the search process goes from one solution to another (with each solution having a tabu-list), the process generates additional nodes. Certain solutions that appear promising may not be used as a take-off point immediately but are retained for future use. If at a certain point during the search process the current solution does not appear promising as a take-off point, the search process can go back in the tabu-tree to another node (solution) that had been retained before and take off again, but now in a different direction.

C.6 Local Search: Genetic Algorithms

Genetic algorithms are more general and abstract than simulated annealing and tabu-search. Simulated annealing and tabu-search may, in a certain way, be viewed as special cases of genetic algorithms.

Genetic algorithms, when applied to scheduling, view sequences or schedules as *individuals* or members of a *population*. Each individual is characterized by its *fitness*. The fitness of an individual is measured by the associated value of the objective function. The procedure works iteratively, and each iteration is referred to as a *generation*. The population of one generation consists of individuals surviving from the previous generation plus the new schedules or *children* from the previous generation. The population size usually remains constant from one generation to the next. The children are generated through reproduction and mutation of individuals that were part of the previous generation (the *parents*). Individuals are sometimes also referred to as *chromosomes*. In a multi-machine environment a chromosome may consist of sub-chromosomes, each one containing the information regarding the job sequence on a machine. A mutation in a parent chromosome may be equivalent to an adjacent pairwise interchange in the corresponding sequence. In each generation the most fit individuals reproduce while the least fit die. The birth, death and reproduction processes that determine the composition of the next generation can be complex, and usually depend on the fitness levels of the individuals of the current generation.

A genetic algorithm, as a search process, differs in one important aspect from simulated annealing and tabu-search. At each iterative step a number of different plans or schedules are generated and carried over to the next step. In simulated annealing and tabu-search only a single plan or schedule is carried over from one iteration to the next. Hence simulated annealing and tabu-search may be regarded as special cases of genetic algorithms with a population size equal to 1. This diversification scheme is an important characteristic of genetic algorithms. In a genetic algorithm a neighbourhood is not based on just one plan or schedule, but rather on a collection of plans or schedules. The design of the neighbourhood of a current population of schedules is therefore based on more general techniques than those used in simulated annealing and tabu-search. A new schedule can be constructed by combining parts of different schedules within the population. For example, in a job shop scheduling problem a new schedule can be generated by combining the sequence of operations on one machine in one parent schedule with a sequence of operations on another machine in another parent schedule. This is often referred to as the *cross-over* effect.

A very simplified version of a genetic algorithm can be described as follows.

Algorithm C.6.1 (Genetic algorithm).
Step 1.

    Set $k = 1$.

    Select ℓ initial sequences $\mathcal{S}_{1,1}, \ldots, \mathcal{S}_{1,\ell}$ using some heuristic.

Step 2.

    Select the two best schedules among $\mathcal{S}_{k,1}, \ldots, \mathcal{S}_{k,\ell}$ and call these $\mathcal{S}_k^+$ and $\mathcal{S}_k^{++}$.

    Select the two worst schedules among $\mathcal{S}_{k,1}, \ldots, \mathcal{S}_{k,\ell}$ and call these $\mathcal{S}_k^-$ and $\mathcal{S}_k^{--}$.

    Select two neighbors $\mathcal{S}^*$ and $\mathcal{S}^{**}$ from the neighbourhood of $\mathcal{S}_k^+$ and $\mathcal{S}_k^{++}$.

    Replace $\mathcal{S}_k^-$ and $\mathcal{S}_k^{--}$ with $\mathcal{S}^*$ and $\mathcal{S}^{**}$;

    Keep all other schedules the same and go to Step 3.

Step 3.

    Increment k by 1.

    If $k = N$ then STOP, otherwise go to Step 2.

Exercises

The use of genetic algorithms has its advantages and disadvantages. One advantage is that they can be applied to a problem without having to know much about the structural properties of the problem. They can be very easily coded and they often give fairly good solutions. However, the amount of computation time needed to obtain such a solution can be relatively long in comparison with the more rigorous problem-specific approaches.

C.7 Discussion

For many scheduling problems one can design procedures that combine several of the techniques presented in this chapter. The following three step approach has proven useful for solving scheduling problems in practice. It combines composite dispatching rules with simulated annealing or tabu-search.

Step 1. Compute the values of a number of statistics such as due date tightness, setup time severity, and so on.

Step 2. Determine the scaling parameters of a composite dispatching rule based on the outcome of Step 1 and apply the composite dispatching rule to the scheduling instance.

Step 3. Use the schedule developed in Step 2 as an initial solution for a local search procedure, and let the local search procedure try to improve on the schedule.

This three step framework would only be useful if the routine would be used frequently (e.g., a new instance of the same problem has to be solved every day). The reason is that the empirical procedure that determines the functions that map values of the job statistics into appropriate values for scaling parameters constitutes a major investment of time. Such an investment pays off only when the routine is used frequently.

It is also clear how beam search can be combined with other heuristics. For instance, it may be possible to use composite dispatching rules within the beam search process in order to evaluate nodes. Also, the final solution obtained with beam search can be fed into a local search procedure in order to obtain further improvement.

Exercises

C.1. What does the ATCS rule reduce to
 (a) if both K_1 and K_2 go to ∞,
 (b) if K_1 is very close to zero and $K_2 = 1$,
 (c) and if K_2 is very close to zero and $K_1 = 1$?

C.2. Consider the instance in Example C.5.4. Apply the tabu-search technique once more, starting out with the same initial sequence, under the following conditions.

(a) Keep the length of the tabu-list equal to one, i.e., only the pair of jobs which was swapped during the last move cannot be swapped again. Apply the technique for four iterations and determine whether the optimal sequence is reached.

(b) Keep the length of the tabu-list equal to three, i.e., the pairs of jobs which were swapped during the last three moves cannot be swapped again. Apply the technique for four iterations and determine whether the optimal sequence is reached.

C.3. Consider the following basic mutations that can be applied to a sequence:
  (i) An insertion (a job is selected and put elsewhere in the sequence).
  (ii) A pairwise interchange of two adjacent jobs.
  (iii) A pairwise interchange of two nonadjacent jobs.
  (iv) A sequence interchange of two adjacent subsequences of jobs.
  (v) A sequence interchange of two nonadjacent subsequences of jobs.
  (vi) A reversal of a subsequence of jobs.

Some of these mutations are special cases of others and some mutations can be achieved through repeated applications of others. Taking this into account explain how these six types of mutations are related to one another.

Comments and References

An excellent treatise of general purpose procedures and heuristic techniques for scheduling problems is the text by Morton and Pentico (1993); they cover most of the techniques described in this chapter.

Many of the basic dispatching rules originated in the fifties and early sixties. For example, Smith (1956) introduced the WSPT rule, Jackson (1955) introduced the EDD rule, Hu (1961) the CP rule. Conway (1965a, 1965b) was one of the first to make a comparative study of dispatching rules. A fairly complete list of the most common dispatching rules is given by Panwalkar and Iskander (1977) and a detailed description of composite dispatching rules by Bhaskaran and Pinedo (1992). Special examples of composite dispatching rules are the COVERT rule developed by Carroll (1965) and the ATC rule developed by Vepsalainen and Morton (1987). Ow and Morton (1989) describe a rule for scheduling problems with earliness and tardiness penalties. The ATCS rule is due to Lee, Bhaskaran and Pinedo (1997).

The application of filtered beam search to scheduling was first proposed by Ow and Morton (1988).

A great deal of research has been done on the applications of simulated annealing, tabu-search and genetic algorithms on scheduling. For general overviews of the application of local search techniques to scheduling, see Storer, Wu and Vaccari (1992), Aarts, van Laarhoven, Lenstra, and Ulder (1994), Anderson, Glass and Potts (1995), Vaessens, Aarts and Lenstra (1996), and Aarts and Lenstra (1997). Studies on simulated annealing can be found in Kirkpatrick, Gelatt and Vecchi (1983), Matsuo, Suh and Sullivan (1988) and Van Laarhoven, Aarts and Lenstra (1992). For tabu-search, see Glover (1990), Dell'Amico and Trubian (1991) Nowicki and Smutnicki (1996), Taillard (1990,1994). For genetic algorithms applied to scheduling, see

Comments and References

Lawton (1992), Della Croce, Tadei and Volta (1992), Bean (1994), Pesch (1994), Corne and Ross (1997), and Kreipl (2000).

The three step procedure described in the discussion section is due to Lee, Bhaskaran and Pinedo (1997). There are many other approaches, besides those covered in this chapter, that have proven useful in scheduling. Some of these approaches are based on pricing techniques, i.e., there are costs involved in assigning jobs to specific machines during specific time periods. One such pricing approach is discussed in Roundy, Maxwell, Herer, Tayur and Getzler (1991). Other approaches are based on decomposition techniques; see, for example, Sidney and Steiner (1986), Monma and Sidney (1987), Uzsoy (1993), Ovacik and Uzsoy (1997).

Appendix D

Constraint Programming Methods

D.1 Introduction 437
D.2 Constraint Satisfaction 438
D.3 Constraint Programming 439
D.4 OPL: An Example of a Constraint
    Programming Language 441
D.5 Constraint Programming vs. Mathematical
    Programming 444

D.1 Introduction

In contrast to mathematical programming, which has its roots in the Operations Research community, constraint programming has its origins in the Artificial Intelligence and Computer Science communities. Constraint programming can be traced back to the constraint satisfaction problems studied in the 1970's. These problems require a search for feasible solutions that satisfy all given constraints. To facilitate the search for solutions to such problems various special purpose languages have been developed, e.g., Prolog. However, during the last decade of the twentieth century, constraint programming has not only been used for solving feasibility problems, but also for solving optimization problems. Several approaches have been developed that facilitate the application of constraint programming to optimization problems. One such approach is via the Optimization Programming Language (OPL), which was designed for modeling and solving optimization problems through both constraint programming techniques and mathematical programming procedures.

D.2 Constraint Satisfaction

To describe the constraint programming framework, it is necessary to first define the constraint satisfaction problem. In order to be consistent with the mathematical programming material presented in Appendix A, it is advantageous to present the constraint satisfaction problem using mathematical programming terminology. Assume n decision variables $x_1, \ldots, x_n$ and let $\mathcal{D}_j$ denote the set of allowable values for decision variable x_j. This set is typically referred to as the *domain* of the variable x_j. Decision variables can take integer values, real values, as well as set elements.

Formally, a constraint is a mathematical relation that implies a subset $\mathcal{S}$ of the set $\mathcal{D}_1 \times \mathcal{D}_2 \times \cdots \times \mathcal{D}_n$ such that if $(x_1, \ldots, x_n) \in \mathcal{S}$, then the constraint is said to be satisfied. One can also define a mathematical function f such that

$$f(x_1, \ldots, x_n) = 1$$

if and only if the constraint is satisfied. Using this notation, the Constraint Satisfaction Problem (CSP) can be defined as follows:

$$f_i(x_1, \ldots, x_n) = 1, \quad i = 1, \ldots, m$$
$$x_j \in \mathcal{D}_j \quad j = 1, \ldots, n$$

Since the problem is only a feasibility problem, no objective function has to be defined. The constraints in the constraint set may be of various different types: they may be linear, nonlinear, logical combinations of other constraints, cardinality constraints, higher order constraints or global constraints.

One basic difference between the constraints in a mathematical programming formulation and the constraints in a constraint satisfaction problem lies in the fact that the constraints in a mathematical programming formulation are typically either linear or nonlinear, whereas the constraints in a constraint programming formulation can be of a more general form.

Constraint Satisfaction is typically solved via a tree search algorithm. Each node in the search tree corresponds to a partial solution and going from one node to another is done by assigning a value to a variable. At each node, the following operations have to be performed:

> WHILE not solved AND not infeasible DO
>   consistency checking (domain reduction)
>   IF a dead-end is detected THEN
>   try to escape from dead-end (backtrack)
>   ELSE
>   select variable
>   assign value to variable
>   ENDIF
> ENDWHILE

The selection of the next variable and the assignment of its value is done by variable selection heuristics and value assignment heuristics. In job shop scheduling the variable corresponds to an operation and the value corresponds to its starting time. After a value is assigned to a variable, inconsistent values of unassigned variables are deleted. The process of removing inconsistent values is often referred to as consistency checking or domain reduction. One well-known technique of consistency checking is *constraint propagation*. For a variable x, the *current domain* $\delta(x)$ is the set of values for which no inconsistency can be found with the available consistency checking techniques. If, after removing inconsistent values from the current domains, a current domain has become empty, a so-called *dead-end* has been reached. In such a case, one or more assignments of variables have to be undone (i.e., some form of backtracking has to be done) and alternatives have to be tried out. An instance is solved if every variable is assigned a value and an instance is shown to be infeasible if for a variable in the root of the tree no values are remaining to be tried.

If constraint satisfaction is applied to the job shop problem in Chapter 5, then the problem is to verify whether there exists a feasible job shop schedule with a makespan $C_{\max}$ that is less than a given value z^*.

Domain reduction in job shop scheduling boils down to the following: Given the partial schedule already constructed, each operation yet to be scheduled has an earliest possible starting time and a latest possible completion time. Whenever the starting time and completion time of an operation is fixed, some form of checking has to be done on how the newly scheduled (fixed) operation affects the earliest possible starting times and latest possible completion times of all the operations remaining to be scheduled. The earliest possible starting time of a yet to be scheduled operation may now have to be set a little bit later while the latest possible completion time of that operation may now have to be set a little bit earlier.

Note that the branch-and-bound approach described in Section B.4 constructs a schedule going forward in time. With constraint satisfaction an actual makespan is specified; this allows the construction of the schedule to proceed forward as well as backward in time.

D.3 Constraint Programming

Originally, constraint satisfaction was used only to find feasible solutions for problems. However, the constraint satisfaction structure, when embedded in a more elaborate framework, can be applied to optimization (e.g., minimization) problems as well. An optimization problem may be formulated as follows:

$$\text{minimize} \quad g(x_1, \ldots, x_n)$$

subject to

$$f_i(x_1, \ldots, x_n) = 1, \qquad i = 1, \ldots, m$$

$$x_j \in \mathcal{D}_j \qquad j = 1, \ldots, n$$

The standard search procedure for finding the optimal solution is to first find a feasible solution to the Constraint Satisfaction Problem, while ignoring the objective function. Let $y_1, \ldots, y_n$ represent such a feasible solution. Let $z^* = g(y_1, \ldots, y_n)$ and add the constraint

$$g(x_1, \ldots, x_n) < z^*$$

to the constraint set and solve this modified Constraint Satisfaction Problem. The additional constraint forces the new feasible solution to have a better objective value than the current one. Constraint propagation may cause the domains of the decision variables to be narrowed, thus reducing the size of the search space. As the search goes on, new solutions must have progressively better objective values. The algorithm terminates when no feasible solution is found; when this happens, the last feasible point found is the optimal solution.

A more sophisticated and efficient search procedure, often referred to as *dichotomic* search, requires at the outset a good lower bound L on the objective $g(x_1, \ldots, x_n)$. The procedure must also find an initial feasible solution that represents an upper bound U on the objective function. The dichotomic search procedure essentially performs a binary search on the objective function. The procedure computes the midpoint

$$M = \frac{(U + L)}{2}$$

of the two bounds and then solves the constraint satisfaction problem with the added constraint

$$g(x_1, \ldots, x_n) < M$$

If it finds a feasible solution, then the new upper bound is set equal to M and the lower bound is kept the same; a new midpoint is computed and the search continues. If it does not find a feasible solution, then the lower bound is set equal to M and the upper bound is kept the same; a new midpoint is computed and the search continues.

Dichotomic search is effective when the lower bound is strong, because the computation time that is needed to show that a constraint satisfaction problem does not have a feasible solution usually tends to be long.

Constraint programming can be described as a two-level architecture that at one level has a list of constraints (sometimes referred to as a constraint store) and at the other level a programming language component that specifies the organization and the sequence in which the basic operations are executed. One fundamental feature of the programming language component is its ability to specify a search procedure.

D.4 OPL: An Example of a Constraint Programming Language

The Optimization Programming Language (OPL) was developed during the last decade of the twentieth century to deal with optimization problems through a combination of constraint programming and mathematical programming techniques.

This section describes two constraint programs for scheduling problems. The first program is designed for the job shop scheduling problem with the makespan objective (as described in Chapter 5), and the second program is designed for an automotive assembly line which has a paint shop that is subject to sequence dependent setup costs (as described in Chapter 6).

Example D.4.1 (An OPL Program for a Job Shop). An OPL program for the job shop problem with the makespan objective can be written as follows:

```
01. INT NBMACHINES = ...;
02. RANGE MACHINES 1..NBMACHINES;
03. INT NBJOBS = ...;
04. RANGE JOBS 1..NBJOBS;
05. INT NBOPERATIONS = ...;
06. RANGE OPERATIONS 1..NBOPERATIONS;
07. MACHINES RESOURCE[JOBS,OPERATIONS]= ...;
08. INT+ DURATION[JOBS,OPERATIONS] = ...;
09. INT TOTALDURATION
       =SUM(J IN JOBS, O IN OPERATIONS)DURATION[J,O];
10. SCHEDULEHORIZON=TOTALDURATION;
11. ACTIVITY OPERATION[J IN JOBS, O IN OPERATIONS](DURATION[J,O]);
12. ACTIVITY MAKESPAN(0);
13. UNARYRESOURCE TOOL[MACHINES];
14. MINIMIZE
15.     MAKESPAN.END
16. SUBJECT TO {
17.     FORALL(J IN JOBS)
18.         OPERATION[J,NBOPERATIONS] PRECEDES MAKESPAN;
19.     FORALL(J IN JOBS)
20.         FORALL(O IN 1,..NBOPERATIONS-1)
21.             OPERATION[J,O] PRECEDES OPERATION[J,O+1];
22.     FORALL(J IN JOBS)
23.         FORALL(O IN OPERATIONS)
24.             OPERATION[J,O] REQUIRES TOOL[RESOURCE[J,O]];
25. };
```

The first six lines define the input data of the problem, consisting of the number of machines, the number of jobs, and the number of operations. The array RESOURCE contains input data that consists of the identity and the characteristics of the machine needed for processing a particular operation of a job. The array DURATION specifies the time required for processing each operation of a job. The keyword ACTIVITY implicitly indicates a decision variable. An activity consists of three elements: a start time, an end time and a duration. In the declaration given here, the durations of each activity are given as data. When an activity is declared, the constraint

$$\text{OPERATION}[\text{J},\text{O}].\text{START} + \text{OPERATION}[\text{J},\text{O}].\text{DURATION} = \text{OPERATION}[\text{J},\text{O}].\text{END}$$

is automatically included in the program. The makespan activity is a dummy activity with zero processing time that will be the very last operation in the schedule. UNARYRESOURCE implies also a decision variable; decisions have to be made which activities have to be assigned to which resources at any given time.

The remaining part of the program states the problem: first the objective and then all the constraints. The statement PRECEDES is a keyword of the language. So one of the constraints is internally immediately translated into

$$\text{OPERATION}[\text{J},\text{O}].\text{END} <= \text{OPERATION}[\text{J},\text{O}+1].\text{START}$$

The word REQUIRES is also a keyword of the language. Declaring a set of requirements causes the creation of a set of so-called *disjunctive* constraints (see Appendix A). For a given machine I described by a resource TOOL[I], let OPERATION[J1,O1] and OPERATION[J2,O2] denote two operations that require machine I, i.e., the data given includes

$$\text{RESOURCE}[\text{J1},\text{O1}]=\text{RESOURCE}[\text{J2},\text{O2}]=\text{I}$$

The following disjunctive constraint is now created automatically by the system, ensuring that the two operations cannot occupy machine I at the same time:

$$\text{OPERATION}[\text{J1},\text{O1}].\text{END}<= \text{OPERATION}[\text{J2},\text{O2}].\text{START}$$
$$\text{or}$$
$$\text{OPERATION}[\text{J2},\text{O2}].\text{END}<= \text{OPERATION}[\text{J1},\text{O1}].\text{START}$$

These disjunctive constraints imply that OPERATION[J1,O1] either precedes or follows OPERATION[J2,O2].

The second example of a constraint programming application concerns a paced assembly system in the automotive industry (see Chapter 6).

Example D.4.2 (An OPL Program for a Paced Assembly Line). A set of cars in an automotive plant has to be painted. Each car has a designated color, and the cars have to be sequenced in such a way that the cost incurred because of paint changes in the sequence is minimized. Car j is given an initial position x_j in the sequence and a global interval G such that if a car

D.4 OPL: An Example of a Constraint Programming Language   443

is painted before its initial position x_j or after the position $x_j + G$, a penalty cost is incurred. This problem is a simplified version of the paint shop problem solved at DaimlerChrysler using contraint programming.

An OPL program for the automotive assembly line with a paint shop that is subject to sequence dependent setup costs can be written as follows:

```
01. INT NCARS = ...;
02. ENUM COLORS= ...;
03. RANGE CARRNG 1..NCARS;
04. CARRNG WHENCAR[CARRNG]= ...;
05. COLORS COLOR[CARRNG]= ...;
06. INT+ OKINTERVAL=...;
07. INT SWAPCOST = ...;
08. ASSERT FORALL (J IN 1..NCARS-1) WHENCAR[J]<=WHENCAR[J+1];
09. SETOF(INT) WHENCOLOR[C IN COLORS]
        = {J | J IN CARRNG:COLOR[J]=C};
10. VAR CARRNG POSITION[CARRNG];
11. VAR CARRNG WHICHCAR[CARRNG];
12. VAR INT COLORCHANGES IN CARD(COLORS) -1 .. NCARS-1;
13. VAR INT PENALTYCOST IN 0..NCARS*NCARS;
14. MINIMIZE SWAPCOST*COLORCHANGES + PENALTYCOST
15. SUBJECT TO {
16.     COLORCHANGES = SUM (J IN 1 .. NCARS-1)
17.         (COLOR[WHICHCAR[J]] <> COLOR[WHICHCAR[J+1]]);
18.     PENALTYCOST=SUM (J IN CARRNG)
19.     (MAXL(WHENCAR[J] - POSITION[J], 0)+
        (MAXL(POSITION[J] - (WHENCAR[J] + OKINTERVAL),0));
20.         ALLDIFFERENT(POSITION);
21.         ALLDIFFERENT(WHICHCAR);
22.         FORALL (J IN CARRNG) {
23.             WHICHCAR[POSITION[J]]=J;
24.             POSITION[WHICHCAR[J]]=J;
25.         }
26.         FORALL (C IN COLORS) {
27.             FORALL (K IN WHENCOLOR[C]:K <>
                WHENCOLOR[C].LAST()) {
28.                 POSITION[K] <
                    POSITION[WHENCOLOR[C].NEXT(K)];
29.             }
30.         }
31. };
```

The first two lines of the code specify the number of cars and the set of paint colors. The third line creates a range of positions in which each car has to be placed. The fourth line declares the array WHENCAR which correspond to the values x_j. Without loss of generality these values can be given in increasing order and this is verified in line 8. The fifth line specifies the color of each car. Line 6 specifies the global interval G. Line 7 specifies the cost of having to change the color between two consecutive cars. The ninth line determines for each color the set of cars that have to be painted that color.

The model for solving the problem begins at line 10. The array POSITION contains the positions of each car and the array WHICHCAR indicates which car is in which position. Line 12 declares an integer variable that counts the number of times that the color is changed, the lower bound being the number of colors minus 1 and the upper bound being the number of cars minus 1. The penalty cost is declared in line 13 as an integer decision variable. Line 14 specifies the objective to be minimized. The number of color changes is computed in lines 16 and 17 by comparing the colors of adjacent cars. The penalty costs are computed in lines 18 and 19. The OPL function MAXL(A,B) takes the maximum of the two arguments. The penalty is computed by penalizing both the earliness and the tardiness and assessing no penalty when a car is positioned in the appropriate interval, i.e., when

$$\text{WHENCAR}[\text{I}] \leq \text{POSITION}[\text{I}] \leq \text{WHENCAR}[\text{I}] + \text{OKINTERVAL}$$

Lines 20 and 21 contain two *global* constraints. The constraint ALLDIFFERENT(POSITION) specifies that each element of the POSITION array should have a different value. Because the array POSITION has NCARS elements, each with an integer value between 1 and NCARS, the constraint is equivalent to saying that the array POSITION contains an assignment of each car to a unique position. (An equivalent representation using mathematical programming techniques would require NCARS × NCARS binary variables and 2 ×NCARS constraints.) The constraints in lines 23 and 24 specify that the arrays POSITION and WHICHCAR are inverses of one another. The constraints lines 26 to 28 help reduce the search space. For each color, the cars of that color are considered in the order of the values in the WHENCAR array. It is easy to see that cars of the same color should be ordered in the same order in which they appear in the input data. The constraints are written for each of the cars of each color. The expression WHENCOLOR[C].NEXT(K) indicates the car that follows car K in the set WHENCOLOR[C].

D.5 Constraint Programming vs. Mathematical Programming

Constraint programming, as a modelling tool, is more flexible than mathematical programming. The objective functions as well as the constraints can be of a more general form, since the decision variables may represent, besides integer values and real values, also set elements and even subsets of sets.

D.5 Constraint Programming vs. Mathematical Programming

The ways in which searches for optimal solutions can be conducted in mathematical programming and in constraint programming have similarities as well as differences. One similarity between constraint programming and the branch-and-bound method applied to integer programming is based on the fact that both approaches typically search through a tree of which each one of the nodes represent a partial solution of a schedule. Both approaches have to deal with issues such as how to examine the search space effectively and how to evaluate all the alternatives as efficiently as possible (i.e., how to prune the tree).

One of the differences between the dichotomic search used in constraint programming and a branch-and-bound procedure applied to mixed-integer programming is that the dichotomic search stresses a search for feasible solutions, whereas a branch-and-bound procedure emphasizes improvements on the lower bound.

Constraint programming techniques and mathematical programming techniques can be embedded within one framework. Such an integrated approach has already proven useful for a variety of scheduling problems.

Consider, for example, the crew scheduling problems described in Chapters 11 and 12. These problems are usually of the set partitioning type and the solution techniques used are typically based on column generation. A column in the crew scheduling problem represents a round trip, which is a feasible combination of various flight legs that can be handled by one crew. Since the potential number of columns is rather huge, it is important to be able to start the optimization with an initial subset of columns that is appropriate.

The solution strategy uses constraint programming in two ways. First, it is used when generating the initial subset of columns. Second, it is used as a subproblem algorithm for generating new columns during the optimization process. Thus, the master problem must find a set of round trips that covers every flight at a minimum cost. The main program alternates between solving the linear programming relaxation of the set covering problem and solving a column generation problem that generates new columns for the master problem. Once the columns are satisfactory, the columns are fixed and the set covering problem is solved to find the integer optimal solution.

The constraint programming subproblem is interesting. Constraint programming is used to enumerate the potential round trips. In the initialization stage a set of round trips are generated that are guaranteed to cover every flight leg. In the column generation phase the constraint program is used twice. First, the optimal round trips with the current set of dual multipliers are determined. After this has been determined, a search is done for all round trips with a reduced cost of at least half of the optimal round trip. This provides a large set of entering columns and eliminates one of the major weaknesses of column generation, namely the large number of iterations needed to improve the objective value in the master problem.

Constraint programming turns out to be also suitable for the preprocessing stage. A constraint programming routine can detect columns that are infeasible and reduce this way the number of columns to start out with.

Exercises

D.1. Apply the constraint programming method described in Section 5.5 to the instance presented in Exercise 5.5.

D.2. Adapt the OPL program presented in Example D.4.1 so that it is applicable to the instance in Exercise 5.5. Run the program using the OPL software on the CD-ROM and compare the outcome of your program with the result obtained in Exercise D.1.

Comments and References

There is an enormous literature on constraint satisfaction and on constraint programming. Even the part of the literature that focuses on scheduling is extensive. For general treatises on constraint programming, see Van Hentenryck (1989), Van Hentenryck and Lustig (1999), Hooker (2000), and Lustig and Puget (2001). The presentation in this appendix is mainly based on that in Lustig and Puget (2001).

The book by Baptiste, Le Pape and Nuijten (2001) focuses exclusively on scheduling. A number of papers have focused on the application of constraint programming to scheduling; see, for example, Nuijten and Aarts (1996), Jain and Grossmann (2001), and Lustig and Puget (2001). Various papers considered the integration of constraint programming techniques with column generation; see, for example, Grönkvist (2002).

Appendix E

Selected Scheduling Systems

E.1 Introduction 447
E.2 Generic Systems..............................447
E.3 Application-Specific Systems 448
E.4 Academic Prototypes 448

E.1 Introduction

Over the last three decades hundreds of scheduling systems have been developed. These developments have taken place in industry and academia in a number of countries. An up-to-date list or annotated bibliography of all systems most likely does not exist. However, several survey papers have been written describing a number of systems.

In this appendix a distinction is made between commercial generic systems, industrial systems that are application-specific and academic prototypes (research systems). The considerations in the design and the development of the various classes of systems are usually quite different. In this appendix systems are *not* classified according to the approach used for generating schedules, i.e., whether it is based on constraint programming or on certain types of algorithms. The reason for not providing such a classification is that many scheduling systems are based on a hybrid architecture and are therefore hard to classify.

E.2 Generic Systems

Commercial generic systems are designed for implementation in a wide variety of situations with only minor customization. The software houses that develop generic systems are usually not associated with a single client company. However, they may focus on a specific application or a specific industry. A sample of such systems is given in Table E.1.

Table E.1. Commercial Generic Systems

SYSTEM	COMPANY	APPLICATION AREA
MS Project	Microsoft	Project Management
Primavera	Primavera Systems	Project Management
TESS	Taylor Scheduling Software	Job Shop Scheduling
Aspen MIMI	Aspentech	Planning and Scheduling
APO	SAP	Supply Chain Management
i2 Supply Chain Management	i2 Technologies	Supply Chain Management
NAK	Willis Software	Timetabling
FTScheduler	FriarTuck	Tournament Scheduling
TSSS	Galactix Software	Tournament Scheduling
MultiRail	MultiModal Applied Systems	Railway Scheduling
Carmen Tail Assignment	Carmen Systems	Fleet Scheduling
ScheduleSoft	ScheduleSoft Corp.	Workforce Scheduling
TeleCenter System	TCS Management Group	Workforce Scheduling
Totalview	IEX Corp.	Workforce Scheduling
ALTITUDE PBS	Ad Opt	Crew Scheduling

Some of these systems have a much wider applicability and market than others. For example, MS Project, Primavera, and NAK are for the mass market; TESS can be applied in any job shop environment. However, systems such as MultiRail and Carmen Tail Assignment are tailored to a specific industry.

E.3 Application-Specific Systems

Application-specific systems are designed for either a single installation or a single type of installation. The algorithms embedded in these systems are typically quite elaborate. Many application-specific systems in industry have been developed in collaboration with an academic partner. A sample of application-specific systems is given in Table E.2 (with references).

There are, of course, many more interesting application-specific systems that have been developed in industry. Especially in the airline industry and in the automotive industry many systems have been developed. However, the companies in these industries that develop the systems do not have the tendency to report on their system development efforts in the literature.

E.4 Academic Prototypes

Academic prototypes are usually developed for research and teaching. The programmers are typically graduate students who work part time on the system. The design of these systems is often completely different from the design

Table E.2. Application-Specific Systems

SYSTEM	COMPANY	APPLICATION
DAS	Daewoo Shipbuilding	Project Management
BPSS	International Paper	Job Shop Scheduling
Giotto	Alcan	Job Shop Scheduling
Jobplan	Siemens	Job Shop Scheduling
MacMerl	Pittsburgh Plate & Glass	FAS Sequencing
LMS	IBM	Supply Chain Management
GATES	TWA	Airport Gates Scheduling
DISSY	Bremen Public Transport Authority	Personnel Scheduling

Table E.3. Academic Prototypes

SYSTEM	INSTITUTION	APPLICATION AREA
CIRCA-II	University of Maryland	Project Scheduling
LEKIN	New York University	Job Shop Scheduling
LISA	Universität Magdeburg	Job Shop Scheduling
OPIS	Carnegie-Mellon University	Job Shop Scheduling
QRP	Clemson University	Job Shop Scheduling
SONIA	Université de Paris - Sud	Job Shop Scheduling
TOSCA	University of Edinburgh	Job Shop Scheduling
TTA	Universidad Catolica de Chile	Job Shop Scheduling
ASAP	University of Nottingham	Timetabling
ConBatt	Humboldt University	Timetabling

of commercial systems. Some of the academic prototypes have had a significant impact on the design and architecture of certain commercial systems. A sample of academic prototypes is given in Table E.3.

Most of the academic prototypes listed are job shop scheduling systems. One exception is the ASAP system which is a timetabling system.

Comments and References

Several reviews and survey papers have been written with regard to scheduling systems, see Adelsberger and Kanet (1991), Smith (1992), Arguello (1994) and Fraser (1995).

The DAS project is described by Lee, Lee, Park, Hong, and Lee (1997). The BPSS system is dicussed by Adler, Fraiman, Kobacker, Pinedo, Plotnicoff and Wu (1993). The Giotto System by Alcan is described in a slide show that is included in the CD-ROM that comes with this book. The Jobplan system developed at Siemens is discussed by Kanet and Sridharan (1990). The MacMerl system developed at PPG is described by Hsu, Prietula, Thompson and Ow (1993). The Logistics Management System (LMS), developed for IBM's Burlington plant, is covered by Sullivan and Fordyce (1990). Brazile and Swigger (1988) developed the GATES system for Trans World Airways (TWA).

For a description of the CIRCA-II system, see Atkins, Abdelzaher, Shin and Durfee (1999). The LEKIN system is discussed in Chapter 5 of this book and the software is included in the accompanying CD-ROM. The LISA system is described by Bräsel, Dornheim, Kutz, Mörig and Rössling (2003). The OPIS system was developed by Smith (1994), the QRP system by Adelsberger and Kanet (1991), the SONIA system by Collinot, LePape and Pinoteau (1988), and the TOSCA system by Beck (1993). The TTA system is covered by Nussbaum and Parra (1993). The ASAP system was developed by Burke, Elliman, Ford, and Weare (1996). A description of the ConBaTT system can be found on the website of Hans-Joachim Goltz at the Research Institute for Computer Architecture and Software Technology in Berlin.

Appendix F

The Lekin System User's Guide

F.1 Introduction 451
F.2 Linking External Algorithms 451

F.1 Introduction

The educational version of the LEKIN scheduling system is a teaching tool for job shop scheduling. The system has been designed for use in either a regular Windows or a Windows NT environment.

Installation on a Windows 98 PC is straightforward. The user has to click on the "start" button and open "a:setup.exe" in the run menu.

Installation on a network server in a Windows NT environment may require some (minor) system adjustments. The program will attempt to write in the directory of the network server (which is usually read-only). The program can then be installed in one of the following two ways. The system administrator can create a public directory on the network server where the program can write. Or, a user can create a new directory on a local drive and write a link routine that connects the new directory to the network server.

F.2 Linking External Algorithms

Linking an external program is done by clicking on "Tools" and selecting "Options". A window appears with a button for a New Folder and a button for a New Item. Clicking on New folder creates a new submenu. Clicking on a New Item creates a placeholder for a new algorithm. After the user has clicked on a New Item, he has to enter all the data with respect to the New Item. Under "Executable" he has to enter the full path to the executable file of the program. The development of the code can be done in any environment under Win3.2. In what follows we give examples of the format of a file that contains the workstation information and the format of a file that contains the information pertaining to the job. After a new algorithm has been added, it is included as one of the heuristics under the schedule option menu.

Example F.2.1 (File Containing Workstation Information). The file shown in Table F.1 contains the information pertaining to various workstations. The first workstation consists of two machines in parallel.

Table F.1. Table for Example F.2.1.

Actual line	Comments
Flexible:	Defines "flexible" environment. If the word is "single", then this is an Uni-m/c environment.
Workstation: Wks000	Defines the first workstation. Wks000 is the name of the workstation.
Setup: A;B;1 B;A;2	Setup matrix for this workstation. The setup time between a status A operation and a status B operation is 1, whereas if you switch the order to B,A the setup time becomes 1. All other setup times are 0.
Machine: Wks000.000	Defines a machine with name Wks000.000
Machine: Wks000.001	Defines another machine with name Wks000.001
Workstation: Wks001	Defines the next workstation named Wks001
Setup:	Even when there are no setups the system requires a line for setups.
Machine: Wks001.000	
.....	(more machines in Wks001)

Example F.2.2 (File Containing Information Pertaining to Jobs). The file shown in Table F.2 contains the data pertaining to a job with multiple operations and arbitrary routes.

The following example describes the format of the ouput file.

Example F.2.3 (Format of the Output File). The output file has a very easy format. Only the sequences of the operations on the various machines have to be specified. It is not necessary to provide starting times and completion times in the output.

The following example contains a program written in C++ that schedules a number of jobs on a single machine. This example illustrates how the input data is read and how the output data is written.

F.2 Linking External Algorithms

Table F.2. Table for Example F.2.2.

Actual line		Comments
Shop:	Job	Defines the job shop environment. "Flow" would indicate a flow shop, and "single" would indicate a single workstation or machine.
Job:	Job000	Defines the first job, named Job000.
Release:	0	Release date of Job000
Due:	8	Due date of Job000
Weight:	4	Weight of Job000
Oper:	Wks000;2;B	The first operation on Job000 route. It requires 2 time units at Wks000 and a machine setting B.
Oper:	Wks004;1;A	The second operation on Job000 route.
.....		(More operations of Job000)
Job:	Job001	Defines the second job, named Job001.
Release:	0	Release date of Job001
Due:	10	Due date of Job001
Weight:	2	Weight of Job001
Oper:	Wks001;3;C	The first operation on Job001 route. It requires 3 time units at Wks001 and a machine setting C.
Oper:	Wks003;1;A	The second operation on Job001 route.
.....		(More operations of Job001)

Table F.3. Table for Example F.2.3.

Actual line		Comments
Schedule:	SB for Cmax	Provides the name "SB for Cmax" for the first schedule in the file.
Machine:	Wks000.000	This is the proper name for the machine Wks000.000 at the workstation Wks000.
Oper:	Job01	This is the first operation scheduled on the machine above. Since recirculation is not allowed the pair (Job01, Wks000) uniquely defines the operation.
Oper:	Job06	Second operation on this machine.
......		More operations on this machine.
Machine:	Wks000.001	This is the second machine of workstation Wks000.
Oper:	Job06	This is the first operation scheduled on machine Wks000.001
......		More operations on this machine.
Schedule:	SB for Lmax	provides the name "SB for Lmax" for the second schedule in the file.
....		Data with regard to the second schedule.

Example F.2.4 (Scheduling Jobs on a Single Machine). In this example we illustrate how the input files and the output file are used in an actual program. In this program a set of jobs are scheduled on a single machine according to the WSPT rule.

```cpp
#include <iostream.h>
#include <fstream.h>
#include <string.h>
#include <stdio.h>
#include <stdlib.h>
// We need a data structure to store information about job.
struct Tjob
{
int id;
int release;
int due;
int weight;
int proc;
double wp; //   weight divided by processing time
};
Tjob jobArray[100];
int jobCount;
// a string buffer.
char buffer[1024];
// ---------------------------------------------------------------
// For the single machine setting, we do not have to read
// the machine file. But we will do it here anyhow just to
// verify that it actually represents a single machine.
void ReadMch()
{
// No need to use the qualified path.  Just "_user.mch".
ifstream Fmch("_user.mch", ios::nocreate);
// Check the first line in the file.
// If it is not "Single:", too bad.
Fmch.getline(buffer, 1024);
if (strcmp(buffer, "Single:"))
{
cout <<  "we do not support flexible workstations!\n";
exit(1);
}
// Now we skip several lines. There are two ways to skip:
// Getline or ignore.  Getline allows you to check what you are skipping.
Fmch.getline(buffer, 1024);
// buffer = "Workstation:  Wks000",
// but we do not care.
Fmch.ignore(1024, '\n'); // skip "Setup:"
Fmch.ignore(1024, '\n'); // skip "Machine:"
// We do not need the availability time  or the starting status for the
```

F.2 Linking External Algorithms

```
// machine, but we will read it just to show how it is done.
Fmch.ignore(20);  // skip "Release:"
int avail: Fmch >> avail;
Fmch.ignore(20)   // skip "Status:"
// Counting spaces is not a good idea, so just read till the first character.
Fmch.eatwhite();
char status=Fmch.get();
// Now the rest of the file must contain only white-space characters.
Fmch.eatwhite();
if (!Fmch.eof())
{
cout << "The file must contain at least two workstations!\n";
exit(1);
}
// -----------------------------------------------------------
// With the job file it is less easy; a stream of jobs have to be read.
void readJob()
{
ifstream Fjob("_user.job", ios::nocreate);
Fjob >> buffer; // buffer = "Shop:", ignore
Fjob >> buffer; // check if single machine
if (strcmp(buffer, "Single"))
{
cout << "This is not a single machine file!\n";
exit(1);
}
while(1)
{
Fjob >> buffer; } // buffer = "Job:"
if (strcmp(buffer, "Job:")) // if not, must be the end of the file
break;
Fjob >> buffer;   // buffer = "Job###", ignore
jobarray[jobCount].id=jobCount;
Fjob >> buffer;   // buffer = "release:'
Fjob >> jobArray[jobCount].release;
Fjob >> buffer;   // buffer = "due:'
Fjob >> jobArray[jobCount].due;
Fjob >> buffer;   // buffer = "weight:'
Fjob >> jobArray[jobCount].weight;
Fjob >> buffer;   // buffer = "Oper:'
Fjob >> buffer;   // buffer = "Wks000;#;A" and we need the #
char* ss = strchr(buffer, ';');
if (!ss) break;
if (sscanf(ss+1, "%d", & jobArray[jobCount].proc) < 1) break;
jobArray[jobCount].wp=
double(jobArray[jobCount].weight/jobArray[jobCount].proc);
jobcount++;
}
if (jobCount == 0 )
```

```
{
cout << "No jobs defined!\n";
exit(1);
}
}
// -----------------------------------------------------------
// Compare function for sorting
int compare(const void* j1, const void* j2)
{
TJob* jb1= (TJob*)j1;
TJob* jb2= (TJob*)j2;
double a = jb1 ->wp - jb2->wp;
return a<0  ? -1 : a>0 ? 1: 0;
}
// Since this is just a single machine,
// we can implement any rule by sorting on the job array.
// We use that C standard qsort function.
void SortJobs()
{ qsort(jobArray, jobCount, sizeof(TJob), compare); }
// Output the schedule file.
void WriteSeq()
{
ofstream Fsch("_user.seq");
Fsch << "Schedule: WSPT rule\n"; // schedule name
Fsch << "Machine: Wks000.000\n"; // Name of the first and last machine
// Now enumerate the operations.
for (int i=0; i<jobCount; i++)
Fsch << "Oper: Job" << JobArray[i].id << "\n";
}
// -----------------------------------------------------------
int main (int argc. char* argv[])
{
// We have to have exactly 2 command line segments:
// objective function and time limit.
if (argc !=3)
{
cout << "illegal call!\n";
exit(1)
}
// Check the objective function.
// The WSPT rule is for the total weighted completion time.
// Do not bother to use sscanf
if (strcmp(argv[1], "3"))
{
cout << "The only objective supported is
total weighted   completion time.\n';
exit(1);
}
ReadMch();
```

```
ReadJob();
SortJobs();
WriteSeq();
cout << "Success\n";
return 0;
}
```

Comments and References

The LEKIN system is due to Asadathorn (1997) and Feldman and Pinedo (1998). The general purpose routine of the shifting bottleneck type that is embedded in the system is also due to Asadathorn (1997). The local search routines that are applicable to the flow shop and job shop are due to Kreipl (2000). The more specialized SB-LS routine for the flexible flow shop is due to Yang, Kreipl and Pinedo (2000).

References

E.H.L. Aarts and J.K. Lenstra (1997) *Local Search in Combinatorial Optimization*, J. Wiley, New York.

E.H.L. Aarts, P.J.M. van Laarhoven, J.K. Lenstra, and N.L.J. Ulder (1994) "A Computational Study of Local Search Algorithms for Job Shop Scheduling", *ORSA Journal of Computing*, Vol. 6, pp. 118–125.

I.N.K. Abadie, N.G. Hall and C. Sriskandarajah (2000) "Minimizing Cycle Time in a Blocking Flow Shop", *Operations Research*, Vol. 48, pp. 177–180.

T.S. Abdul-Razaq, C.N. Potts, and L.N. Van Wassenhove (1990) "A Survey of Algorithms for the Single Machine Total Weighted Tardiness Scheduling Problem", *Discrete Applied Mathematics*, Vol. 26, pp. 235–253.

J. Adams, E. Balas and D. Zawack (1988) "The Shifting Bottleneck Procedure for Job Shop Scheduling", *Management Science*, Vol. 34, pp. 391–401.

H.H. Adelsberger and J.J. Kanet (1991) "The LEITSTAND - A New Tool for Computer-Integrated-Manufacturing", *Production and Inventory Management Journal*, Vol. 32, pp. 43–48.

L. Adler, N.M. Fraiman, E. Kobacker, M.L. Pinedo, J.C. Plotnicoff and T.-P. Wu (1993) "BPSS: A Scheduling System for the Packaging Industry", *Operations Research*, Vol. 41, pp. 641–648.

A. Aggoun and A. Vazacopoulos (2004) "Solving Sports Scheduling and Timetabling Problems with Constraint Programming", in *Economics, Management and Optimization in Sports*, S. Butenko, J. Gil-Lafuente, and P. Pardalos (Eds.), Springer Verlag, New York.

R. Akkiraju, P. Keskinocak, S. Murthy, F. Wu (1998) "A New Decision Support System for Paper Manufacturing", in *Proceedings of the Sixth International Workshop on Project Management and Scheduling (1998)*, pp. 147–150, Bogazici University Printing Office, Istanbul, Turkey.

R. Akkiraju, P. Keskinocak, S. Murthy, F. Wu (2001) "An Agent based Approach to Multi Machine Scheduling", *Journal of Applied Intelligence,* Vol. 14, pp. 135–144.

K.M. Alguire and C.P. Gomes (1996) "ROMAN – An Application of Advanced Technology to Outage Management", in *Proceedings of the Sixth Annual Dual-Use Technology and Applications Conference,* pp. 290–295, IEEE Computer Society Press, Los Alamitos, California.

A. Anagnostopoulos, L. Michel, P. Van Hentenryck, and Y. Vergados (2003) "A Simulated Annealing Approach to the Traveling Tournament Problem", in *Proceedings CPAIOR'03.*

E.J. Anderson, C.A. Glass and C.N. Potts (1997) "Machine Scheduling", Chapter 11 in *Local Search in Combinatorial Optimization,* E.H.L. Aarts and J.K. Lenstra (Eds.), J. Wiley, New York.

D. Applegate and W. Cook (1991) "A Computational Study of the Job-Shop Scheduling Problem", *ORSA Journal on Computing,* Vol. 3, pp. 149–156.

M. Arguello (1994) "Review of Scheduling Software", Technology Transfer 93091822A-XFR, SEMATECH, Austin, Texas.

R. Armstrong, S. Gu and L. Lei (1996) "A Greedy Algorithm to Determine the Number of Transporters in a Cyclic Electroplating Process', *IIE Transactions,* Vol. 28, pp. 347–355.

B.C. Arntzen, G.G. Brown, T.P. Harrison and L.L. Trafton (1995) "Global Supply Chain Management at Digital Equipment Corporation", *Interfaces,* Vol. 25, pp. 69–93.

N. Asadathorn (1997) "Scheduling of Assembly Type of Manufacturing Systems: Algorithms and Systems Developments", Ph.D Dissertation, Department of Industrial Engineering, New Jersey Institute of Technology, Newark, NJ.

H. Atabakhsh (1991) "A Survey of Constraint Based Scheduling Systems Using an Artificial Intelligence Approach", *Artificial Intelligence in Engineering,* Vol. 6, No. 2, pp. 58–73.

E.M. Atkins, T.F. Abdelzaher, K.G. Shin, and E.H. Durfee (1999) "Planning and Resource Allocation for Hard Real-Time, Fault-Tolerant Plan Execution", in *Proceedings of the Third International Conference on Artificial Intelligence,* Seattle, Washington, pp. 244–251.

H. Aytug, S. Bhattacharyya, G.J. Koehler and J.L. Snowdon (1994) "A Review of Machine Learning in Scheduling", *IEEE Transactions on Engineering Management,* Vol. 41, pp. 165–171.

T.P. Bagchi (1999) *Multiobjective Scheduling by Genetic Algorithms,* Kluwer Academic Publishers, Boston.

References

K.R. Baker (1974) *Introduction to Sequencing and Scheduling*, John Wiley, NY.

K.R. Baker (1975) "A Comparative Survey of Flowshop Algorithms", *Operations Research,* Vol. 23, pp. 62–73.

K.R. Baker (1995) "Elements of Sequencing and Scheduling", Kenneth R. Baker, Tuck School of Business Administration, Dartmouth College, Hanover, NH.

E. Balas (1970) "Project Scheduling with Resource Constraints', in *Applications of Mathematical Programming Techniques,* E.M.L. Beale (Ed.), pp. 187–200, English University Press, London.

E. Balas, J.K. Lenstra, and A. Vazacopoulos (1995) "The One-Machine Problem with Delayed Precedence Constraints and its Use in Job Shop Scheduling", *Management Science,* Vol. 41, pp. 94–109.

Ph. Baptiste, C.L. Le Pape, and W. Nuijten (2001) *Constraint-Based Scheduling,* Kluwer Academic Publishers, Boston.

G. Barbarosoglu and D. Ozgur (1999) "Hierarchical Design of an Integrated Production and 2-Echelon Distribution System", *European Journal of Operational Research,* Vol. 118, pp. 464–484.

J.R. Barker and G.B. McMahon (1985) "Scheduling the General Job-shop", *Management Science,* Vol. 31, pp. 594–598.

C. Barnhart, P. Belobaba and A. Odoni (2003) "Applications of Operations Research in the Air Transport Industry", *Transportation Science,* Vol. 37, pp. 368–391.

C. Barnhart, F. Lu, and R. Shenoi (1998) "Integrated Airline Schedule Planning", Chapter 13 in *Operations Research in the Airline Industry,* G. Yu (Ed.), Kluwer Academic Publishers, Boston, pp. 384–403.

J.J. Bartholdi III, J.B. Orlin, and H.D. Ratliff (1980) "Cyclic Scheduling via Integer Programs with Circular Ones", *Operations Research,* Vol. 28, pp. 1074–1085.

T. Bartsch, A. Drexl and S. Kröger (2004) "Scheduling the Professional Soccer Leagues of Austria and Germany", *Computers and Operations Research,* to appear.

C. Basnet and J.H. Mize (1994) "Scheduling and Control of Flexible Manufacturing Systems: a Critical Review", *International Journal of Computer Integrated Manufacturing,* Vol. 7, pp. 340–355.

J. Bean (1994) "Genetic Algorithms and Random Keys for Sequencing and Optimization", *ORSA Journal of Computing,* Vol. 6, pp. 154–160.

J. Bean, J. Birge, J. Mittenthal and C. Noon (1991) "Matchup Scheduling with Multiple Resources, Release Dates and Disruptions", *Operations Research,* Vol. 39, pp. 470–483.

H. Beck (1993) "The Management of Job Shop Scheduling Constraints in TOSCA", in *Intelligent Dynamic Scheduling for Manufacturing Systems,* L. Interrante (Ed.), Proceedings of Workshop Sponsored by the National Science Foundation, the University of Alabama in Huntsville and Carnegie Mellon University, held in Cocoa Beach, January, 1993.

D.P. Bertsekas (1987) *Dynamic Programming: Deterministic and Stochastic Models,* Prentice Hall, New Jersey.

K. Bhaskaran and M. Pinedo (1992) "Dispatching", Chapter 83 in *Handbook of Industrial Engineering,* G. Salvendy (Ed.), pp. 2184–2198, J. Wiley, New York.

L. Bianco, S. Ricciardelli, G. Rinaldi and A. Sassano (1988) "Scheduling Tasks with Sequence-Dependent Processing Times", *Naval Research Logistics Quarterly,* Vol. 35, pp. 177–184.

Chr. Bierwirth and D.C. Mattfeld (1999) "Production Scheduling and Rescheduling with Genetic Algorithms", *Evolutionary Computation,* Vol. 7, pp. 1–17.

R.E. Bixby and E.K. Lee (1998) "Solving a Truck Dispatching Problem Using Branch-and-Cut", *Operations Research,* Vol. 46, pp. 355–367.

J.T. Blake and M.W. Carter (1997) "Surgical Process Scheduling: A Structured Review", *Journal of the Society for Health Systems,* Vol. 5, No. 3, pp. 17–30.

J.T. Blake and M.W. Carter (2002) "A Goal-Programming Approach to Resource Allocation in Acute-Care Hospitals", *European Journal of Operational Research,* Vol. 140, pp. 541–561.

J.T. Blake and J. Donald (2002) "Using Integer Programming to Allocate Operating Room Time at Mount Sinai Hospital", *Interfaces,* Vol. 32, No. 2, pp. 63–73.

J. Blazewicz, W. Cellary, R. Slowinski and J. Weglarz (1986), *Scheduling under Resource Constraints - Deterministic Models, Annals of Operations Research,* Vol. 7, Baltzer, Basel.

J. Blazewicz, M. Dror, and J. Weglarz (1991) "Mathematical Programming Formulations for Machine Scheduling: A Survey", *European Journal of Operational Research,* Vol. 51, pp. 283–300.

J. Blazewicz, K. Ecker, G. Schmidt and J. Weglarz (1993) *Scheduling in Computer and Manufacturing Systems,* Springer Verlag, Berlin.

References

L. Bodin, B. Golden, A. Assad, and M. Ball (1983) "Routing and Scheduling of Vehicles and Crews: The State of the Art", *Computers and Operations Research,* Vol. 10, pp. 63–211.

S. Bollapragada, M. Bussieck, and S. Mallik (2004) "Scheduling Commercial Videotapes in Broadcast Television", *Operations Research,* pp. ....

S. Bollapragada, H. Cheng, M. Phillips, M. Garbiras, M. Scholes, T. Gibbs, and M. Humphreville (2002) "NBC's Optimization Systems Increase Revenues and Productivity", *Interfaces,* Vol. 32, No. 1, pp. 47–60.

S. Bollapragada and M. Garbiras (2004), "Scheduling Commercials on Broadcast Television", *Operations Research,* Vol. 52, pp. 337–345.

G. Booch (1994) *Object-Oriented Analysis and Design with Applications (second edition),* Benjamin Cummings.

K.I. Bouzina and H. Emmons (1996) "Interval Scheduling on Identical Machines", *Journal of Global Optimization,* Vol. 9, pp. 379–393.

D.J. Bowersox and D.J. Closs (1996) *Logistical Management, the Integrated Supply Chain Process,* McGraw-Hill, New York.

H. Bräsel, L. Dornheim, S. Kutz, M. Mörig, and I. Rössling (2003) "LiSA - A Library of Scheduling Algorithms", Fakultät für Mathematik, Otto-von-Guericke Universität Magdeburg, Preprint Nr. 34 (2003).

H. Braun (2001) "Optimization in Supply Chain Management", *SAP Insider,* SAP AG., Vol. 2.

H. Braun and C. Groenewald (2000) "New Advances in SAP APO to Optimize the Supply Chain", *SAP Insider,* SAP AG. Vol. 1.

R.P. Brazile and K.M. Swigger (1988) "GATES: An Airline Assignment and Tracking System", *IEEE Expert,* Vol. 3, pp. 33–39.

D. Brelaz (1979) "New Methods to Color the Vertices of a Graph", *Communications of the ACM,* Vol. 22, pp. 251–256.

A. Brown and Z.A. Lomnicki (1966) "Some Applications of the Branch and Bound Algorithm to the Machine Sequencing Problem", *Operational Research Quarterly,* Vol. 17, pp. 173–186.

D.E. Brown and W.T. Scherer (Eds.) (1995) "Intelligent Scheduling Systems", Kluwer Academic Publishers, Boston.

G.G. Brown, G.W. Graves and D. Ronen (1987) "Scheduling Ocean Transportation of Crude Oil", *Management Science,* Vol. 33, pp. 335–346.

P. Brucker (2004) *Scheduling Algorithms,* (Fourth Edition), Springer Verlag, Berlin.

P. Brucker, A. Drexl, R. Möhring, K. Neumann, and E. Pesch (1999) "Resource Constrained Project Scheduling: Notation, Classification, Models, and Methods," *European Journal of Operational Research*, Vol. 112, pp. 3–41.

W. Brüggemann (1995) *Ausgewahlte Probleme der Produktionsplanung.* Physica-Verlag (Springer Verlag), Heidelberg.

W.J. Burgess and R.E. Busby (1992) "Personnel Scheduling", Chapter 81 in *Handbook of Industrial Engineering*, G. Salvendy (Ed.), pp. 2155–2169, J. Wiley, New York.

E. Burke and M.W. Carter (Eds.) (1998) *The Practice and Theory of Automated Timetabling II,* Selected Papers from the 2nd International Conference on the Practice and Theory of Automated Timetabling (held August 1997 in Toronto), Lecture Notes in Computer Science No. 1408, Springer Verlag, Berlin.

E. Burke and P. De Causmaecker (Eds.) (2003) *The Practice and Theory of Automated Timetabling IV,* Selected Papers from the 4th International Conference on the Practice and Theory of Automated Timetabling (held August 2002 in Gent, Belgium), Lecture Notes in Computer Science No. 2749, Springer Verlag, Berlin.

E. Burke, P. De Causmaecker, G. Vanden Berghe and H. Van Landeghem (2004) "The State of the Art of Nurse Rostering", *Journal of Scheduling*, Vol. 7, pp. 441–499.

E. Burke, D.G. Elliman, P.H. Ford, and R.F. Weare (1996) "Exam Timetabling in British Universities: A Survey", in *The Practice and Theory of Automated Timetabling*, E. Burke and P. Ross (Eds.), Lecture Notes in Computer Science, Vol. 1153, Springer Verlag, Berlin.

E. Burke and W. Erben (Eds.) (2001) *The Practice and Theory of Automated Timetabling III,* Selected Papers from the 3rd International Conference on the Practice and Theory of Automated Timetabling (held August 2000 in Konstanz, Germany), Lecture Notes in Computer Science, Vol. 2079, Springer Verlag, Berlin.

E. Burke and P. Ross (Eds.) (1996) *The Practice and Theory of Automated Timetabling,* Selected Papers from the 1st International Conference on the Practice and Theory of Automated Timetabling (held August 1995 in Edinburgh), Lecture Notes in Computer Science, Vol. 1153, Springer Verlag, Berlin.

L. Burns and C.F. Daganzo (1987) "Assembly Line Job Sequencing Principles", *International Journal of Production Research*, Vol. 25, pp. 71–99.

R.N. Burns and M.W. Carter (1985) "Work Force Size and Single Shift Schedules with Variable Demands", *Management Science*, Vol. 31, pp. 599–608.

References

R.N. Burns and G.J. Koop (1987) "A Modular Approach to Optimal Multiple Shift Manpower Scheduling", *Operations Research*, Vol. 35, pp. 100–110.

S. Butenko, J. Gil-Lafuente, and P. Pardalos (2004) *Economics, Management and Optimization in Sports*, Springer Verlag, Berlin.

H.G. Campbell, R.A. Dudek and M.L. Smith (1970) "A Heuristic Algorithm for the n Job m Machine Sequencing Problem", *Management Science*, Vol. 16, pp. B630–B637.

A. Caprara, M. Fischetti and P. Toth (2002) "Modeling and Solving the Train Timetabling Problem" *Operations Research*, Vol. 50, pp. 851–861.

M. Carey and D. Lockwood (1995) "A Model, Algorithms and Strategy for Train Pathing", *Journal of the Operational Research Society*, Vol. 46, pp. 988–1005.

J. Carlier (1982) "The One-Machine Sequencing Problem", *European Journal of Operational Research*, Vol. 11, pp. 42–47.

J. Carlier and E. Pinson (1989) "An Algorithm for Solving the Job Shop Problem", *Management Science*, Vol. 35, pp. 164–176.

J.J. Carreño (1990) "Economic Lot Scheduling for Multiple Products on Parallel Identical Processors", *Management Science*, Vol. 36, pp. 348–358.

D.C. Carroll (1965) "Heuristic Sequencing of Single and Multiple Component Jobs", *Ph.D Thesis*, Sloan School of Management, M.I.T., Cambridge, MA.

M.W. Carter (1986) "A Survey of Practical Applications of Examination Timetabling Algorithms", *Operations Research*, Vol. 34, pp. 193–202.

M.W. Carter and C.A. Tovey (1992) "When is the Classroom Assignment Problem Hard?" *Operations Research*, Vol. 40, pp. S28–S39.

T. Cayirli and E. Veral (2004) "Outpatient Scheduling in Health Care: A Review of the Literature", *Production and Operations Management*, Vol. 12, pp. 519–549.

X. Chao and M. Pinedo (1992) "A Parametric Adjustment Method for Dispatching", Technical Report, Department of Industrial Engineering and Operations Research, Columbia University, New York, NY.

X. Chao and M. Pinedo (1996) "Lot Sizing and Scheduling on Machines in Parallel and in Series", Technical Report, Department of Industrial Engineering and Operations Research, Columbia University, New York, NY.

C.-C. Cheng and S.F. Smith (1997) "Applying Constraint Satisfaction Techniques to Job Shop Scheduling", *Annals of Operations Research*, Vol. 70, pp. 327–357.

S. Chopra and P. Meindl (2001) *Supply Chain Management - Strategy, Planning, and Operation*, Prentice-Hall, Upper Saddle River, New Jersey.

Ph. Chrétienne, E.G. Coffman, Jr., J.K. Lenstra, and Z. Liu (Eds.) (1995) *Scheduling Theory and its Applications,* J. Wiley, New York.

M. Christiansen (1999) "Decomposition of a Combined Inventory and Time Constrained Ship Routing Problem", *Transportation Science,* Vol. 33, pp. 3–16.

M. Christiansen, K. Fagerholt and D. Ronen (2004) "Ship Routing and Scheduling – Status and Perspective", *Transportation Science,* Vol. 38, pp. 1–18.

A. Collinot, C. LePape and G. Pinoteau (1988) "SONIA: A Knowledge-Based Scheduling System", *Artificial Intelligence in Engineering,* Vol. 2, pp. 86–94.

R.W. Conway (1965a) "Priority Dispatching and Work-In-Process Inventory in a Job Shop", *Journal of Industrial Engineering,* Vol. 16, pp. 123–130.

R.W. Conway (1965b) "Priority Dispatching and Job Lateness in a Job Shop", *Journal of Industrial Engineering,* Vol. 16, pp. 228–237.

R.W. Conway, W.L. Maxwell, L.W. Miller (1967) *Theory of Scheduling,* Addison-Wesley, Reading, MA.

J.-F. Cordeau, G. Stojkovic, F. Soumis, and J. Desrosiers (2001) "Benders Decomposition for Simultaneous Aircraft Routing and Crew Scheduling", *Transportation Science,* Vol. 35, pp. 375–388.

D.W. Corne and P.M. Ross (1997) "Practical Issues and Advances in Evolutionary Scheduling", in *Genetic Algorithms in Engineering Applications,* Dasgupta and Michalewicz (Eds.), Springer Verlag, Berlin.

Y. Crama, V. Kats, J. Van de Klundert and E. Levner (2000) "Cyclic Scheduling in Robotic Flowshops", *Annals of Operations Research,* Vol. 96, pp. 97–124.

Y. Crama, W.J. Kolen, and A.G. Oerlemans (1990) "Throughput Rate Optimization in the Automated Assembly of Printed Circuit Boards", *Annals of Operations Research,* Vol. 26, pp. 455–480.

W.B. Crowston, M. Wagner, and J.F. Williams (1973) "Economic Lot Size Determination in Multi-Stage Assembly Systems", *Management Science,* Vol. 19, pp. 517–527.

F.H. Cullen, J.J. Jarvis and H.D. Ratliff (1981) "Set Partitioning Based Heuristics for Interactive Routing", *Networks,* Vol. 11, pp. 125–144.

J.R. Daduna, I. Branco, and J.M. Pinto Paixao (1995) *Computer Aided Scheduling of Public Transport.* Lecture Notes in Economics and Mathematical Systems No. 430, Springer Verlag, Berlin.

C.F. Daganzo (1989) "The Crane Scheduling Problem", *Transportation Research,* Vol. 23B, pp. 159–175.

References

D.G. Dannenbring (1977) "An Evaluation of Flowshop Sequencing Heuristics", *Management Science,* Vol. 23, pp. 1174–1182.

S. Dauzère-Pérès and J.-B. Lasserre (1993) "A Modified Shifting Bottleneck Procedure for Job-Shop Scheduling', *International Journal of Production Research,* Vol. 31, pp. 923–932.

S. Dauzère-Pérès and J.-B. Lasserre (1994) *An Integrated Approach in Production Planning and Scheduling,* Lecture Notes in Economics and Mathematical Systems, Vol. 411, Springer Verlag, Berlin.

M. Dawande, H.N. Geismar, S.P. Sethi and C. Sriskandarajah (2004) "Sequencing and Scheduling in Robotic Cells: Recent Developments", under review in *Journal of Scheduling.*

K. De Bontridder (2001) "Integrating Purchase and Production Planning – Using Local Search in Supply Chain Optimization", PhD Thesis, Technische Universiteit Eindhoven, the Netherlands.

M. Dell'Amico and M. Trubian (1991) "Applying Tabu-Search to the Job Shop Scheduling Problem", *Annals of Operations Research,* Vol. 41, pp. 231–252.

F. Della Croce, R. Tadei and G. Volta (1992) "A Genetic Algorithm for the Job Shop Problem", *Computers and Operations Research,* Vol. 22, pp. 15–24.

E.L. Demeulemeester and W.S. Herroelen (2002) *Project Scheduling: A Research Handbook,* Kluwer Academic Publishers, Boston, MA.

M.A.H. Dempster, J.K. Lenstra and A.H.G. Rinnooy Kan (Eds.) (1982) *Deterministic and Stochastic Scheduling,* Reidel, Dordrecht.

E.V. Denardo (1982) *Dynamic Programming: Models and Applications,* Prentice-Hall, NJ.

J. Desrosiers, A. Lasry, D. McInnes, M.M. Solomon, and F. Soumis (2000) "ALTITUDE: The Airline Operations Management System at Air Transat", *Interfaces,* Vol. 30, No. 2, pp. 41–53.

G. Desaulniers, J. Desrosiers, Y. Dumas, M. M. Solomon, and F. Soumis (1997) "Daily Aircraft Routing and Scheduling", *Management Science,* Vol. 43, pp. 841–855.

M. Desrochers and J.-M. Rousseau (1992) *Computer Aided Scheduling of Public Transport.* Lecture Notes in Economics and Mathematical Systems No. 386, Springer Verlag, Berlin.

D. de Werra (1988) "Some Models of Graphs for Scheduling Sports Competitions", *Discrete Applied Mathematics,* Vol. 21, pp. 47–65.

C. Dhaenens-Flipo and G. Finke (2001) "An Integrated Model for an Industrial Production-Distribution Problem", *IIE Transactions,* Vol. 33, pp. 705–715.

L. Di Gaspero (2003) *Local Search Techniques for Scheduling Problems: Algorithms and Software Tools*, PhD Thesis Series, Computer Science 2003/2, University of Udine, Udine, Italy.

G. Dobson (1987) "The Economic Lot-Scheduling Problem: Achieving Feasibility Using Time-Varying Lot Sizes", *Operations Research*, Vol. 35, pp. 764 771.

G. Dobson (1992) "The Cyclic Lot Scheduling Problem with Sequence-Dependent Setups", *Operations Research,* Vol. 40, pp. 736–749.

V.R. Dondeti and H. Emmons (1992) "Fixed Job Scheduling with Two Types of Processors", *Operations Research,* Vol. 40, pp. S76–S85.

A. Drexl and A. Kimms (1997) "Lot Sizing and Scheduling - Survey and Extensions", *European Journal of Operational Research,* Vol. 99, pp. 221–235.

A. Drexl and A. Kimms (2001) "Sequencing JIT Mixed-Model Assembly Lines Under Station-Load and Part-Usage Constraints", *Management Science,* Vol. 47, pp. 480–491.

A. Elkamel and A. Mohindra (1999) "A Rolling Horizon Heuristic for Reactive Scheduling of Batch Process Operations", *Engineering Optimization,* Vol. 31, pp. 763–792.

S.E. Elmaghraby (1967) "On the Expected Duration of PERT Type Networks", *Management Science,* Vol. 13, pp. 299–306.

H. Emmons (1969) "One-Machine Sequencing to Minimize Certain Functions of Job Tardiness", *Operations Research,* Vol. 17, pp. 701–715.

H. Emmons (1985) "Work–Force Scheduling with Cyclic Requirements and Constraints on Days Off, Weekends Off, and Work Stretch", *IIE Transactions,* Vol. 17, pp. 8–16.

H. Emmons and R.N. Burns (1991) "Off-Day Scheduling with Hierarchical Worker Categories", *Operations Research,* Vol. 39, pp. 484–495.

A. Feldman and M. Pinedo (1998) "The Design and Implementation of an Educational Scheduling System", Technical Report, Operations Management Area, Stern School of Business, New York University, New York.

A. Feldman (1999) *Scheduling Algorithms and Systems,* PhD Thesis, Department of Industrial Engineering and Operations Research, Columbia University, New York.

M.L. Fisher (1976) "A Dual Algorithm for the One-Machine Scheduling Problem", *Mathematical Programming,* Vol. 11, pp. 229–251.

M.L. Fisher (1981) "The Lagrangean Relaxation Method for Solving Integer Programming Problems", *Management Science,* Vol. 27, pp. 1–18.

References

M.L. Fisher and M.B. Rosenwein (1989) "An Interactive Optimization System for Bulk-Cargo Ship Scheduling", *Naval Research Logistics,* Vol. 36, pp. 27–42.

R. Fourer (1998) "Predictions for Web Technologies in Optimization", *INFORMS Journal on Computing,* Vol. 10, pp. 388–389.

R. Fourer and J.-P. Goux (2001) "Optimization as an Internet Resource", *Interfaces,* Vol. 31, No. 2, pp. 130–150.

M.S. Fox and S.F. Smith (1984) "ISIS – A Knowledge-Based System for Factory Scheduling", *Expert Systems,* Vol. 1, pp. 25–49.

J. Fraser (1995) "Finite Scheduling and Manufacturing Synchronization: Tools for Real Productivity", *IIE Solutions,* Vol. 27, No. 9 pp. 44–55.

S. French (1982) *Sequencing and Scheduling: an Introduction to the Mathematics of the Job Shop,* Horwood, Chichester.

D.R. Fulkerson (1962) "Expected Critical Path Lengths in PERT Type Networks', *Operations Research,* Vol. 10, pp. 808–817.

G. Gallego (1988) *Real-Time Scheduling of Several Products on a Single Facility with Setups and Backorders,* Ph.D. Thesis, ORIE Department, Cornell University, Ithaca, NY.

G. Gallego and M. Queyranne (1995) "Inventory Coordination and Pricing Decisions: Analysis of a Simple Class of Heuristics", Chapter 6 in *Mathematical Programming and Modelling Techniques in Practice,* A. Sciomachen (Ed.), J. Wiley, New York.

G. Gallego and R. Roundy (1992) "The Economic Lot Scheduling Problem with Finite Backorder Costs", *Naval Research Logistics,* Vol. 39, pp. 729–739.

G. Gallego and D.X. Shaw (1997) "Complexity of the ELSP with General Cyclic Schedules", *IIE Transactions,* Vol. 29, pp. 109–113.

J.P. Garcia-Sabater (2001) "The Problem of JIT Dynamic Sequencing. A Model and a Parametric Procedure", Proceedings of the ORP3 conference, Université Paris Dauphine, September 2001, Paris, France.

M.R. Garey and D.S. Johnson (1979) *Computers and Intractability - A Guide to the Theory of NP-Completeness,* W.H. Freeman and Company, San Francisco.

M.R. Garey, D.S. Johnson, B.B. Simons and R.E. Tarjan (1981) "Scheduling Unit-Time Tasks with Arbitrary Release Times and Deadlines", *SIAM Journal of Computing,* Vol. 10, pp. 256–269.

M. Gawande (1996) "Workforce Scheduling Problems with Side Constraints", presented at the semiannual INFORMS meeting in Washington DC (May, 1996).

J. Gaylord (1987) *Factory Information Systems,* Marcel Dekker, NY.

Y. Ge and Y. Yih (1995) "Crane Scheduling with Time Windows in Circuit Board Production Lines", *International Journal of Production Research*, Vol. 33, pp. 1187–1199.

B. Giffler and G.L. Thompson (1960) "Algorithms for Solving Production Scheduling Problems", *Operations Research*, Vol. 8, pp. 487–503.

P.C. Gilmore and R.E. Gomory (1964) "Sequencing a One-State Variable Machine: a Solvable Case of the Travelling Salesman Problem", *Operations Research*, Vol. 12, pp. 655–679.

C.R. Glassey and M. Mizrach (1986) "A Decision Support System for Assigning Classes to Rooms", *Interfaces,* Vol. 16, No. 5, pp. 92–100.

F. Glover (1990) "Tabu Search: A Tutorial", *Interfaces,* Vol. 20, No. 4, pp. 74–94.

C.P. Gomes, D. Smith, and S. Westfold (1996) "Synthesis of Schedulers for Planned Shutdowns of Power Plants", in *Proceedings of the Eleventh Knowledge-Based Software Engineering Conference,* pp. 12–29, IEEE Computer Society Press, Los Alamitos, California.

K. Gosselin and M. Truchon (1986) "Allocation of Classrooms by Linear Programming", *Journal of the Operational Research Society*, Vol. 37, pp. 561–569.

S.C. Graves (1981) "A Review of Production Scheduling", *Operations Research*, Vol.29, pp. 646–676.

S.C. Graves, H. C. Meal, D. Stefek and A. H. Zeghmi (1983) "Scheduling of Reentrant Flow Shops", *Journal of Operations Management,* Vol. 3, pp. 197–207.

M. Grönkvist (2002) "Using Constraint Propagation to Accelerate Column Generation in Aircraft Scheduling", Technical Report 2002, Computing Science Department, Chalmers University of Technology, Gothenburg, Sweden.

J.N.D. Gupta (1972) "Heuristic Algorithms for the Multistage Flow Shop Problem", *AIIE Transactions*, Vol. 4, pp. 11–18.

K. Haase (1994) *Lotsizing and Scheduling for Production Planning.* Lecture Notes in Economics and Mathematical Systems No. 408, Springer Verlag, Berlin.

K. Hadavi (1998) "Supply Chain Planning Integration: Design Consideration or Afterthought?" *APS Magazine,* Vol. 1, No. 2, MFG Publishing, Inc., Beverly, MA.

K. Hadavi and K. Voigt (1987) "An Integrated Planning and Scheduling Environment", in *Proceedings of AI in Manufacturing Conference,* Long Beach, California.

K. Hägele, C.O. Dúnlaing, and S. Riis (2001) "The Complexity of Scheduling TV Commercials", *Electronic Notes in Theoretical Computer Science*, Vol. 40.

URL: http://www.elsevier.nl/locate/entcs/volume40.html

N.G. Hall, H. Kamoun, and C. Sriskandarajah (1997) "Scheduling in Robotic Cells: Classification, Two and Three Machine Cells", *Operations Research*, Vol. 45, pp. 421–439.

N.G. Hall, W. Kubiak and S.P. Sethi (1991) "Earliness-Tardiness Scheduling Problems, II: Weighted Deviation of Completion Times about a Common Due Date", *Operations Research*, Vol. 39, pp. 847–856.

N.G. Hall, W.-P. Liu and J.B. Sidney (1998) "Scheduling in Broadcast Networks", *Networks*, Vol. 32, pp. 233–253.

N.G. Hall and M.E. Posner (1991) "Earliness-Tardiness Scheduling Problems, I: Weighted Deviation of Completion Times about a Common Due Date", *Operations Research*, Vol. 39, pp. 836–846.

N.G. Hall and C.N. Potts (2003) "Supply Chain Scheduling: Batching and Delivery", *Operations Research*, Vol. 51, pp. 566–584.

N.G. Hall and C. Sriskandarajah (1996) "A Survey of Machine Scheduling Problems with Blocking and No-Wait in Process", *Operations Research*, Vol. 44, pp. 421–439.

J.-P. Hamiez and J.-K. Hao (2001) "Solving the Sports League Scheduling Problem with Tabu Search", in *Local Search for Planning and Scheduling*, A. Nareyek (Ed.), Lecture Notes in Computer Science, Vol. 2148, pp. 24–36, Springer Verlag, Berlin.

A.N. Haq (1991) "An Integrated Production-Inventory-Distribution Model for Manufacturing of Urea: A Case", *International Journal of Production Economics*, Vol. 39, pp. 39–49.

F.W. Harris (1915) *Operations and Cost*, Factory Management Series, Shaw, Chicago.

A.C. Hax and H.C. Meal (1975) *Hierarchical Integration of Production Planning and Scheduling,*, M.A. Geisler (Ed.), TIMS Studies in the Management Sciences, Vol. 1: Logistics, North Holland, Amsterdam.

M. Henz (2001) "Scheduling a Major College Basketball Conference-Revisited", *Operations Research*, Vol. 49, pp. 163–168.

M. Henz, T. Müller, and S. Thiel (2004) "Global Constraints for Round Robin Tournament Scheduling", *European Journal of Operational Research*, Vol. 153, pp. 92–101.

J. Herrmann, C.-Y. Lee and J. Snowdon (1993) "A Classification of Static Scheduling Problems", in *Complexity in Numerical Optimization*, P.M. Pardalos (Ed.), World Scientific, pp. 203 – 253.

T.J. Hodgson and G.W. McDonald (1981a) "Interactive Scheduling of a Generalized Flow Shop. Part I: Success Through Evolutionary Development", *Interfaces*, Vol. 11, No. 2, pp. 42–47.

T.J. Hodgson and G.W. McDonald (1981b) "Interactive Scheduling of a Generalized Flow Shop. Part II: Development of Computer Programs and Files", *Interfaces*, Vol. 11, No. 3, pp. 83–88.

T.J. Hodgson and G.W. McDonald (1981c) "Interactive Scheduling of a Generalized Flow Shop. Part III: Quantifying User Objectives to Create Better Schedules", *Interfaces*, Vol. 11, No. 4, pp. 35–41.

K.L. Hoffman and M. Padberg (1993) "Solving Airline Crew Scheduling Problems by Branch-and-Cut", *Management Science*, Vol. 39, pp. 657–682.

D.J. Hoitomt, P.B. Luh and K.R. Pattipati (1993) "A Practical Approach to Job Shop Scheduling Problems", *IEEE Transactions on Robotics and Automation*, Vol. 9, pp. 1–13.

J.N. Hooker (2000) *Logic-Based Methods for Optimization: Combining Optimization and Constraint Satisfaction*, J. Wiley, New York.

J.H. Horen (1980) "Scheduling of Network Television Programs', *Management Science*, Vol. 26, pp. 354–370.

W.-L. Hsu, M. Prietula, G. Thompson and P.S. Ow (1993) "A Mixed-Initiative Scheduling Workbench: Integrating AI, OR and HCI", *Decision Support Systems*, Vol. 9, pp. 245–257.

T.C. Hu (1961) "Parallel Sequencing and Assembly Line Problems", *Operations Research*, Vol. 9, pp. 841–848.

R. Hung and H. Emmons (1993) "Multiple-Shift Workforce Scheduling under the 3-4 Compressed Workweek with a Hierarchical Workforce", *IIE Transactions*, Vol. 25, pp. 82–89.

L. Interrante (Ed.) (1993) *Intelligent Dynamic Scheduling for Manufacturing Systems*, Proceedings of a Workshop Sponsored by National Science Foundation, the University of Alabama in Huntsville and Carnegie Mellon University, held at Cocoa Beach, January, 1993.

J.R. Jackson (1955) "Scheduling a Production Line to Minimize Maximum Tardiness", Research Report 43, Management Science Research Project, University of California, Los Angeles.

J.R. Jackson (1956) "An Extension of Johnson's Results on Job Lot Scheduling", *Naval Research Logistics Quarterly*, Vol. 3, pp. 201–203.

V. Jain and I.E. Grossmann (2001) "Algorithms for Hybrid MILP/CP Models for a Class of Optimization Problems", *Informs Journal on Computing*, Vol. 13, pp. 258–276.

S.M. Johnson (1954) "Optimal Two and Three-Stage Production Schedules with Setup Times Included" *Naval Research Logistics Quarterly*, Vol. 1, pp. 61–67.

P. Jones and R. Inman (1989) "When is the Economic Lot Scheduling Problem Easy?" *IIE Transactions*, Vol. 21, pp. 11–20.

P. Jones and R. Inman (1996) "Product Grouping for Batch Processes", *International Journal of Production Research*, Vol. 31, pp. 3095–3105.

J.J. Kanet and H.H. Adelsberger (1987) "Expert Systems in Production Scheduling", *European Journal of Operational Research*, Vol. 29, pp. 51–59.

J.J. Kanet and V. Sridharan (1990) "The Electronic Leitstand: A New Tool for Shop Scheduling", *Manufacturing Review*, Vol. 3, pp. 161–170.

K.R. Karwan and J.R. Sweigart (Eds.) (1989) *Proceedings of the Third International Conference on Expert Systems and the Leading Edge in Production and Operations Management*, Conference held on Hilton Head Island, South Carolina, 1989, sponsored by Management Science Department, College of Business Administration, University of South Carolina.

K.G. Kempf (1989) "Intelligent Interfaces for Computer Integrated Manufacturing", in *Proceedings of the Third International Conference on Expert Systems and the Leading Edge in Production and Operations Management*, K.R. Karwan and J.R. Sweigart (Eds.), pp. 269–280, Management Science Department, College of Business Administration, University of South Carolina.

H. Kerzner (1994) *Project Management: A Systems Approach to Planning, Scheduling and Controlling (Fifth Edition)*, Van Nostrand Reinhold, New York.

P. Keskinocak, R. Goodwin, F. Wu, R. Akkiraju and S. Murthy (2001) "Decision Support for Managing an Electronic Supply Chain", *Electronic Commerce Research Journal*, Vol. 1, pp. 15–31.

P. Keskinocak, F. Wu, R. Goodwin, S. Murthy, R. Akkiraju, S. Kumaran, and A. Derebail (2002) "Scheduling Solutions for the Paper Industry", *Operations Research*, Vol. 50, pp. 249–259.

P. Keskinocak and S. Tayur (1998) "Scheduling of Time-Shared Jet Aircraft", *Transportation Science*, Vol. 32, pp. 277–294.

A. Kimms (1997) *Multi-Level Lot Sizing and Scheduling*, Physica Verlag, Heidelberg.

S. Kirkpatrick, C.D. Gelatt and M. P. Vecchi (1983) "Optimization by Simulated Annealing", *Science,* Vol. 220, pp. 671–680.

D. Kjenstad (1998) *Coordinated Supply Chain Scheduling,* PhD Thesis and NTNU Report 1998:24, Department of Production and Quality Engineering, Norwegian University of Science and Technology (NTNU), Trondheim, Norway.

R. Kolish (1995) *Project Scheduling under Resource Constraints,* Physica-Verlag (Springer Verlag), Heidelberg.

S. Kreipl (1998) *Ablaufplanung bei alternativen Arbeitsplanen,* Ph.D thesis, Universität Passau, Germany.

S. Kreipl (2000) "A Large Step Random Walk for Minimizing Total Weighted Tardiness in a Job Shop", *Journal of Scheduling,* Vol. 3, pp. 125–138.

S. Kreipl and M. Pinedo (2004) "Planning and Scheduling in Supply Chains – An Overview of Issues in Practice", *Production and Operations Management,* Vol. 13, pp. 77–92.

A. Kusiak (1990) *Intelligent Manufacturing Systems,* Prentice–Hall, Englewood Cliffs, New Jersey.

A. Kusiak (Ed.) (1992) *Intelligent Design and Manufacturing,* J. Wiley, New York.

A. Kusiak and M. Chen (1988) "Expert Systems for Planning and Scheduling Manufacturing Systems", *European Journal of Operational Research,* Vol. 34, pp. 113–130.

G. Laporte and S. Desroches (1984) "Examination Timetabling by Computer", *Computers and Operations Research,* Vol. 11, pp. 351–360.

E.L. Lawler, J.K. Lenstra and A.H.G. Rinnooy Kan (1982) "Recent Developments in Deterministic Sequencing and Scheduling: A Survey", in *Deterministic and Stochastic Scheduling,* Dempster, Lenstra and Rinnooy Kan (Eds.), pp. 35–74, Reidel, Dordrecht.

E.L. Lawler, J.K. Lenstra, A.H.G. Rinnooy Kan and D. Shmoys (1993) "Sequencing and Scheduling: Algorithms and Complexity", in *Handbooks in Operations Research and Management Science, Vol. 4: Logistics of Production and Inventory,* S.C. Graves, A. H. G. Rinnooy Kan and P. Zipkin, (Eds.), pp. 445–522, North-Holland, New York.

G. Lawton (1992) "Genetic Algorithms for Schedule Optimization", *AI Expert,* May Issue, pp. 23–27.

J.K. Lee, K.J. Lee, H.K. Park, J.S. Hong, and J.S. Lee (1997) "Developing Scheduling Systems for Daewoo Shipbuilding: DAS Project", *European Journal of Operational Research,* Vol. 97, pp. 380–395.

C.Y. Lee and L. Lei (Eds.) (1997) *Deterministic Sequencing and Scheduling,* Annals of Operations Research, Vol. 70, P. Hammer (Ed.), Baltzer, Basel.

C.Y. Lee, L. Lei and M. Pinedo (1997) "Current Trends in Deterministic Scheduling", Chapter 1 in *Annals of Operations Research,* Vol. 70, C.Y. Lee and L. Lei (Eds.), pp. 1–41. Baltzer, Basel.

C.-Y. Lee, R. Uzsoy, and L.A. Martin-Vega (1992) "Efficient Algorithms for Scheduling Semiconductor Burn-In Operations", *Operations Research,* Vol. 40, pp. 764–995.

Y.H. Lee, K. Bhaskaran and M.L. Pinedo (1997) "A Heuristic to Minimize the Total Weighted Tardiness with Sequence Dependent Setups", *IIE Transactions,* Vol. 29, pp. 45–52.

P. Lefrancois, M.H. Jobin and B. Montreuil (1992) "An Object-Oriented Knowledge Representation in Real-Time Scheduling", in *New Directions for Operations Research in Manufacturing,* G. Fandel, Th. Gulledge and A. Jones (Eds.), Springer Verlag, pp. 262–279.

V.J. Leon and S.D. Wu (1994) "Robustness Measures and Robust Scheduling for Job Shops", *IIE Transactions,* Vol. 26, pp. 32–43.

V.J. Leon, S.D. Wu and R.H. Storer (1994) "A Game Theoretic Approach for Job Shops in the Presence of Random Disruptions", *International Journal of Production Research,* Vol. 32, pp. 1451–1476.

L. Lettovsky, E.L. Johnson, G.L. Nemhauser (2000) "Airline Crew Recovery", *Transportation Science,* Vol. 34, pp. 337–348.

J. Y.-T. Leung (Ed.) (2004) *Handbook of Scheduling,* Chapman & Hall/CRC, Boca Raton, Florida.

J. Y.-T. Leung (2004) "Some Basic Scheduling Algorithms", Chapter 3 in *Handbook of Scheduling,* Chapman & Hall/CRC, Boca Raton, Florida.

J. Y.-T. Leung, H. Li and M. Pinedo (2005) "Scheduling Orders for Multiple Product Types with Due Date Related Objectives", *European Journal of Operational Research,* to appear.

J. Y.-T. Leung, H. Li and M. Pinedo (2005) "Order Scheduling in an Environment with Dedicated Resources in Parallel", *Journal of Scheduling,* Vol. 8.

J. Liu, Y. Jiang and Z. Zhou (2002) "Cyclic Scheduling of a Single Hoist in Extended Electroplating Lines: A Comprehensive Integer Programming Solution", *IIE Transactions,* Vol. 34, pp. 347–355.

J. Liu and B.L. MacCarthy (1996) "The Classification of FMS Scheduling Problems", *International Journal of Production Research,* Vol. 34, pp. 647–656.

J. Liu and B.L. MacCarthy (1997) "A Global MILP Model for FMS Scheduling", to appear in *European Journal of Operational Research,* Vol. 100, No. 3.

Z.A. Lomnicki (1965) "A Branch and Bound Algorithm for the Exact Solution of the Three-Machine Scheduling Problem", *Operational Research Quarterly,* Vol. 16, pp. 89–100.

P.B. Luh, and D.J. Hoitomt (1993) "Scheduling of Manufacturing Systems Using the Lagrangean Relaxation Technique", *IEEE Transactions on Automatic Control, Special Issue: Meeting the Challenge of Computer Science in the Industrial Applications of Control,* Vol. 38, pp. 1066–1080.

I.J. Lustig and J.-F. Puget (2001) "Program Does Not Equal Program: Constraint Programming and Its Relationship to Mathematical Programming" *Interfaces,* Vol. 31, No. 6, pp. 29–53.

B.L. MacCarthy and J. Liu (1993) "A New Classification Scheme for Flexible Manufacturing Systems", *International Journal of Production Research,* Vol. 31, pp. 299–309.

R.E. Marsten and F. Shepardson (1981) "Exact Solution of Crew Scheduling Problems Using the Set Partitioning Model: Recent Successful Applications", *Networks,* Vol. 11, pp. 165–177.

C.U. Martel (1982a) "Scheduling Uniform Machines with Release Times, Deadlines and Due times" in *Deterministic and Stochastic Scheduling,* M.A. Dempster, J.K. Lenstra and A.H.G. Rinnooy Kan (Eds.), pp. 89–99, Reidel, Dordrecht.

C.U. Martel (1982b) "Preemptive Scheduling with Release Times, Deadlines and Due Times", *Journal of the Association of Computing Machinery,* Vol. 29, pp. 812–829.

C. Martin, D. Jones and P. Keskinocak (2003) "Optimizing On-Demand Aircraft Schedules for Fractional Aircraft Operators", *Interfaces,* Vol. 33, No. 5, pp. 22–35.

J. Martin (1993) *Principles of Object-Oriented Analysis and Design,* Prentice-Hall, Englewood Cliffs, NJ.

H. Matsuo (1990) "Cyclic Sequencing Problems in the Two-machine Permutation Flow Shop: Complexity, Worst-Case and Average Case Analysis", *Naval Research Logistics,* Vol. 37, pp. 679–694.

H. Matsuo, C.J. Suh and R.S. Sullivan (1988) "A Controlled Search Simulated Annealing Method for the General Job Shop Scheduling Problem", Working Paper 03-44-88, Graduate School of Business, University of Texas, Austin.

W.L. Maxwell (1964) "The Scheduling of Economic Lot Sizes", *Naval Research Logistics Quarterly,* Vol. 11, pp. 89–124.

K. McAloon, C. Tretkoff, and G. Wetzel (1997) "Sports League Scheduling", in *Proceedings of the Third ILOG Optimization Suite International Users' Conference*, Paris, France.

S.T. McCormick, M.L. Pinedo, S. Shenker and B. Wolf (1989) "Sequencing in an Assembly Line with Blocking to Minimize Cycle Time", *Operations Research*, Vol. 37, pp. 925–936.

S.T. McCormick, M.L. Pinedo, S. Shenker and B. Wolf (1990) "Transient Behavior in A Flexible Assembly System", *The International Journal of Flexible Manufacturing Systems*, Vol. 3, pp. 27–44.

G.B. McMahon and M. Florian (1975) "On Scheduling with Ready Times and Due dates to Minimize Maximum Lateness", *Operations Research*, Vol. 23, pp. 475–482.

R. McNaughton (1959) "Scheduling with Deadlines and Loss Functions", *Management Science*, Vol. 6, pp. 1–12.

S.V. Mehta and R. Uzsoy (1999) "Predictable Scheduling of a Single Machine subject to Breakdowns", *International Journal of Computer Integrated Manufacturing*, Vol. 12, pp. 15–38.

T.C. Miller (2002) *Hierarchical Operations and Supply Chain Planning*, Springer Verlag, London, UK.

R. Miyashiro and T. Matsui (2003) "Round-Robin Tournaments with a Small Number of Breaks", Department of Mathematical Informatics Technical Report METR 2003-29, Graduate School of Information Science and Technology, the University of Tokyo, Tokyo, Japan.

J.G. Moder and C.R. Philips (1970) *Project Management with CPM and PERT*. Van Nostrand Reinhold, New York.

Y. Monden (1983) *Toyota Production System: Practical Approach to Production Management*, Institute of Industrial Engineers, Norcross, Georgia.

C.L. Monma and J.B. Sidney (1987) "Optimal Sequencing via Modular Decomposition: Characterizations of sequencing Functions", *Mathematics of Operations Research*, Vol. 12, pp. 22–31.

J.M. Moore (1968) "An n Job, One Machine Sequencing Algorithm for Minimizing the Number of Late Jobs", *Management Science*, Vol. 15, pp. 102–109.

T.E. Morton and D. Pentico (1993) *Heuristic Scheduling Systems*, John Wiley, NY.

J.A. Muckstadt and R. Roundy (1993) "Analysis of Multi-Stage Production Systems", Chapter 2 in *Handbooks in Operations Research and Management Science, Vol. 4: Logistics of Production and Inventory*, S.C. Graves, A.H.G. Rinnooy Kan (Eds.), pp. 59–132, North Holland, New York.

J.M. Mulvey (1982) "A Classroom/Time Assignment Model", *European Journal of Operational Research,* Vol. 9, pp. 64–70.

S. Murthy, R. Akkiraju, R. Goodwin, P. Keskinocak, J. Rachlin, F. Wu, S. Kumaran, J. Yeh, R. Fuhrer, A. Aggarwal, M. Sturzenbecker, R. Jayaraman and R. Daigle (1999) "Cooperative Multi-Objective Decision-Support for the Paper Industry", *Interfaces,* Vol. 29, No. 5, pp. 5–30.

E.J. Muth (1979) "The Reversibility Property of a Production Line", *Management Science,* Vol. 25, pp. 152–158.

R. Nanda and J. Browne (1992) *Introduction to Employee Scheduling,* Van Nostrand Reinhold, NY.

A. Nareyek (Ed.) (2001) *Local Search for Planning and Scheduling,* Revised Papers of ECAI 2000 Workshop in Berlin, Germany, Lecture Notes in Computer Science, No. 2148, Springer Verlag, Berlin.

G.L. Nemhauser and M. Trick (1998) "Scheduling a Major College Basketball Conference", *Operations Research,* Vol. 46, pp. 1–8.

G.L. Nemhauser and L.A. Wolsey (1999) *Integer and Combinatorial Optimization,* Wiley-Interscience, New York.

K. Neumann, C. Schwindt, and J. Zimmermann (2001) *Project Scheduling with Time Windows and Scarce Resources,* Lecture Notes in Economics and Mathematical Systems No. 508, Springer Verlag, Berlin.

S.J. Noronha and V.V.S. Sarma (1991) "Knowledge-Based Approaches for Scheduling Problems: A Survey", *IEEE Transactions on Knowledge and Data Engineering,* Vol. 3, pp. 160–171.

E. Nowicki and C. Smutnicki (1996) "A Fast Taboo Search algorithm for the Job Shop Problem", *Management Science,* Vol. 42, pp. 797–813.

E. Nowicki and S. Zdrzalka (1986) "A Note on Minimizing Maximum Lateness in a One-Machine Sequencing Problem with Release Dates", *European Journal of Operational Research,* Vol. 23, pp. 266–267.

W.P.M. Nuijten and E.H.L. Aarts (1996) "A Computational Study of Constraint Satisfaction for Multiple Capacitated Job Shop Scheduling", *European Journal of Operational Research,* Vol. 90, pp. 269–284.

M. Nussbaum and E.A. Parra (1993) "A Production Scheduling System", *ORSA Journal on Computing,* Vol. 5, pp. 168–181.

M.D. Oliff (Ed.) (1988) *Expert Systems and Intelligent Manufacturing, Proceedings of the Second International Conference on Expert Systems and the Leading Edge in Production Planning and Control,* held May 1988 in Charleston, South Carolina, Elsevier, New York.

M.D. Oliff and E.E. Burch (1985) "Multiproduct Production Scheduling at Owens-Corning Fiberglas", *Interfaces*, Vol. 15, No. 5, pp. 25–34.

I.M. Ovacik and R. Uzsoy (1997) *Decomposition Methods for Complex Factory Scheduling Problems*, Kluwer Academic Publishers, Boston.

P.S. Ow (1985) "Focused Scheduling in Proportionate Flowshops", *Management Science*, Vol. 31, pp. 852–869.

P.S. Ow and T.E. Morton (1988) "Filtered Beam Search in Scheduling", *International Journal of Production Research*, Vol. 26, pp. 297–307.

P.S. Ow and T.E. Morton (1989) "The Single Machine Early/Tardy Problem", *Management Science*, Vol. 35, pp. 177–191.

M. Padberg (1995) *Linear Optimization and Extensions*, Springer Verlag, Berlin.

D.S. Palmer (1965) "Sequencing Jobs Through a Multi-Stage Process in the Minimum Total Time - A Quick Method of Obtaining a Near Optimum", *Operational Research Quarterly*, Vol. 16, pp. 101–107.

S.S. Panwalkar and W. Iskander (1977) "A Survey of Scheduling Rules", *Operations Research*, Vol. 25, pp. 45–61.

C.H. Papadimitriou and P.C. Kannelakis (1980) "Flowshop Scheduling with Limited Temporary Storage", *Journal of the Association of Computing Machinery*, Vol. 27, pp. 533–549.

C.H. Papadimitriou and K. Steiglitz (1982) *Combinatorial Optimization: Algorithms and Complexity*, Prentice-Hall, NJ.

S. Park, N. Raman and M.J. Shaw (1997) "Adaptive Scheduling in Dynamic Flexible Manufacturing Systems: A Dynamic Rule Selection Approach," *IEEE Transactions on Robotics and Automation*, Vol. 13, pp. 486–502.

R.G. Parker (1995) *Deterministic Scheduling Theory*, Chapman & Hall, London.

R.G. Parker and R.L. Rardin (1988) *Discrete Optimization*, Academic Press, San Diego.

J.H. Patterson (1984) "A Comparison of Exact Approaches for Solving the Multiple Constrained Resources Project Scheduling Problem", *Management Science*, Vol. 30, pp. 854–867.

A.N. Perakis and W.M. Bremer (1992) "An Operational Tanker Scheduling Optimization System: Background, Current Practice and Model Formulation", *Maritime Policy and Management*, Vol. 19, pp. 177–187.

PERT (1958) "Program Evaluation Research Task", Phase I Summary Report, Special Projects Office, Bureau of Ordnance, Department of the Navy, Washington, D.C.

E. Pesch (1994) *Learning in Automated Manufacturing*, Physica-Verlag (Springer Verlag), Heidelberg.

J.R. Pimentel (1990) *Communication Networks for Manufacturing*, Prentice-Hall, Englewood Cliffs, NJ.

M. Pinedo (1982) "Minimizing the Expected Makespan in Stochastic Flow Shops", *Operations Research*, Vol. 30, pp. 148–162.

M. Pinedo (1985) "A Note on Stochastic Shop Models in which Jobs have the Same Processing Requirements on each Machine", *Management Science*, Vol. 31, pp. 840–845.

M. Pinedo (1995) *Scheduling – Theory, Algorithms, and Systems*, Prentice-Hall, Upper Saddle River, New Jersey.

M. Pinedo (2002) *Scheduling – Theory, Algorithms, and Systems*, (Second Edition), Prentice-Hall, Upper Saddle River, New Jersey.

M. Pinedo and X. Chao (1999) *Operations Scheduling with Applications in Manufacturing and Services*, Irwin/McGraw-Hill, Burr Ridge, Illinois.

M. Pinedo, R. Samroengraja and P.C. Yen (1994) "Design Issues with Regard to Scheduling Systems in Manufacturing", in *Control and Dynamic Systems*, C. Leondes (Ed.), Vol. 60, pp. 203–238, Academic Press, San Diego, California.

M. Pinedo, S. Seshadri and G. Shantikumar (1999) "Call Centers in Financial Services: Strategies, Technologies and Operations", Chapter 18 in *Creating Value in Financial Services*, E.L. Melnick, P.R. Nayyar, M. Pinedo and S. Seshadri (Eds.), pp. 357–388, Kluwer, Boston.

M. Pinedo and M. Singer (1999) "A Shifting Bottleneck Heuristic for Minimizing the Total Weighted Tardiness in a Job Shop", *Naval Research Logistics*, Vol. 46, pp. 1–18.

M. Pinedo, B. Wolf and S.T. McCormick (1986) "Sequencing in a Flexible Assembly Line with Blocking to Minimize Cycle Time", in *Proceedings of the Second ORSA/TIMS Conference on Flexible Manufacturing Systems*, K. Stecke and R. Suri (Eds.), Elsevier, Amsterdam, pp. 499–508.

M. Pinedo and B. P.-C. Yen (1997) "On the Design and Development of Object-oriented Scheduling Systems", in *Annals of Operations Research*, Vol. 70, C.-Y. Lee and L. Lei (Eds.), pp. 359–378, Baltzer, Basel.

E. Pinson (1995) "The Job Shop Scheduling Problem: A Concise Survey and Some Recent Developments", in *Scheduling Theory and its Applications*, Ph. Chrétienne, E.G. Coffman, Jr., J.K. Lenstra, and Z. Liu (Eds.), John Wiley, New york, pp. 177–293.

M.E. Posner (1985) "Minimizing Weighted Completion Times with Deadlines", *Operations Research*, Vol. 33, pp. 562–574.

C.N. Potts (1980) "Analysis of a Heuristic for One Machine Sequencing with Release Dates and Delivery Times", *Operations Research,* Vol. 28, pp. 1436–1441.

C.N. Potts and L.N. van Wassenhove (1982) "A Decomposition Algorithm for the Single Machine Total Tardiness Problem", *Operations Research Letters,* Vol. 1, pp. 177–181.

C.N. Potts and L.N. van Wassenhove (1985) "A Branch and Bound Algorithm for the Total Weighted Tardiness Problem", *Operations Research,* Vol. 33, pp. 363–377.

C.N. Potts and L.N. van Wassenhove (1987) "Dynamic Programming and Decomposition Approaches for the Single Machine Total Tardiness Problem", *European Journal of Operational Research,* Vol. 32, pp. 405–414.

C.N. Potts and L.N. van Wassenhove (1988) "Algorithms for Scheduling a Single Machine to Minimize the Weighted Number of Late Jobs", *Management Science,* Vol. 34, pp. 843–858.

R.M.V. Rachamadugu (1987) "A Note on the Weighted Tardiness Problem", *Operations Research,* Vol. 35, pp. 450–452.

R. Rachamadugu and K.E. Stecke (1994), "Classification and review of FMS scheduling procedures", *Production Planning and Control,* Vol. 5, pp. 2–20.

J. Rachlin, R. Goodwin, S. Murthy, R. Akkiraju, F. Wu, S. Kumaran, and R. Das (2001) "A-Teams: An Agent Architecture for Optimization and Decision Support", Working Paper, IBM Supply Chain Optimization Solutions, Atlanta, GA.

S.K. Reddy, J.E. Aronson, and A. Stam (1998) "SPOT: Scheduling Programs Optimally for Television", *Management Science,* Vol. 44, pp. 83–102.

J.C. Régin (1999) "Sports Scheduling and Constraint Programming", presentation at *INFORMS* meeting in Cincinnati, Ohio.

A.H.G. Rinnooy Kan (1976) *Machine Scheduling Problems: Classification, Complexity and Computations,* Nijhoff, The Hague.

F.A. Rodammer and K.P. White (1988) "A Recent Survey of Production Scheduling", *IEEE Transactions on Systems, Man and Cybernetics,* Vol. 18, pp. 841–851.

R. Roundy (1992) "Cyclic Schedules for Job Shops with Identical Jobs" *Mathematics of Operations Research,* Vol. 17, pp. 842–865.

R. Roundy, W. Maxwell, Y. Herer, S. Tayur and A. Getzler (1991) "A Price-Directed Approach to Real-Time Scheduling of Production Operations", *IIE Transactions,* Vol. 23, pp. 449–462.

B. Roy and B. Sussmann (1964) "Les Problemes d'Ordonnancement avec Constraintes Disjonctives", Note DS No. 9 bis, SEMA, Montrouge.

I. Sabuncuoglu and M. Bayiz (2000) "Analysis of Reactive Scheduling Problems in a Job Shop Environment", *European Journal of Operational Research*, Vol. 126, pp. 567–586.

N.M. Sadeh, D.W. Hildum and D. Kjenstad (2003) "Agent-Based e-Supply Chain Decision Support", *Journal of Organizational Computing and Electronic Commerce*, Vol. 13, No. 3.

S. Sahni and Y. Cho (1979) "Scheduling Independent Tasks with Due Times on a Uniform Processor System", *Journal of the Association of Computing Machinery*, Vol. 27, pp. 550–563.

S.C. Sarin, S. Ahn and A.B. Bishop (1988) "An Improved Branching Scheme for the Branch and Bound Procedure of Scheduling n Jobs on m Machines to Minimize Total Weighted Flow Time", *International Journal of Production Research*, Vol. 26, pp. 1183–1191.

M. Sasieni (1986) "A Note on PERT Times", *Management Science*, Vol. 30, pp. 1652–1653.

J. Sauer (1993) "Dynamic Scheduling Knowledge for Meta-Scheduling", in *Proceedings of the 6th International Conference on Industrial Engineering Applications of Artificial Intelligence and Expert Systems (IEA/AIE)*, Edinburgh, Scotland.

A. Schaerf (1999) "Scheduling Sport Tournaments Using Constraint Logic Programming", *Constraints*, Vol. 4, pp. 43–65.

A.-W. Scheer (1988) *CIM - Computer Steered Industry*, Springer Verlag, New York.

A. Scholl (1998) *Balancing and Sequencing of Assembly Lines (2nd Edition)* Physica-Verlag (Springer Verlag), Heidelberg.

J. Schönberger, D.C. Mattfeld and H. Kopfer (2004) "Memetic Algorithm Timetabling for Non-Commercial Sport Leagues", *European Journal of Operational Research*, Vol. 153, pp. 102–116.

J.A.M. Schreuder (1992) "Combinatorial Aspects of Construction of Competition Dutch Professional Football Leagues", *Discrete Applied Mathematics*, Vol. 35, pp. 301–312.

A. Schrijver (2003) *Combinatorial Optimization - Polyhedra and Efficiency (Vols. A, B, and C)*, Springer Verlag, Berlin.

J.F. Shapiro (2001) *Modeling the Supply Chain*, Duxbury, Pacific Grove, California.

M.J.P. Shaw (1988) "Knowledge-Based Scheduling in Flexible Manufacturing Systems: An Integration of Pattern-Directed Inference and Heuristic Search", *International Journal of Production Research,* Vol. 6, pp. 821–844.

M.J.P. Shaw, S. Park and N. Raman (1992) "Intelligent Scheduling with Machine Learning Capabilities: The Induction of Scheduling Knowledge", *IIE Transactions on Design and Manufacturing,* Vol. 24, pp. 156–168.

M.J.P. Shaw and A.B. Whinston (1989) "An Artificial Intelligence Approach to the Scheduling of Flexible Manufacturing Systems", *IIE Transactions,* Vol. 21, pp. 170–183.

J. Shepherd and L. Lapide (1998) "Supply Chain Optimization: Just the Facts", AMR Research, Inc.

J.B. Sidney and G. Steiner (1986) "Optimal Sequencing by Modular Decomposition: Polynomial Algorithms", *Operations Research,* Vol. 34, pp. 606–612.

D. Simchi-Levi, P. Kaminsky, and E. Simchi-Levi (2000) *Designing and Managing the Supply Chain,* McGraw-Hill/Irwin, Burr Ridge, IL.

M. Singer and M. Pinedo (1998) "A Computational Study of Branch and Bound Techniques for Minimizing the Total Weighted Tardiness in Job Shops", *IIE Transactions Scheduling and Logistics,* Vol. 30, pp. 109–118.

M.L. Smith, S.S. Panwalkar and R.A. Dudek (1975) "Flow Shop Sequencing with Ordered Processing Time Matrices", *Management Science,* Vol. 21, pp. 544–549.

M.L. Smith, S.S. Panwalkar and R.A. Dudek (1976) "Flow Shop Sequencing Problem with Ordered Processing Time Matrices: A General Case", *Naval Research Logistics Quarterly,* Vol. 23, pp. 481–486.

S.F. Smith (1992) "Knowledge-based Production Management: Approaches, Results and Prospects", *Production Planning and Control,* Vol. 3, pp. 350–380.

S.F. Smith (1994) "OPIS: a Methodology and Architecture for Reactive Scheduling", in *Intelligent Scheduling,* M. Zweben and M. Fox (Eds.), Morgan and Kaufmann, San Mateo, California.

S.F. Smith, M.S. Fox and P.S. Ow (1986) "Constructing and Maintaining Detailed Production Plans: Investigations into the Development of Knowledge-Based Factory Scheduling Systems", *AI Magazine,* Vol. 7, pp. 45–61.

S.F. Smith and O. Lassila (1994) "Configurable Systems for Reactive Production Management", in *Knowledge-Based Reactive Scheduling (B-15),* E. Szelke and R.M. Kerr (Eds.), Elsevier Science, North–Holland, pp. 93-106.

S.F. Smith, N. Muscettola, D.C. Matthys, P.S. Ow and J.Y. Potvin (1990) "OPIS: An Opportunistic Factory Scheduling System", *Proceedings of the*

Third International Conference on Industrial and Expert Systems, (IEA/AIE 90), Charleston, South Carolina.

W.E. Smith (1956) "Various Optimizers for Single Stage Production", *Naval Research Logistics Quarterly*, Vol. 3, pp. 59–66.

J.J. Solberg (1989) "Production Planning and Scheduling in CIM", *Information Processing*, Vol. 89, pp. 919–925.

H. Stadtler and C. Kilger (Eds.) (2002) *Supply Chain Management and Advanced Planning - Concepts, Models, Software, and Case Studies (2nd Edition)* Springer, Berlin.

G. Stojkovic, F. Soumis, J. Desrosiers, and M. Solomon (2002) "An Optimization Model for a Real-Time Flight Scheduling Problem", *Transportation Research Part A: Policy and Practice*, Vol. 36, pp. 779–788.

M. Stojkovic, F. Soumis, and J. Desrosiers (1998) "The Operational Airline Crew Scheduling Problem", *Transportation Science*, Vol. 32, pp. 232–245.

R.H. Storer, S.D. Wu and R. Vaccari (1992) "New Search Spaces for Sequencing Problems with Application to Job Shop Scheduling", *Management Science*, Vol. 38, pp. 1495–1509.

K. Strobel (2001) "mySAP Supply Chain Management: Managing the Supply Chain beyond Corporate Borders", *SAP Insider*, SAP AG., Vol. 2.

D.R. Sule (1996) "Industrial Scheduling", PWS Publishing Co. (ITP), Boston.

G. Sullivan and K. Fordyce (1990) "IBM Burlington's Logistics Management System", *Interfaces*, Vol. 20, pp. 43–64.

C.S. Sung and S.H. Yoon (1998) "Minimizing Total Weighted Completion Time at a Pre-Assembly Stage Composed of Two Feeding Machines", *International Journal of Production Economics*, Vol. 54, pp. 247–255.

W. Szwarc (1971) "Elimination Methods in the $m \times n$ Sequencing Problem", *Naval Research Logistics Quarterly*, Vol. 18, pp. 295–305.

W. Szwarc (1973) "Optimal Elimination Methods in the $m \times n$ Flow Shop Scheduling Problem", *Operations Research*, Vol. 21, pp. 1250–1259.

W. Szwarc (1978) "Dominance Conditions for the Three-Machine Flow-Shop Problem", *Operations Research*, Vol. 26, pp. 203–206.

E. Taillard (1990) "Some Efficient Heuristic Methods for the Flow Shop Sequencing Problem', *European Journal of Operational Research*, Vol. 47, pp. 65–74.

E. Taillard (1994) "Parallel Taboo Search Techniques for the Job Shop Scheduling Problem", *ORSA Journal of Computing*, Vol. 6, pp. 108–117.

F.B. Talbot (1982) "Resource Constrained Project Scheduling with Time-Resource Trade-Offs: The Nonpreemptive Case", *Management Science*, Vol. 28, pp. 1197–1210.

C.S. Tang and E.V. Denardo (1988) "Models Arising from a Flexible Manufacturing Machine, Part I: Minimization of the Number of Tool Switches," *Operations Research*, Vol. 36, pp. 767–777.

J.M. Tien and A. Kamiyama (1982) "On Manpower Scheduling Algorithms", *SIAM Review*, Vol. 24, pp. 275–287.

M. Trick (2001) "A Schedule-and-Break Approach to Sports Scheduling" in *Practice and Theory of Automated Timetabling III*, E.K. Burke and W. Erben (Eds.), Selected Papers of Third International Conference PATAT 2000 (Konstanz, Germany), Lecture Notes in Computer Science No 2079, pp. 242–253, Springer Verlag, Berlin.

R. Uzsoy (1993) "Decomposition Methods for Scheduling Complex Dynamic Job Shops", in *Proceedings of the 1993 NSF Design and Manufacturing Systems Conference*, Vol. 2, pp. 1253–1258, Society of Manufacturing Engineers, Dearborn, MI

R. Uzsoy, C.-Y. Lee and L.A. Martin-Vega (1992) "A Review of Production Planning and Scheduling Models in the Semiconductor Industry, Part I: System Characteristics, Performance Evaluation and Production Planning", *IIE Transactions*, Vol. 24, pp. 47–61.

R.J.M. Vaessens, E.H.L. Aarts, and J.K. Lenstra (1996) "Job Shop Scheduling by Local Search", *INFORMS Journal on Computing*, Vol. 8, pp. 302–317.

N.J. Vandaele and M. Lambrecht (2001) "Planning and Scheduling in an Assemble-to-Order Environment: Spicer-Off-Highway Products Division", Technical Report, Antwerp University, Belgium.

S.L. Van de Velde (1991) *Machine Scheduling and Lagrangean Relaxation*, Ph.D thesis, Technische Universiteit Eindhoven, the Netherlands.

P. Van Hentenryck (1989) *Constraint Satisfaction in Logic Programming*, MIT Press, Cambridge, MA.

P. Van Hentenryck and I. Lustig (1999) *The OPL Optimization Programming Language*, MIT Press, Cambridge, MA.

P.J.M. Van Laarhoven, E.H.L. Aarts and J.K. Lenstra (1992) "Job Shop Scheduling by Simulated Annealing", *Operations Research*, Vol. 40, pp. 113–125.

A. Vepsalainen and T.E. Morton (1987) "Priority Rules and Lead Time Estimation for Job Shop Scheduling with Weighted Tardiness Costs", *Management Science*, Vol. 33, pp. 1036–1047.

G.E. Vieira, J.W. Herrmann, and E. Lin (2003) "Rescheduling Manufacturing Systems: A Framework of Strategies, Policies, and Methods", *Journal of Scheduling*, Vol. 6, pp. 39–62.

S. Voss and J.R. Daduna (2001) *Computer Aided Scheduling of Public Transport*. Lecture Notes in Economics and Mathematical Systems No. 505, Springer Verlag, Berlin.

M.R. Walker and J.S. Sayer (1959) "Project Planning and Scheduling", Report 6959, E.I. du Pont de Nemours & Co., Inc., Wilmington, Del.

S. Webster (2000) "Frameworks for Adaptable Scheduling Algorithms", *Journal of Scheduling*, Vol. 3, pp. 21–50.

L.M. Wein (1988) "Scheduling Semi-Conductor Wafer Fabrication", *IEEE Transactions on Semiconductor Manufacturing*, Vol. 1, pp. 115–129.

L.M. Wein and P.B. Chevelier (1992) "A Broader View of the Job-Shop Scheduling Problem", *Management Science*, Vol. 38, pp. 1018–1033.

M. Widmer and A. Hertz (1989) "A New Heuristic Method for the Flow Shop Sequencing Heuristic", *European Journal of Operational Research*, Vol. 41, 186–193.

J.D. Wiest and F.K. Levy (1977) *A Management Guide to PERT/CPM*. Prentice-Hall, Englewood Cliffs.

N.H.M. Wilson (1999) *Computer Aided Scheduling of Public Transport*. Lecture Notes in Economics and Mathematical Systems No. 471, Springer Verlag, Berlin.

R.H. Wilson (1934) "A Scientific Routine for Stock Control," *Harvard Business Review*, Vol. 13, pp. 116–128.

D.A. Wismer (1972) "Solution of Flowshop Scheduling Problem with No Intermediate Queues", *Operations Research*, Vol. 20, pp. 689–697.

R.J. Wittrock (1985) "Scheduling Algorithms for Flexible Flow Lines", *IBM Journal of Research and Development*, Vol. 29, pp. 401–412.

R.J. Wittrock (1988) "An Adaptable Scheduling Algorithm for Flexible Flow Lines", *Operations Research*, Vol. 36, pp. 445–453.

R.J. Wittrock (1990) "Scheduling Parallel Machines with Major and Minor Setup Times", *International Journal of Flexible Manufacturing Systems*, Vol. 2, pp. 329–341.

I.W. Woerner and E. Biefeld (1993) "HYPERTEXT-BASED Design of a User Interface for Scheduling", in *Proceedings of the AIAA Computing in Aerospace 9*, San Diego, California.

L.A. Wolsey (1998) *Integer Programming*, John Wiley, New York.

A. Wren and J.R. Daduna (1988) *Computer Aided Scheduling of Public Transport.* Lecture Notes in Economics and Mathematical Systems No. 308, Springer Verlag, Berlin.

S.D. Wu, E.S. Byeon, and R.H. Storer (1999) "A Graph-Theoretic Decomposition of Job Shop Scheduling Problems to Achieve Schedule Robustness", *Operations Research,* Vol. 47, pp. 113–124.

S.D. Wu, R.H. Storer, and P.C. Chang (1991) "A Rescheduling Procedure for Manufacturing Systems under Random Disruptions", in *New Directions for Operations Research in Manufacturing,* T. Gulledge and A. Jones (Eds.), Springer Verlag, Berlin.

Y. Yang, S. Kreipl, and M. Pinedo (2000) "Heuristics for Minimizing the Total Weighted Tardiness in Flexible Flow Shops", *Journal of Scheduling,* Vol. 3, pp. 89–108.

C.A. Yano and A. Bolat (1989) "Survey, Development and Applications of Algorithms for Sequencing Paced Assembly Lines", *Journal of Manufacturing and Operations Management,* Vol. 2, pp. 172–198.

P.C. Yen (1995) *On the Architecture of an Object-Oriented Scheduling System,* Ph.D thesis, Department of Industrial Engineering and Operations Research, Columbia University, New York, New York.

P.C. Yen (1997) "Interactive Scheduling Agents on the Internet", in *Proceedings of the Hawaii International Conference on System Science (HICSS-30),* Hawaii.

Y. Yih (1994) "An Algorithm for Hoist Scheduling Problems", *International Journal of Production Research,* Vol. 32, pp. 501-516.

Y. Yih (1990) "Trace-driven Knowledge Acquisition (TDKA) for Rule-Based Real Time Scheduling Systems", *Journal of Intelligent Manufacturing,* Vol. 1, pp. 217–230.

E. Yourdon (1994) *Object-Oriented Design: An Integrated Approach,* Prentice-Hall, Englewood Cliffs, NJ.

G. Yu (Ed.) (1998) *Operations Research in the Airline Industry,* Kluwer Academic Publishers, Boston.

P.H. Zipkin (2000) *Foundations of Inventory Management,* McGraw-Hill/Irwin, Burr Ridge, IL.

M. Zweben and M. Fox (Eds.) (1994) *Knowledge-Based Scheduling,* Morgan and Kaufmann, San Mateo, California.

Notation

Symbol	Description	Chapter (Section)
A	set of arcs; precedence constraints	4; 5; A
B	set of (disjunctive) arcs	5
c_{ab}^m	cost of moving (transporting) one unit from a to b	2(4); 8(4)
c_{ij}^p	cost to produce one unit of type j in facility i	8(4)
c_{jk}^s	cost of setup for job k after job j	2(4); 7
C_{ij}	completion time of job j on machine i	4; 5; 6
C_j	final completion time of job j in system	4; 5; 6
$C_{\max}$	makespan	4; 5; 9; A; B
d_j	due date	2; 3; 5; 8(5); 9; B; C
D_j	demand rate of items of type j	7; 8(4)
$\mathcal{D}$	domain of variable x_j	D
E_j	earliness of job j	2; 14(2)
G	graph	2; 4; 5
h_j	holding (inventory carrying) cost for one unit of type j	2; 7; 8
L_j	lateness of job j	2; 5; B
$L_{\max}$	maximum lateness	2; 5; B
m	number of machines	5; 6; 7; 9; A; B; C
M	set of machines	5; 9
M_j	set of machines that can process job j	9
n	number of jobs	4; 5; 6; 7; 9; A; B; C
$\mathcal{N}_j$	number of jobs of type j	6(2)
o_{ik}	overload on machine i due to job k	6(4)
O_{ik}	cumulative overload on machine i	6(4)
Q_j	production rate of items of type j	7

Symbol	Description	Chapter (Section)
r_j	release date	2; 3; 5; 9; B; C
R_i	number of parts of type i	6(5)
s_{ijk}	setup time between jobs j and k on machine i	7; 8
S_{ij}	starting time of job j on machine i	5
S_j	starting time of job j	4; 5
$\mathcal{S}$	set of schedules	7; 11
$\mathcal{S}_k$	schedule k	C(5)
t	a point in time (an epoch); slot	8; 9; 10; A(4); B(4)
T_j	tardiness of job j	2; 5; C
u^l	idle time after items in position l of the sequence	7(4)
U	source node	4(4); 5
V	sink node	4(4); 5
$\mathcal{V}$	node in branching tree	B(4)
w_j	weight	2; 3; 8(5); 9; A; C
W_i	number of operators in workforce pool i	4; 9; 12
W	number of operators in entire workforce	4; 9; 12
$\mathcal{W}$	workload	6(4)
X_{ij}	departure (exit) time of job j from machine i	6(2)
x	decision variable	7; 8; 9; 11; A; D
y_k	production frequency	7(4)
α	weight	8(5)
Δ	difference	6(5)
μ_j	expected processing time of job j	4(3)
ν_{it}	potential utilization of machine i at time t	9(3)
π	profit	9; 11
ρ_j	utilization factor (D_j/Q_j)	7
ρ^l	feasible row price	12(6)
$\sigma_{(...)}$	slack	5(5)
σ_j^2	variance of the processing time of job j	4(3)
τ	duration; run time	7; 8
ϕ	flexibility	5(5)
ψ	penalty	6(3); 8(4)
Ω	set of operations	B(4)

Subscripts

Symbol	Description	Chapter
cp	critical path	4
i (h)	machine i (h) or facility i (h)	
j (k)	job or activity j (k)	
l	stage	8
t	point in time (epoch)	8; 9

Superscripts

Symbol	Description	Chapter
l	position in sequence	7
m	moving (transportation)	8
p	production	8
s	setup	7; 8

Name Index

A
Aarts, E.H.L., 113, 434, 446
Abadie, I.N.K., 139
Abdelzaher, T.F., 450
Abdul-Razaq, T.S., 112
Adams, J., 112
Adelsberger, H.H., 343, 449, 450
Adler, L., 450
Aggarwal, A., 201, 379
Aggoun, A., 251
Ahn, S., 411
Akkiraju, R., 201, 368, 379
Alguire, K.M., 79
Anagnostopoulos, A., 251
Anderson, E.J., 434
Applegate, D., 112, 412
Arguello, M., 449
Armstrong, R., 379
Arntzen, B.C., 201
Aronson, J.E., 252
Asadathorn, N., 113, 457
Assad, A., 315
Atabakhsh, H., 18, 343
Atkins, E.M., 450
Aytug, H., 368

B
Bagchi, T.P., 17
Baker, K.R., 16, 17, 79, 112
Balas, E., 79, 112
Ball, M., 315

Baptiste, Ph., 17, 113, 446
Barbarosoglu, G., 201
Barker, J.R., 112, 412
Barnhart, C., 287
Bartholdi III, J.J., 315
Bartsch, T., 252
Basnet, C., 130
Bayiz, M., 368
Bean, J., 139, 435
Beck, H., 450
Belobaba, P., 287
Bertsekas, D.P., 411
Bhaskaran, K., 112, 434, 435
Bhattacharyya, S., 368
Bianco, L., 139
Biefeld, E., 343
Bierwirth, Chr., 368
Birge, J., 139
Bishop, A.B., 411
Bixby, R.E., 412
Blake, J.T., 379
Blazewicz, J., 16, 394
Bodin, L., 315
Bolat, A., 139
Bollapragada, S., 252
Booch, G., 369
Bouzina, K.I., 227
Bowersox, D.J., 201
Branco, I., 17, 48, 287
Bräsel, H., 450

Braun, H., 201
Brazile, R.P., 450
Brelaz, D., 228
Bremer, W.M., 287
Brown, A., 112, 412
Brown, D.E., 18
Brown, G.G., 201, 287
Browne, J., 17, 48, 314
Brucker, P., 16, 17, 35, 79
Brüggemann, W., 17, 169
Burch, E.E., 169
Burgess, W., 18, 314
Burke, E., 17, 18, 48, 228, 450
Burns, L., 139
Burns, R.N., 314
Busby, R.E., 18, 314
Bussieck, M., 252
Butenko, S., 17
Byeon, E.S., 368

C

Campbell, H.G., 112
Caprara, A., 287
Carey, M., 287
Carlier, J., 112, 412
Carreño, J.J., 169
Carroll, D.C., 112, 434
Carter, M.W., 17, 48, 228, 314, 379
Cayirli, T., 379
Cellary, W., 16
Chang, P.C., 368
Chao, X., 169, 368
Chen, M., 343
Cheng, C.-C., 113
Cheng, H., 252
Chevelier, P.B., 113
Cho, Y., 228
Chopra, S., 201
Chrétienne, P., 17
Christiansen, M., 48, 287
Closs, D.J., 201
Coffman, E.G., 17
Collinot, A., 369, 450
Conway, R.W., 16, 35, 111, 434
Cook, W., 112, 412
Cordeau, J.-F., 287

Corne, D.W., 435
Crama, Y., 139, 379
Crowston, W.B., 169
Cullen, F.H., 315

D

Daduna, J.R., 17, 48, 287
Daganzo, C.F., 139, 379
Daigle, R., 201, 379
Dannenbring, D.G., 112
Das, R., 201, 379
Dauzère-Pérès, S., 17, 112
Dawande, M., 379
De Bontridder, K., 201
De Causmaecker, P., 17, 18, 48, 228
Della Croce, F., 113, 434
Dell'Amico, M., 113, 434
Dempster, M.A.H., 17, 48
Demeulemeester, E.L., 16, 79
Denardo, E.V., 139, 411
Derebail, A., 201
Desaulniers, G., 287
Desrochers, M., 17, 48, 287
Desroches, S., 228
Desrosiers, J., 287, 315
de Werra, D., 251
Dhaenens-Flipo, C., 201
Di Gaspero, L., 228
Dobson, G., 169
Donald, J., 379
Dondeti, V.R., 228
Dornheim, L., 450
Drexl, A., 16, 17, 35, 79, 139, 169, 252
Dror, M., 394
Dudek, R.A., 112
Dumas, Y., 287
Dúnlaing, C.O., 252
Durfee, E.H., 450

E

Ecker, K., 16
Elkamel, A., 368
Elliman, D.G., 450
Elmaghraby, S.E., 79
Emmons, H., 227, 314
Erben, W., 17, 48, 228

F

Fagerholt, K., 48
Feldman, A., 113, 369, 457
Finke, G., 201
Fischetti, M., 287
Fisher, M.L., 112, 287, 411
Florian, M., 112, 412
Ford, P.H., 450
Fordyce, K., 450
Fourer, R., 379
Fox, M.S., 18, 343, 369
Fraiman, N.M., 450
Fraser, J., 449
French, S., 16
Fuhrer, R., 201, 379
Fulkerson, D.R., 79

G

Gallego, G., 169
Garbiras, M., 252
Garcia-Sabater, J.P., 139
Garey, M.R., 228
Gawande, M., 315
Gaylord, J., 343
Ge, Y., 379
Geismar, H.N., 379
Gelatt, C.D., 434
Getzler, A., 435
Gibbs, T., 252
Giffler, B., 412
Gil-Lafuente, J., 17
Gilmore, P.C., 139
Glass, C.A., 434
Glassey, C.R., 228
Glover, F., 434
Golden, B., 315
Goltz, H.-J., 450
Goodwin, R., 201, 379
Gosselin, K., 228
Gomes, C.P., 79
Gomory, R.E., 139
Goux, J.-P., 379
Graves, G.W., 287
Graves, S.C., 17, 113, 169
Groenewald, C., 201
Grönkvist, M., 446

Grossmann, I.E., 446
Gu, S., 379
Gupta, J.N.D., 112

H

Haase, K., 17, 169
Hadavi, K., 201
Hägele, K., 252
Hall, N.G., 139, 201, 228, 252
Hamiez, J.-P., 251
Hao, J.-K., 251
Haq, A.N., 201
Harris, F.W., 169
Harrison, T.P., 201
Hax, A.C., 201
Henz, M., 251, 252
Herer, Y., 435
Herrmann, J.W., 35, 368
Herroelen, W.S., 16, 79
Hertz, A., 112
Hildum, D.W., 201, 379
Hodgson, T.J., 113
Hoffman, K.L., 315, 412
Hoitomt, D.J., 112, 412
Hong, J.S., 450
Hooker, J.N., 446
Horen, J.B., 252
Hsu, W.-L., 450
Hu, T.C., 111, 434
Humphreville M., 252
Hung, R., 314

I

Inman, R., 169
Interrante, L., 343
Iskander, W., 112, 434

J

Jackson, J.R., 111, 112, 434
Jain, V., 446
Jarvis, J.J., 315
Jayaraman, R., 201, 379
Jiang, Y., 379
Jobin, M.H., 343
Johnson, D.S., 228
Johnson, E.L., 315
Johnson, S.M., 112
Jones, D., 287
Jones, Ph., 169

K

Kaminsky, P., 201
Kamiyama, A., 17, 314
Kamoun, H., 139
Kanet, J.J., 343, 449, 450
Kannelakis, P.C., 112
Karwan, K.R., 343
Kats, V., 379
Kempf, K.G., 343
Kerzner, H., 16
Keskinocak, P., 201, 287, 368, 379
Kilger, C., 17, 201
Kimms, A., 17, 139, 169
Kirkpatrick, S., 434
Kjenstad, D., 201, 379
Kobacker, E., 450
Koehler, G.J., 368
Kolen, W.J., 139
Kolish, R., 16, 79
Koop, G.J., 314
Kopfer, H., 252
Kreipl, S., 112, 113, 201, 435, 457
Kröger, S., 252
Kubiak, W., 228
Kumaran, S., 201, 379
Kusiak, A., 17, 343
Kutz, S., 450

L

Lambrecht, M., 201
Lapide, L., 201
Laporte, G., 228
Lasry, A., 287
Lasserre, J.-B., 17, 112
Lassila, O., 369
Lawler, E.L., 16, 17, 35
Lawton, G., 435
Lee, C.-Y., 17, 35, 113, 379
Lee, E.K., 412
Lee, J.K., 450
Lee, J.S., 450
Lee, K.J., 450
Lee, Y.H., 112, 435
Lefrancois, P., 343
Lei, L., 17, 379
Lenstra, J.K., 16, 17, 35, 48, 112, 434

Leon, V.J., 368
Le Pape, C.L., 17, 113, 369, 446, 450
Lettovsky, L., 315
Leung, J. Y.-T., 18, 379
Levner, E., 379
Levy, F.K., 79
Li, H., 379
Lin, E., 368
Liu, J., 35, 139, 379
Liu, W.-P., 252
Liu, Z., 17
Lockwood, D., 287
Lomnicki, Z.A., 112, 412
Lu, F., 287
Luh, P.B., 112, 412
Lustig, I.J., 201, 446

M

MacCarthy, B.L., 35, 139
Mallik, S., 252
Marsten, R.E., 315
Martel, C.U., 228
Martin, C., 287
Martin, J., 369
Martin-Vega, L.A., 113
Matsui, T., 251
Matsuo, H., 113, 139, 434
Mattfeld, D.C., 252, 368
Matthys, D.C., 369
Maxwell, W.L., 16, 35, 169, 435
McAloon, K., 251
McCormick, S.T., 139
McDonald, G.W., 113
McInnes, D., 287
McMahon, G.B., 112, 412
McNaughton, R., 228
Meal, H.C., 113, 201
Mehta, S.V., 368
Meindl, P., 201
Michel, L., 251
Miller, L.W., 16, 35
Miller, T.C., 17, 201
Mittenthal, J., 139
Miyashiro, R., 251
Mize, J.H., 139
Mizrach, M., 228

Moder, J.G., 16, 79
Mohindra, A., 368
Möhring, R., 16, 35, 79
Monden, Y., 140
Monma, C.L., 435
Montreuil, B., 343
Moore, J.M., 228
Mörig, M., 450
Morton, T.E., 16, 79, 112, 228, 434
Muckstadt, J.A., 201
Müller, T., 251
Mulvey, J.M., 228
Murthy, S., 201, 368, 379
Muscettola, N., 369
Muth, E.J., 112

N

Nanda, R., 17, 48, 314
Nareyek, A., 17
Nemhauser, G.L., 251, 252, 315, 394
Neumann, K., 16, 35, 79
Noon, C., 139
Noronha, S.J., 18, 343
Nowicki, E., 112, 113, 434
Nuijten, W.P.M., 17, 113, 446
Nussbaum, M., 450

O

Odoni, A., 287
Oerlemans, A.G., 139
Oliff, M.D., 169, 343
Orlin, J.B., 315
Ovacik, I., 17, 112, 435
Ow, P.-S., 112, 228, 343, 369, 434, 450
Ozgur, D., 201

P

Padberg, M., 315, 394, 412
Palmer, D.S., 112
Panwalkar, S.S., 112, 434
Papadimitriou, C.H., 112, 394
Pardalos, P., 17
Park, H.K., 450
Park, S., 368
Parker, R.G., 17, 18, 48, 394
Parra, E.A., 450
Patterson, J.H., 79

Pattipati, K.R., 112, 412
Pentico, D., 16, 17, 79, 434
Perakis, A.N., 287
Pesch, E., 16, 35, 79, 368, 435
Philips, C.R., 16, 79
Phillips, M., 252
Pimentel, J.R., 343
Pinedo, M.L., 17, 35, 112, 113, 139, 169, 201, 343, 368, 369, 379, 434, 435, 450, 457
Pinoteau, G., 369, 450
Pinson, E., 112, 412
Pinto Paixao, J.M., 17, 48, 287
Plotnicoff, J.C., 450
Posner, M.E., 228
Potts, C.N., 112, 201, 228, 434
Potvin, J.-Y., 369
Prietula, M., 450
Puget, J.-F., 446

Q

Queyranne, M., 169

R

Rachamadugu, R.M.V., 112, 139
Rachlin, J., 201, 379
Raman, N., 368
Rardin, R.L., 394
Ratliff, H.D., 315
Reddy, S.K., 252
Régin, J.C., 251
Ricciardelli, S., 139
Riis, S., 252
Rinaldi, G., 139
Rinnooy Kan, A.H.G., 16, 17, 35, 48
Rodammer, F.A., 17
Ronen, D., 48, 287
Rosenwein, M.B., 287
Ross, P., 17, 228, 435
Rössling, I., 450
Roundy, R., 139, 169, 201, 435
Rousseau, J.-M., 17, 48, 287
Roy, B., 112

S

Sabuncuoglu, I., 368
Sadeh, N.M., 201, 379

Sahni, S., 228
Samroengraja, R., 343
Sarin, S.C., 411
Sarma, V.V.S., 18, 343
Sasieni, M., 79
Sassano, A., 139
Sauer, J., 369
Sayer, J.S., 79
Schaerf, A., 251
Scheer, A.-W., 343
Scherer, W.T., 18
Scholes, M., 252
Scholl, A., 17, 139
Schönberger, J., 252
Schmidt, G., 16
Schreuder, J.A.M., 251, 252
Schrijver, A., 394
Schwindt, C., 16, 79
Seshadri, S., 379
Sethi, S.P., 228, 379
Shanthikumar, G., 379
Shapiro, J.F., 17, 201
Shaw, D.X., 169
Shaw, M.J.P., 343, 368
Shenker, S., 139
Shenoi, R., 287
Shepardson, F., 315
Shepherd, J., 201
Shin, K.G., 450
Shmoys, D., 16, 17, 35
Sidney, J.B., 252, 435
Simchi-Levi, D., 201
Simchi-Levi, E., 201
Simons, B.B., 228
Singer, M., 112
Slowinski, R., 16
Smith, D., 79
Smith, M.L., 112
Smith, S.F., 18, 113, 343, 368, 369, 449, 450
Smith, W.E., 111, 434
Smutnicki, C., 113, 434
Snowdon, J.L., 35, 368
Solberg, J.J., 343
Solomon, M.M., 287

Soumis, F., 287, 315
Sridharan, V., 450
Sriskandarajah, C., 139, 379
Stadtler, H., 17, 201
Stam, A., 252
Stecke, K.E., 139
Stefek, D., 113
Steiglitz, K., 394
Steiner, G., 435
Stojkovic, G., 287, 315
Storer, R.H., 113, 368, 434
Strobel, K., 201
Sturzenbecker, M., 201, 379
Suh, C.J., 113, 434
Sule, D.R., 17
Sullivan, G., 450
Sullivan, R.S., 113, 434
Sung, C.S., 379
Sussmann, B., 112
Sweigart, J.R., 343
Swigger, K.M., 450
Szwarc, W., 112

T

Tadei, R., 113, 435
Taillard, E., 112, 434
Talbot, F.B., 79
Tang, C.S., 139
Tarjan, R.E., 228
Tayur, S., 287, 435
Thiel, S., 251
Thompson, G.L., 412, 450
Tien, J.M., 17, 314
Toth, P., 287
Tovey, C.A., 228
Trafton, L.L., 201
Tretkoff, C., 251
Trick, M., 251, 252
Trubian, M., 113, 434
Truchon, M., 228

U

Ulder, N.L.J., 434
Uzsoy, R., 17, 112, 113, 368, 435

V

Vaccari, R., 113, 434
Vaessens, R.J.M., 434
Vandaele, N.J., 201

Name Index

Van de Klundert, J., 379
Vanden Berghe, G., 18, 48
Van de Velde, S.L., 394
Van Hentenryck, P., 251, 446
Van Laarhoven, P.J.M., 434
Van Landeghem, H., 18, 48
Van Wassenhove, L.N., 112, 228
Vazacopoulos, A., 112, 201, 251
Vecchi, M.P., 434
Vepsalainen, A., 112, 434
Veral, E., 379
Vergados, Y., 251
Vieira, G.E., 368
Voigt, K., 201
Volta, G., 113, 435
Voss, S., 17, 48, 287

W

Wagner, M., 169
Walker, M.R., 79
Weare, R.F., 450
Webster, S., 369
Weglarz, J., 16, 394
Wein, L.M., 113
Westfold, S., 79
Wetzel, G., 251
Whinston, A.B., 343, 368
White, K.P., 17
Widmer, M., 112
Wiest, J.D., 79

Williams, J.F., 169
Wilson, N.H.M., 17, 48, 287
Wilson, R.H., 169
Wismer, D.A., 139
Wittrock, R.J., 140
Woerner, I.W., 343
Wolf, B., 139
Wolsey, L.A., 394
Wren, A., 17, 48, 287
Wu, F., 201, 368, 379
Wu, S.D., 113, 368, 434
Wu, T.-P., 450

Y

Yang, Y., 112, 113, 457
Yano, C.A., 139
Yeh, J., 201, 379
Yen, P.-C., 343, 369, 379
Yih, Y., 379
Yoon, S.H., 379
Yourdon, E., 369
Yu, G., 17, 48, 287

Z

Zawack, D., 112
Zdrzalka, S., 112
Zeghmi, A.H., 113
Zhou, Z., 379
Zimmermann, J., 79
Zipkin, P.H., 169
Zweben, M., 18

Subject Index

A

Academic prototype, 445-448
Acceptance-Rejection Criterion, 424-431
Active Schedule, 404, 405
Activity, 40, 41
Aircraft, 7, 258-270
Alldifferent constraint, 239, 442-444
Apparent Tardiness Cost (ATC) rule, 83, 328, 416-419
Apparent Tardiness Cost with Setups (ATCS) rule, 352, 353, 419-421
Application-specific system, 336-339, 448, 449
Aspiration Criterion, 428-431
Assignment problem, 237, 243-245, 386
Attribute, 122-127, 325
Automobile Assembly, 5, 122-127, 442-444

B

Backward Procedure, 56, 399
Barrier function method, 387
Batch, 141-170, 177
Beam Search, 421-423
Beam Width, 421-423
Bin Packing, 213-216
Bottleneck, 87-93, 105, 106, 280
Branch-and-bound, 254-270, 400-411
Branch-and-cut, 307, 401

Branch-and-price, 262-271, 400, 401
Branch-cut-and-price, 307, 401, 402
Branching scheme, 256, 265
Break (in tournament schedule), 230-240, 245-248
Break (in shift schedule), 307-311
Break placement, 309-311
B2B, 377, 378
Buffer, 115-140

C

Call center, 307-311
Capacity buckets interface, 331-335
Capacity constrained operation, 123-127
Carlsberg Denmark, 193-197
Carmen Systems, 280-283
Car rental, 6, 207-210
Case-based reasoning, 351-354
Cascading effects, 331-333
Chromatic number, 217, 232
Chromosome, 431-433
Classroom scheduling, 221-224
Clique, 84, 232
College basketball conference, 245-248
Column generation, 262-271, 303-307
Column selection, 262-271, 303-307
Complete graph, 232
Completion time, 21
Completion problem, 250

Composite dispatching rule, 416-421
Concatenation (of procedures), 357, 358
Constraint Programming, 72-76, 93-101, 237-240, 327-330, 437-446
Constraint propagation, 73, 74, 93-102, 237-240, 327-330, 437-446
Constraint satisfaction 73, 74, 93-102, 237-240, 327-330, 437-446
Constraint store, 322-327
Continuous manufacturing, 173-178
Crane scheduling, 372, 373
Crew scheduling, 279-283, 303-307, 374
Critical path, 54-68, 85, 407
Critical Path Method (CPM), 54-69
Critical Path (CP) priority rule, 414-417
Crossover effect, 356, 432
Cut set, 62-67
Cutting plane method, 396, 400-402
Cycle length, 116-122, 127-132
Cycle time; see MPS cycle time; Unit cycle time
Cyclic schedule, 116-122, 127-132
Cyclic Staffing problem, 299-301

D

Database, 320-327, 362-366
Days-off scheduling, 290-296
Decomposition technique, 87-93, 183, 272-279
Degree (of node), 216-221
Delayed precedence constraints, 92
Demand rate, 141-169
Dichotomic search, 256, 265, 440
Disjunctive graph representation, 84-93
Disjunctive programming, 84-87, 391-393
Dispatching rule, 82-84, 106, 177, 190-193, 414-421
Dispatch-list interface, 103, 331-334
Domain, 438-440
Due date range factor, 353, 419-421

Due date, 21, 29-31, 82-84, 122-127, 190-193, 205-213, 351-353, 413-431
Due date tightness factor, 353, 419-421
Dummy job, 52-54
Duration, 40
Dynamic Balancing heuristic, 127-135
Dynamic data, 21, 22
Dynamic dispatching rule, 414
Dynamic programming, 396-400

E

Earliest Due Date first (EDD) rule, 82, 83, 106, 191-193, 415, 416
Earliest Release Date first (ERD) rule, 415, 416
Earliness costs, 32, 350
Economic Lot Size (ELS), 144
Economic Order Quantity (EOQ), 144
Equipment maintenance scheduling, 374
Exam scheduling, 214

F

Facility, 24, 171-197, 320
Factors, 353, 418-421
Filter, 421-424
Finished goods inventory costs, 32, 141-169
First Fit (FF) heuristic, 213-216
First Fit Decreasing (FFD) heuristic, 213-216
Flexible Assembly Systems, 115-140
Flexible Flow Line Loading (FFLL) heuristic, 127-132
Flexible flow shop, 23, 102, 127-132, 136, 162
Flexible job shop, 23, 93, 102, 136
Flexible Manufacturing Systems, 136
Float, 56
Forward procedure, 55, 396-399
Frequency Fixing and Sequencing (FFS), 150-159

G

Gantt chart interface, 103-106

Subject Index

Generic system, 102-109, 336-339, 447, 448
Genetic algorithms, 431-433
Global constraint, 239, 444
Global rule, 415
Goal Chasing method, 132-135
Gradient method, 387
Graph coloring, 216-220, 232
Graph coloring heuristic, 217-219
Grouping and Spacing (GS) heuristic, 122-127, 177

H

Hamiltonian path, 232
Headway, 275-278
Hospital, 7, 43, 289, 374
Hotel, 205-210

I

If-Then rule, 326-329
Induction method, 352-356
Inference engine, 327-330
Information access system, 362
Information coordination system, 362
Information processing system, 362, 363
Integer programming, 69-72, 208, 209, 222, 230-237, 243-245, 254-279, 296-307, 388-391
Integration (of procedures), 357-360, 433
Interior point method, 384
Inventory holding costs, 141-169

J

Job-on-arc, 52-54
Job-on-node, 52-54
Johnson's rule, 111
Just-in-Time (JIT), 31, 133, 177

K

Knowledge base, 322-327

L

Lagrangean multiplier, 154, 387
Lagrangean relaxation, 400, 401
Largest Number of Successors (LNS) rule, 416

Lateness, 29, 87-92, 402-411, 415, 416
Lead-in effect, 244, 245
Learning, 351-356
Least Flexible Job first (LFJ) rule, 415, 416
LEKIN job shop scheduling system, 102-109, 451-457
Library (of algorithms), 357-360
Linear programming, 67, 68, 163, 164, 193-197, 258-272, 296-303, 383-386
Load balancing, 83, 128, 416
Logic rule, 326, 327
Local rule, 414, 415
Local search, 424-433
Longest path; see Critical path
Longest Processing Time First (LPT) rule, 83, 128, 155, 156, 415, 416
Lot size, 141-169

M

Machine criticality, 88
Machine eligibility constraints, 24, 25, 207-213, 415, 416
Machine learning, 351-356
Makespan, 28, 29, 51-79, 83-102, 213-221, 415, 416
Make-To-Order (MTO), 27, 174, 373
Make-To-Stock (MTS), 27, 141, 174, 373
Material handling constraints, 26, 115-136
Material handling system, 26, 115-136, 174
Material Requirements Planning (MRP), 10, 191, 321, 378
Maximum Lateness, 29, 87-92, 402-411, 415, 416
Medium term planning, 5, 142, 176-190
Minimum Part Set (MPS), 116-121, 127-132
Minimum Slack first (MS) rule, 82, 83, 415-421
MPS cycle time, 116-121, 127-132
Multi-step look-back interchange, 425-427
Mutation, 241-243, 424-431

N

Neighbourhood design, 241-243, 424-431
Neighbourhood search, 241-243, 424-431
Network television programs, 243-245
Neural network, 353-356
Nonlinear programming, 68, 69, 150-159, 386-388
Normal distribution, 58-61
NP-hardness, 69, 70, 81-87, 150-162, 205-220, 253-279, 303-307, 388-392, 396-411, 413
Nurse scheduling, 7, 289

O

One step look-back interchange, 425-427
Operator, 216-221, 289-311
operator constraints, 216-221, 289-311
Optimization Programming Language (OPL), 201, 441-444
Order scheduling, 374
Overload, 128-132
Owens-Corning Fiberglas, 162-164

P

Paced assembly system, 122-127, 442-444
Pairwise interchange, 160, 241-243
Paper mill, 5, 173, 174, 323
Parallel machines, 82-84, 159-162, 205-211, 285, 326, 327, 354-356, 385, 421
Parameter adjustment, 352, 353
Penalty function method, 387
Personnel costs, 44, 45, 296-303, 307-311
Polynomial time algorithm, 207-210, 296-302, 383-386, 396
Precedence constraints, 24, 51-76, 92
Preemptions, 27, 28, 42, 224
Processing time, 21, 40
Production rate, 21, 141-164, 184-190
Production synchronization, 373

Profile Fitting (PF) heuristic, 118-121
Program Evaluation and Review Technique (PERT), 58-61
Propagation effects, 73, 74, 93-102, 239, 240, 331-333, 438, 439

R

Reactive scheduling, 346-351
Reconfigurable system, 360-362
Release date, 21, 40, 87-102, 205-213, 354-356, 402-411, 415, 416
Reoptimization, 88, 89, 279, 331-333, 346-351
Rescheduling, 88, 89, 279, 331-333, 346-351
Reservation system, 6, 205-213
Resource constrained project scheduling, 69-72, 82, 207
Resource constraints, see workforce constraints
Robotic cell scheduling, 372, 373
Robustness, 346-351
ROMAN system, 72-76
Room assignment heuristic, 221-224
Rote learning, 352
Rotation schedule, 146-150
Routing, 25, 26, 81-109, 258-272, 303-307
Row prices, 303-307
Run (length), 141-164

S

SAP-APO system, 193-197, 448
Saturation level, 216-221
Scaling parameter, 83, 328, 352, 353, 416-421
Scheduling description language, 378, 379
Scheduling engine, 327-330, 357-360, 363-365, 375-377
Semi-Active schedule, 105
Semiconductor manufacturing, 4, 81, 327, 328
Sequence dependent setup, 26, 31, 44, 107, 108, 141-164, 190-193, 285, 324, 325, 352, 353, 419-421

Service In Random Order (SIRO) rule, 415, 416
Set Covering, 390, 391
Set Packing, 254-258, 390, 391
Set Partitioning, 258-272, 303-307, 390, 391
Setup, 26, 31, 44, 107, 108, 141-164, 190-193, 285, 324, 325, 352, 353, 419-421
Setup time severity factor, 353, 419-421
Shifting bottleneck heuristic, 87-93, 177, 279, 357-360
Shift scheduling, 39, 296-303
Shortest Processing Time first (SPT) rule, 82-84, 414-416
Shortest Queue at Next Operation (SQNO), 416
Shortest Setup Time first (SST) rule, 149, 191-193, 415, 416
Simulated Annealing, 240-243, 424-429
Simplex method, 384
Single machine, 22, 82, 83, 87-93, 102, 125-127, 141-158, 191-193, 348-353
Slack, 56, 82, 83, 94-102, 415-421
Soccer tournament, 6, 7, 229-243
Solid tour, 307, 311
Sport tournament, 6, 7, 229-243, 245-249
Standard deviation, 58-61
Starting time, 22, 40, 54-57, 67, 68, 73, 85, 86, 93-102, 130, 206
Starvation avoidance, 351
Static rule, 414
Storage space constraints, 27, 115-135
Supply chain, 24, 171-198, 371, 377, 378, 448

T
Tabu-list, 429-431
Tabu-search, 429-431
Tabu-tree, 431
Tail assignment, 279-283
Tanker scheduling, 254-258

Tardiness, 29-31, 44, 45, 83, 84, 191-193, 347-350, 352-356, 397-399, 418-424, 430, 431
Throughput diagram interface, 331-336
Throughput (rate), 28, 29, 117, 122, 331-336
Timetabling, 213-221, 230-243, 272-279, 448-449
Time window, 41, 205-213
Tooling constraints, 42, 43, 216-221
Total weighted completion time, 32, 81-84, 388, 389, 392
Total weighted tardiness, 29-31, 83, 84, 191-193
Tournament scheduling, 6, 7, 229-243, 245-248
Train pathing, 272-279
Train timetabling, 272-279
Transportation constraints, 28, 184-190, 275-278
Transportation costs, 32, 186-190, 254-259
Transportation problem, 385, 386
Transportation time, 28, 184-190
Travelling Salesman Problem (TSP), 148, 149
Travelling Tournament Problem, 242, 243
Truck routing, 285, 303-307, 391
Turnaround time, 42, 258-272

U
Underload, 128-132
Unit Cycle Time, 122-127
Unpaced assembly system, 116-121

V
Variance, 58-61

W
Wafer fabrication, 4, 81, 327, 328
Waiting time constraints, 27
Web-based system, 362-365
Weighted number of late jobs, 31, 207-213

Weighted Shortest Processing Time first Rule (WSPT), 82-84, 106, 191, 415-421

Workforce constraints, 43, 213-216, 289-311

Work-In-Process (WIP), 27, 31, 32, 335, 351

X

XPress Mosel (Dash Optimization), 201, 376